Mechanisms and Machine Science

Volume 45

Series editor

Marco Ceccarelli
LARM: Laboratory of Robotics and Mechatronics
DICeM: University of Cassino and South Latium
Via Di Biasio 43, 03043 Cassino (Fr), Italy
e-mail: ceccarelli@unicas.it

More information about this series at http://www.springer.com/series/8779

Lena Zentner · Burkhard Corves
Brian Jensen · Erwin-Christian Lovasz
Editors

Microactuators and Micromechanisms

Proceedings of MAMM-2016, Ilmenau, Germany, October 5–7, 2016

Springer

Editors
Lena Zentner
Department of Mechanism Technology
Technische Universität Ilmenau
Ilmenau, Thüringen
Germany

Brian Jensen
435 CTB
Brigham Young University
Provo, UT
USA

Burkhard Corves
Institut für Getriebetechnik und Maschinendynamik
RWTH Aachen University
Aachen, Nordrhein-Westfalen
Germany

Erwin-Christian Lovasz
Faculty of Mechanical Engineering
Universitatea Politehnica Timişoara
Timişoara
Romania

ISSN 2211-0984 ISSN 2211-0992 (electronic)
Mechanisms and Machine Science
ISBN 978-3-319-45386-6 ISBN 978-3-319-45387-3 (eBook)
DOI 10.1007/978-3-319-45387-3

Library of Congress Control Number: 2016950393

Printed on acid-free paper

This Springer imprint is published by Springer Nature
The registered company is Springer International Publishing AG
The registered company address is: Gewerbestrasse 11, 6330 Cham, Switzerland

Preface

After the successful conferences—MAMM-2010 in Aachen, Germany; MAMM-2012 in Durgapur, India; and MAMM-2014 in Timisoara, Romania—the 4th conference on Microactuators and Micromechanisms was held 2016 in Ilmenau, Germany. This conference was organized by the Mechanism Technology Group, Department of Mechanical Engineering at Technische Universität Ilmenau and under the patronage of IFToMM (International Federation for the Promotion of Mechanism and Machine Science), in particular the Technical Committees of Micromachines and of Linkages and Mechanical Controls.

The aim of the conference is to bring together scientists, industry experts, and students and to provide a special opportunity for know-how exchange and collaboration in various disciplines referring to microsystems technology. This event has presented following topics: microactuators and microassembly, microsensors involving movable solids, micro-opto-mechanical devices, mechanical tools for cell and tissue studies, micromanipulation and microstages, microscale flight and swimming, microrobotics and surgical tools, micron-scale power generation, miniature manufacturing machines, micromechatronics and micromechanisms, biomechanics micro and nano scales, control issues in microsystems and compliant mechanisms, and actuators. In addition to focusing on general topics of interest, the last named topic was introduced new in this conference and found a great resonance. This book has 22 peer-reviewed papers. The conference has exhibited high international participation: the participants were from Germany, Hungary, India, Japan, The Netherlands, Romania, Russia, Serbia, Slovakia, and Spain.

We want to express our special thanks to all authors for the interesting contributions, which had covered the most of the given topics.

The editors would like to express many thanks to IFToMM for support of the conference and for providing assistance to young scientists and engineers through the Young Delegates Program (YDP). In the framework of the conference, three young scientists from three different countries received assistance from the YDP.

Particular thank goes to the International Scientific Committee for their support and cooperation: Burkhard Corves (Germany), Amitabha Ghosh (India), Antoni Gronowicz (Poland), Just L. Herder (Netherlands), Antal Huba (Ungarn), Brian

Jensen (USA), Andrés Kecskeméthy (Germany), Erwin-Christian Lovasz (Romania), Karl-Heinz Modler (Germany), Nenad D. Pavlovic (Serbia), Victor Petuya (Spain), Srikar Vengalattore (Canada), Yao Yan-An (P.R. China), and Lena Zentner (Germany).

We would like to thank the Local Organising Committee of Technische Universität Ilmenau for hosting the scientific event and for the support of all associated activities: Stefan Griebel, Heidi Kirsten, Sebastian Linß, Andrea Schneider, Dirk Wetzlich.

The staff at Springer Publishing, namely Johanna F.A. Pot and Nathalie Jacobs, provided excellent technical and editorial support, for which we thank you.

Ilmenau, Germany — Lena Zentner
Aachen, Germany — Burkhard Corves
Provo, USA — Brian Jensen
Timişoara, Romania — Erwin-Christian Lovasz
October 2016

Contents

Study on Polymer-Made 3DOF Spatial Parallel Manipulator

M. Horie

Abstract In this research, in order to realize the functions equivalent to the functions of the spherical bearing which is frequently used in conventional spatial mechanisms, a polymeric manufactured hinge with Hytrel® (DUPONT Co., Ltd.), which has excellent flexural fatigue resistance, have been proposed. Next, the shape and dimensions of the polymeric manufactured hinge having low stress values at bending deformation have been found by FEM analysis results. In addition, a polymer-made three-degrees-of-freedom(3DOF) spatial parallel manipulator which consists of this polymer-made hinge is designed and developed, and the output displacement characteristics of the manipulator are revealed.

Keywords Polymer-made hinges · Three degrees-of-freedom · Spatial parallel manipulator · Hytrel

1 Introduction

Maintenance of the bearing, including lubrication is time-consuming. In order to dispel the negative side, in the Horie laboratory, the rotation bearing was replaced with a polymer material, and a polymer-made revolute hinge (R hinge) was proposed. Next, design and development of a polymer-made two-degrees-of-freedom pantograph mech. and a polymer-made three-degrees-of-freedom planar parallel manipulator which consists of R hinges was carried out, and their characteristics were revealed (Horie et al. 2002, 2006). In this paper, polymer-made spherical hinges which can represent three degrees-of-freedom (3DOF) as spherical bearings

M. Horie (✉)
Laboratory for Materials and Structures (MSL) and Laboratory for Future Interdisciplinary Research of Science and Technology (FIRST), Tokyo Institute of Technology, Tokyo, Japan
e-mail: mahorie@pi.titech.ac.jp; horie.m.aa@m.titech.ac.jp; mahorie767@gmail.com

L. Zentner et al. (eds.), *Microactuators and Micromechanisms*,
Mechanisms and Machine Science 45, DOI 10.1007/978-3-319-45387-3_1

are proposed. Then, in order to design and manufacture polymer-made spatial mech. consisting of polymer-made spherical hinges, the shape and dimension of the polymer-made spherical hinges are discussed, and characteristics of 3DOF spatial parallel manipulator which consists of S hinges are revealed.

2 Polymer–Made Revolute Hinges

This paper is a study on the spatial mechanism containing a polymer-made S hinge having a function of the spherical pair. However, as the previous studies, the author studied with respect to the two DOF planar pantograph mechanism and the three DOF planar mechanism made of R hinge having a function of the revolute pair. This study is the study of its extension. Therefore, first, characteristics of polymer-made R hinge are described. Next, research contents regarding 2DOF pantograph mechanisms and a 3DOF mechanism made of R hinges are revealed.

2.1 *Fatigue Test Under a Static Tensile Load and Repeated Bending*

Figure 1 shows a polymer-made revolute hinge. A fatigue test was conducted under a static tensile load *W* and bending repeated angle ±45°. Bending is carried out by use of the polymer-made revolute hinge specimen.

In the results, Fig. 2 is obtained as a fatigue test results of the R hinge. From this figure, PP (Polypropylene) has been found that repeated bending motion of ±45° can be conducted more than one million times. From Fig. 2, we discovered that PP (Polypropylene) is a good material for long life materials.

Fig. 1 Revolute hinge specimen

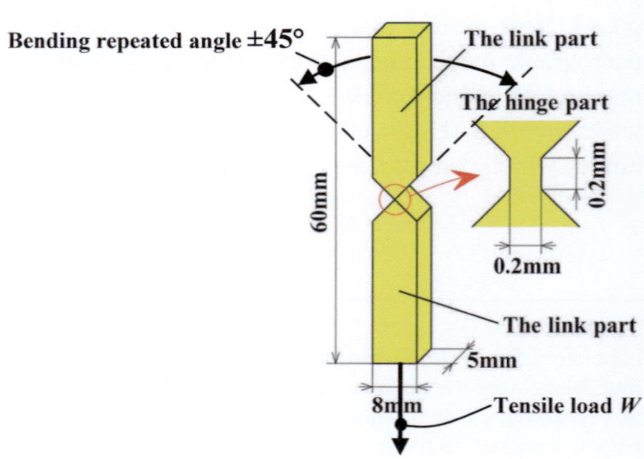

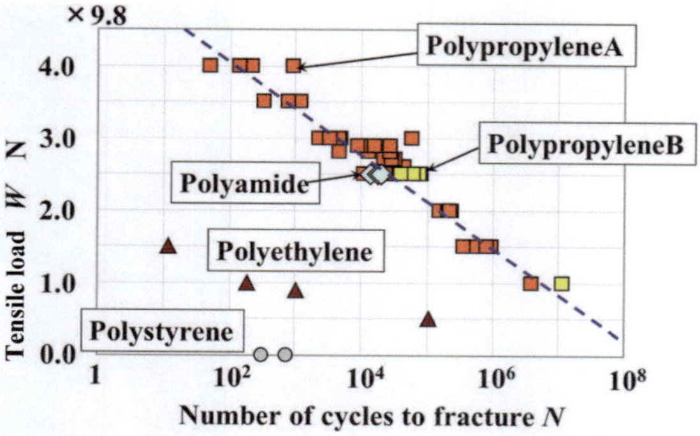

Fig. 2 Results of the fatigue test of the R hinge with static pull load and repeated angular-displacement ±45° bending (Horie et al. 2006)

2.2 *Polymer-Made 2DOF Planar Pantograph Mechanism and Polymer-Made 3DOF Planar Mechanism*

PP (Polypropylene) was discovered to be a good material with a long life from the fatigue test. Therefore, first, the 2DOF planar pantograph mechanism shown in Fig. 3 was proposed and developed (Horie et al. 2006). Figure 1a is Version 1 of the pantograph mechanism. Unfortunately, the direction of the output link changes when two liner actuators move. Well, we thought Version 2 shown in Fig. 3b. This pantograph mechanism has three parallelograms. Therefore, the direction of the output link with a short length does not change the direction when two linear inputs move, three short length links are the same length and parallel; the motion is always parallel, since the leftmost vertical input link moves vertically as usual.

In the case of Fig. 3a, there is one parallelogram in the mechanism. Therefore, the output link orientation changes when the two input links have linear motion, vertical and horizontal. In Fig. 3b, there are three parallelograms in the mechanism.

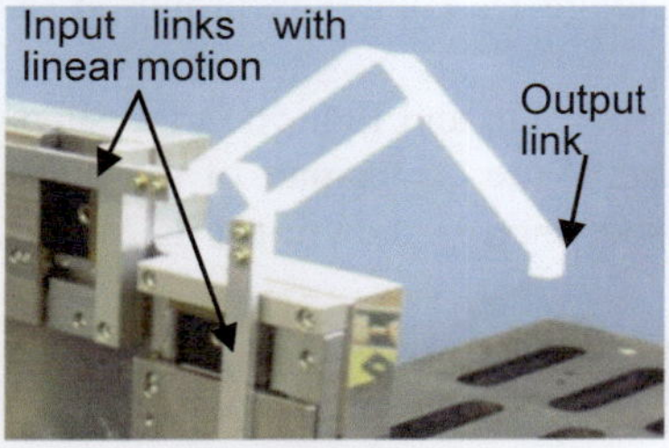

(a) A pantograph mechanism having a parallelogram (Version 1)

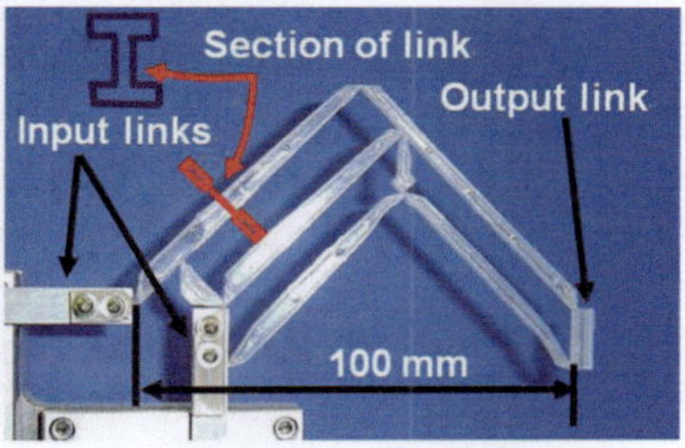

(b) A pantograph mechanism having three parallelograms (Version 2)

Fig. 3 Polymer-made 2DOF pantograph mechanisms (*Material* Polypropylene). **a** A pantograph mechanism having a parallelogram (Version 1). **b** A pantograph mechanism having three parallelograms (Version 2)

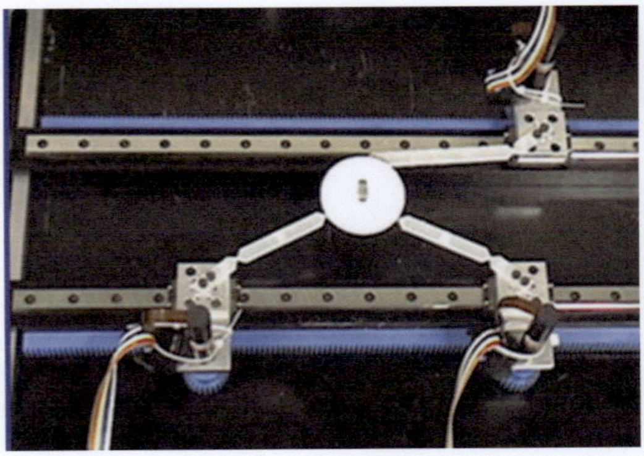

Fig. 4 A polymer-made 3 DOF planar parallel manipulator (*Material* Polypropylene)

Therefore, when two input links move, the short output link oriented in the vertical direction does not change the orientation. Moreover, as the forces acting at the hinge part are minimized, each link whose cross-section is I-shaped and is lightweight; has an optimum mass distribution.

In Fig. 4, there are three actuators on two rails and a 3DOF planar parallel manipulator with hinge joints made of polypropylene (Horie et al. 2002).

In the above mechanisms, polymer-made planar mechanisms are shown. However, typically, in the case of realizing polymer-made spatial mechanisms, it is necessary to propose and make a compact polymer-made hinge which has the function of the ball bearing. We call this polymer-made hinge "polymer-made spherical hinge" or "S hinge". In the next chapter, the polymer-made spherical hinge is discussed.

3 Polymer–Made Spherical Hinges

In this chapter, a polymer-made spherical hinge (S Hinge) is proposed and its bending characteristics are discussed.

Figure 5a is one of example of a hinge specimen, and Fig. 5b is the shape and dimension of the hinge. In Fig. 5a, the shape of the center part of the hinge is a cylinder shape. Half of the spherical hinge specimen is shown in Fig. 5b, with its dimensions.

Next, three types of central parts of the S hinge between two links, as shown in Fig. 6, are considered. Figure 6a through c show respectively a cylindrical type, an hourglass type, and a mountain type.

About S hinges shown in Fig. 6, the stress analysis by use of FEM is tried in the next session.

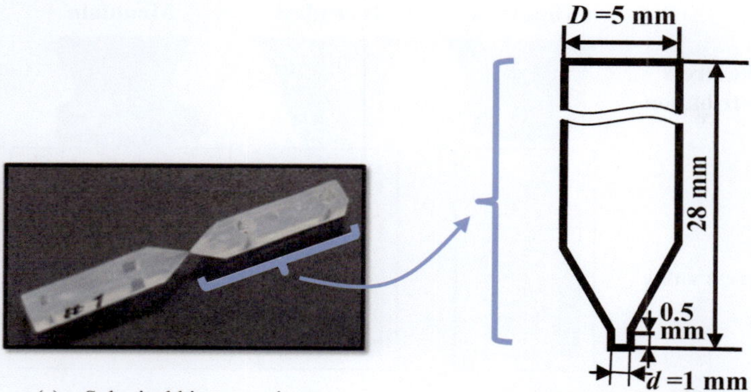

(a) Spherical hinge specimen (an example)

(b) Shape and dimensions of the half of the S-hinge

Fig. 5 Spherical hinge (Center part is cylindrical). **a** Spherical hinge specimen (an example). **b** Shape and dimensions of the half of the S-hinge

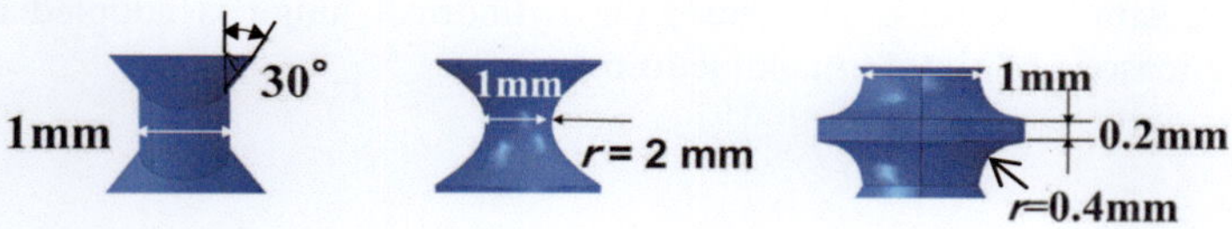

(a) Cylindrical (b) Hourglass (c) Mountain

Fig. 6 Three types of central parts of the S hinge. **a** Cylindrical. **b** Hourglass. **c** Mountain

3.1 *FEM Analysis of Polymer-Made S Hinge*

In order to stress analysis of the central part of the S hinge, the FEM called Abaqus/CAE was used. As the analysis conditions, it gave the following conditions. The material is polypropylene (PP), material density is 0.9 gw/cm^3, Poisson's ratio is 0.4, and the tensile modulus of elasticity is 750 MPa. Moreover, the boundary condition of the analysis is shown in Fig. 7. In Fig. 7, the center of the test piece is fixed. And, as one of the links bends 30°, the displacement vector (16.1658 mm) is added to the end of the left link.

The results of FEM analysis are shown in Fig. 8.

Fig. 7 The boundary condition of FEM analysis for S hinge

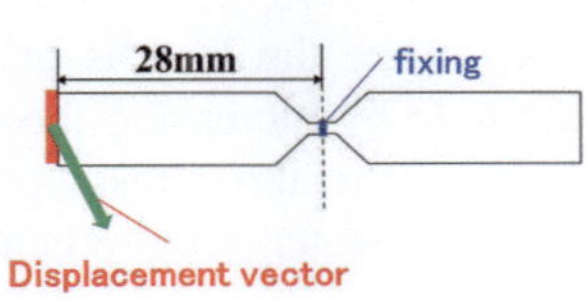

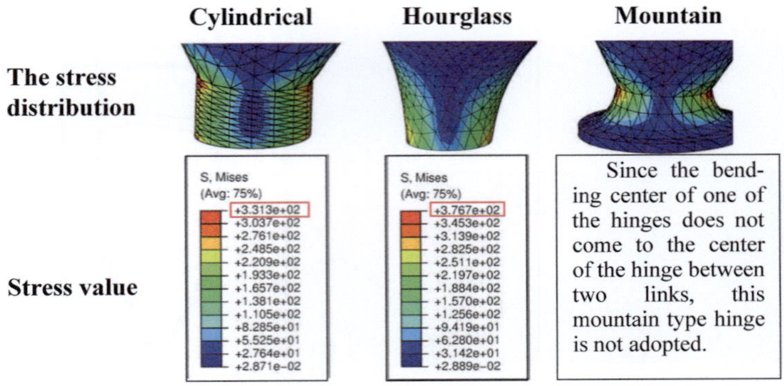

Fig. 8 The FEM analysis results/stress of the S-hinge (*Unit* Pa)

In Fig. 8, since the bending center of one of the hinges does not come to the center of the hinge between two links, the mountain type hinge is not adopted. Moreover, in Fig. 8, maximum stresses of the cylindrical hinge and the hourglass hinge have similar values. Therefore, the cylindrical hinge is adopted as the S hinge, for reasons of simple manufacturing.

3.2 Fatigue Test of S Hinge

A cylindrical hinge from the results of the previous section was fabricated using an injection molding machine. Polypropylene was used as the fatigue test piece material. Using this test piece, the fatigue bending test by tensile load and repeated bending operation is performed. The test piece dimension is already shown in Fig. 5. For the fatigue test, the experimental apparatus shown in Fig. 9 is used.

In Fig. 9c, the rocker of the four-bar linkage has a ±45° and 5 Hz movement. The lower part of the hinge, the dead weight 90 gw, is acting in a downward direction, and the upper part of the hinge is coupled to the rocker of the four-bar linkage. As shown in Fig. 10a, after the fatigue test was repeated 340,000 times, the hinge extend by plastic deformation, having a diameter of $d = 0.5$ mm. If there is a diameter of $d = 1.0$ mm or more, after 9000 times of movement, cracking occurs in the hinge center, which is shown in Fig. 10b, c. Then, in the subsequent repeated bending movement, cracking does not easily proceed.

To improve the strength of the central portion of the polypropylene hinge, in the case where the lengths of the hinges were the same, it was found to be able to improve by increasing the diameter d.

From the above results, in order to obtain a long life of S hinge, a material with strong-repeated fatigue bending properties such as rubber, is need.

(a) A four-bar linkage and specimens

(b) Specimens

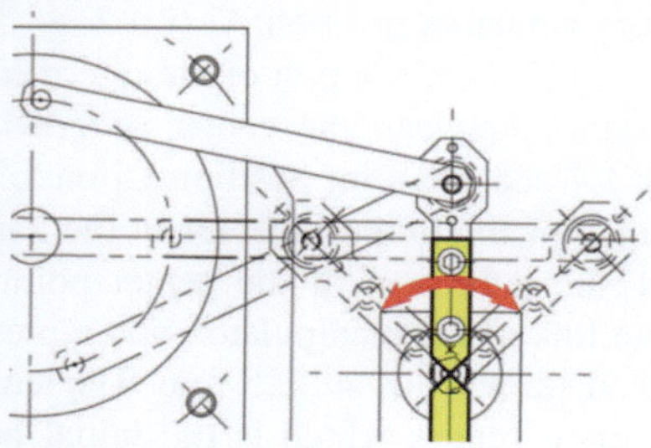

(c) Detailed movement of the four-linkage

Fig. 9 A four-bar linkage, S hinge specimens and detailed movement of the linkage. **a** A four-bar linkage and specimens. **b** Specimens. **c** Detailed movement of the four-linkage

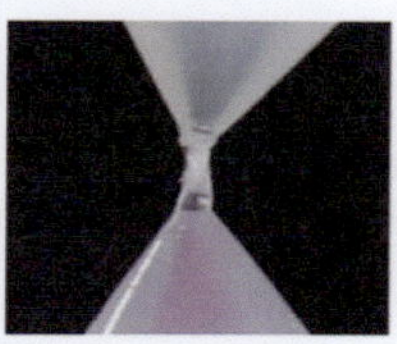

(a) d= 0.5 mm
(After 340000 times)

(b) d=1.0 mm
(After 9000 times)

(c) d=1.5 mm
(After 9000 times)

Fig. 10 Central part view of the S hinge after repeated bending movement. **a** d = 0.5 mm (after 340,000 times), **b** d = 1.0 mm (after 9000 times), **c** d = 1.5 mm (after 9000 times)

In the results from examining a material having such properties, elastomer Hytrel, which has heat resistance, oil resistance, excellent chemical resistance, impact resistance, rebound resilience, and excellent flexural fatigue resistance, was found. Then, this Hytrel/Thermoplastic-polyester-elastomer is used as the S hinge material. In the next session, the 3DOF spatial mechanism using polymer-made S hinges is proposed.

4 Polymer-Made 3DOF Spatial Parallel Manipulator

In the present chapter, there is a description of the design and development of a polymer-made three-degrees-of-freedom(3DOF) spatial parallel manipulator with Hytrel-made S hinges, as described in Chap. 3.

The polymer-made 3DOF parallel manipulator and its mechanism parameters are shown in Fig. 11. In Fig. 11, on the fixed link L1 of the mechanism, there is an equilateral triangle. The center of the equilateral triangle is the origin O of the orthogonal coordinate system O-XYZ. The distance from its center of gravity (center) until the three rotary actuators position Aj (j = 1 ∼ 3) is l_1. The length of the input links L2j (j = 1 ∼ 3) is l_2, the length of the intermediate link/coupler L3j (j = 1 ∼ 3) is l_3, the distance between the center of gravity of the equilateral triangle on the output link L4 and the joint position Cj on L3 is l_4, and the initial position of the center point P of the output link L4 in the Z-axis is Z = Z0.

The squares in Fig. 11 show the loci of the center point P of the equilateral triangle, which is the output link of the manipulator, and represents the limits of the working space of the mechanism when Z = ±25 mm. The length of one side of the square is 100 mm. The square center, which is the initial position of the output point P, is on the Z-axis.

4.1 Mechanism Synthesis

In order to synthesize the mechanism, at the position Z = Z0 mm and Z = Z0 ± 25 mm, the mechanism parameters l_1, l_2, l_3, l_4, and the value of the Z0, each varied by 1 mm, were determined by considering the following conditions.

(1) As the point P passes through the square of the four corners point and the square center point, i.e., 10 points on the square in Fig. 11, mechanism

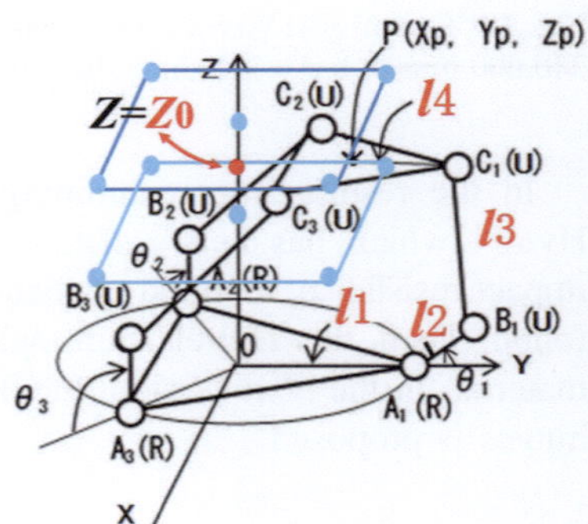

Fig. 11 Mechanism parameters and working space

parameters were determined by using inverse kinematics (Laribil et al. 2008; Lo´pez et al. 2006). Among the results obtained, only the mechanism parameters having three input angles determined by the equation of inverse kinematics were the correct solution.
(2) Reduce the size for miniaturization.
(3) The relative angular displacement of the two links across the hinge portion is in the range of ±45°.
(4) In order to reduce the driving torque of the motor, the ratio of the length of the links L2 and L3, that is, L2/L3 to 0.6 (Arai et al. 2012).

4.2 Mechanism Parameters and Discussion

In the results of the mechanisms synthesis described, in the previous session, mechanism parameters are obtained, as shown in Table 1. By use of these parameters, the 3DOF spatial parallel manipulator is manufactured. Next, in order to consider the relative displacement angles between L2 and L3, L3, and L4 at the S hinge positions, the displacement analysis was carried out. Table 2 shows the analytical results about pitch angles and yaw angles at joints Bj and Cj (j = 1 ~ 3).

From Table 2, it is revealed that the relative displacement angles are within ±45°.

4.3 Displacement Analysis and Joint Force Analysis of the 3DOF Parallel Manipulator

By use of the method of inverse kinematics, the input displacements are obtained when the output displacements are given. Moreover, by using the angular velocity

Table 1 The decided mechanism-parameter

	l_1	l_2	l_3	l_4	Z0
mm	35	60	100	27.5	97

Table 2 Difference angle between Max and Min relative angles between AB and BC (Pitch 2j), BC and CP (Pitch 3j), sj: j-Chain number (j = 1 ~ 3), Yaw angles from Bi to Ci (*Unit* Degree)

Pitch angles 21	Pitch angles 22	Pitch angles 23	Pitch angles 31	Pitch angle 32	Pitch angles 33	Yaw angles s1	Yaw angles s2	Yaw angles s3
79.84	72.94	81.90	83.75	81.72	83.75	60.00	86.16	86.16

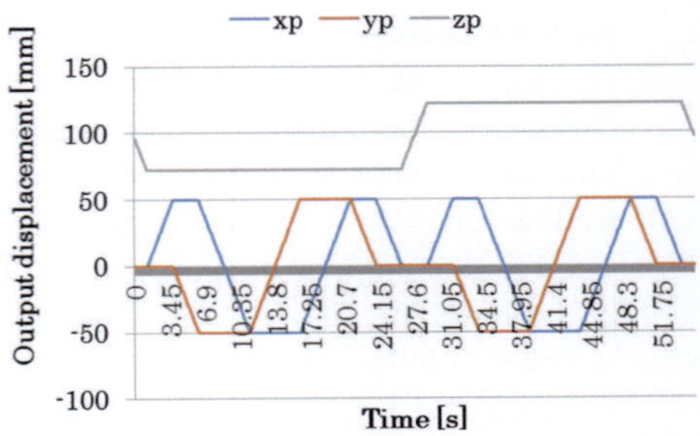

Fig. 12 Output displacements given in the inverse kinematics

Fig. 13 Polymer-made 3DOF parallel manipulator consisting of three RSS chains (There are four coupler links with S-hinges per one SS chain)

and the angular acceleration obtained by differentiating the expression for the angular displacement, each kinematic pair acting force was obtained (Fig. 12).

As a result, as the forces acting on the S hinge per one hinge are small, SS link is assumed to be numerous. In this case, it was decided to use four of the SS-hinge links per one link chain as shown in Fig. 13.

4.4 Polymer-Made 3DOF Parallel Manipulator

In Fig. 14, the polymer-made 3DOF parallel manipulator with stepping motors is shown. The left one is the manipulator without an output link. The right one is the manipulator with an output link. The locus of the output point for measuring is shown in Fig. 15. The output link for measuring is shown in Fig. 16.

As shown in Fig. 15, in order to measure the positioning accuracy and repeatability, four measurement points represented by a hemisphere concave (HC) diameter of 1 cm, were created on the output link.

When the central point P on the output link moves along the circle in Fig. 16, the output link will move while maintaining a horizontal position. And the direction of the output link/square-plate is not changed. The coordinates of four points are measured by use of the three-dimensional measuring machine (Crysta-Apex C7106) made by Mitutoyo Corporation.

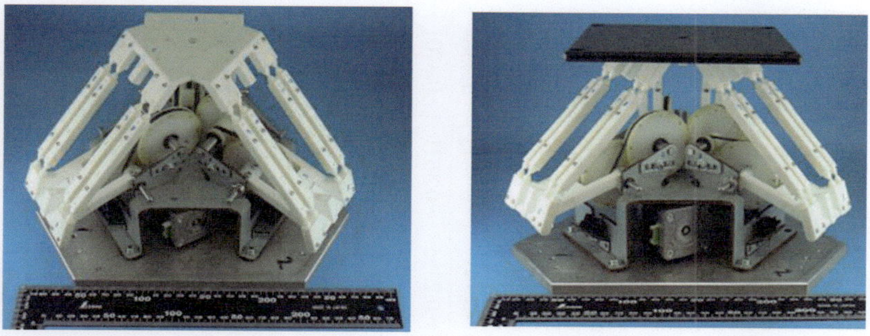

Fig. 14 Polymer-made 3DOF spatial parallel manipulator (*Right* With output table)

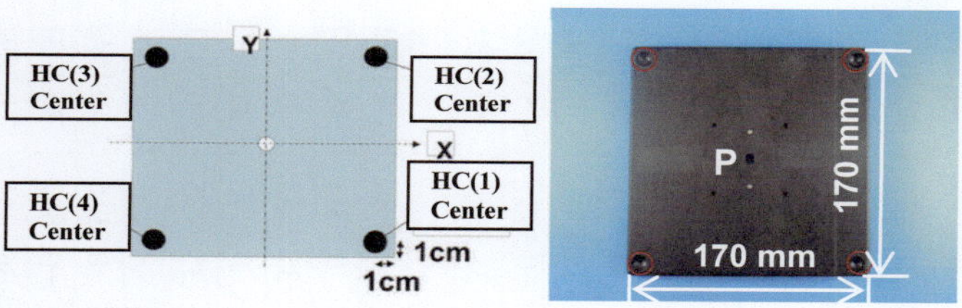

Fig. 15 Output link (table) and measuring points [*Hemisphere Concave* HC (1) ~ (4)]

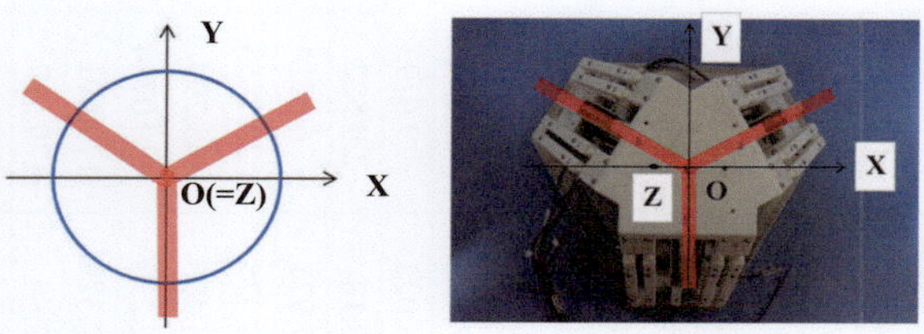

Fig. 16 Locus of the central point of the output link for measuring (*circle*) and rotational motion planes of the input links (*three thick lines*) in the O-XYZ coordinate system

In the results, the positioning accuracy, that is, the coordinates at the measurement point HC (1) ~ (4) are obtained as shown in Table 3, when the order of movement of the output link center (Xp, Yp, Zp) is given as shown in Table 4.

Moreover, the repeatability is obtained, as shown in Table 5.

From Table 3, the positioning accuracy in the Z-axis direction is 2 mm at maximum in the working space 100 * 100 mm^2. This value may be possible to improve, by local possibilities of further reducing the diameter of the hinge. And, from δ_z in Table 5, the reproducibility is found to be 0.082 mm at maximum.

Table 3 Positioning accuracy (Suffix m, measured; d, displace. error; Δ, measured data—start data) (*Unit* mm)

	Xp	Yp	Zp	Xp	Ym	Zm	ΔXm	ΔYm	ΔZm	Xd	Yd	Zd
HC(1)[Start data]	0	0	0	−17.4663	−179.059	195.1906	–	–	–	–	–	–
X−0, Y−0	0	0	−25	−17.7775	−179.042	170.3303	−0.3112	0.0171	−24.8603	−0.3112	0.0171	0.1397
X = 0, Y = 36	0	36	−25	−17.6413	−143.041	171.4225	−0.175	36.0177	−23.7681	−0.175	0.0177	1.2319
X = −36, Y = 0	−36	0	−25	−53.6359	−178.855	171.7004	−36.1696	0.2036	−23.4902	−0.1696	0.2036	1.5098
X = 0, Y = −36	0	−36	−25	−17.8601	−214.666	169.8161	−0.3938	−35.6074	−25.3745	−0.3938	0.3926	−0.3745
X = 36, Y = 0	36	0	−25	18.3609	−179.417	170.3696	35.8272	−0.3582	−24.821	−0.1728	−0.3582	0.179
HC (2) [Start data]	0	0	0	−18.1046	−29.0309	194.3438	–	–	–	–	–	–
X = 0, Y = 0	0	0	−25	−18.2771	−29.0139	169.2848	−0.1725	0.017	−25.059	−0.1725	0.017	−0.059
X = 0, Y = 36	0	36	−25	−18.0185	6.9823	169.587	0.0861	36.0132	−24.7568	0.0861	0.0132	0.2432
X = −36, Y = 0	−36	0	−25	−54.0549	−28.8418	170.4662	−35.9503	0.1891	−23.8776	0.0497	0.1891	1.1224
X = 0, Y = −36	0	−36	−25	−18.4996	−64.6461	170.7002	−0.395	−35.6152	−23.6436	−0.395	0.3848	1.3564
X = 36, Y = 0	36	0	−25	17.6559	−29.394	168.9737	35.7605	−0.3631	−25.3701	−0.2395	−0.3631	−0.3701
HC (3) [Start.data]	0	0	0	−168.022	−29.6793	194.5826	–	–	–	–	–	–
X = 0, Y = 0	0	0	−25	−168.18	−29.522	169.9056	−0.1585	0.1573	−24.677	−0.1585	0.1573	0.323
X = 0, Y = 36	0	36	−25	−167.921	6.6214	170.4642	0.1003	36.3007	−24.1184	0.1003	0.3007	0.8816
X = −36, Y = 0	−36	0	−25	−203.975	−29.2853	169.5892	−35.9532	0.394	−24.9934	0.0468	0.394	0.0066
X = 0, Y = −36	0	−36	−25	−168.406	−65.299	170.7742	−0.3843	−35.6197	−23.8084	−0.3843	0.3803	1.1916
X = 36, Y = 0	36	0	−25	−132.229	−30.0724	171.1818	35.7925	−0.3931	−23.4008	−0.2075	−0.3931	1.5992
HC (4) [Start data]	0	0	–	−167.379	−179.691	195.4131	–	–	–	–	–	–
X = 0, Y = 0	0	0	−25	−167.69	−179.537	170.9313	−0.3116	0.1542	−24.4818	−0.3116	0.1542	0.5182
X = 0, Y = 36	0	36	−25	−167.56	−143.388	172.2647	−0.1807	36.3031	−23.1484	−0.1807	0.3031	1.8516
X = −36, Y = 0	−36	0	−25	−203.565	−179.292	170.8998	−36.1861	0.3991	−24.5133	−0.1861	0.3991	0.4867
X = 0, Y = −36	0	−36	−25	−167.769	−215.316	169.8815	−0.3904	−35.6249	−25.5316	−0.3904	0.3751	−0.5316
X = 36, Y = 0	36	0	−25	−131.561	−180.047	172.374	35.8182	−0.3555	−23.0391	−0.1818	−0.3555	1.9609

Table 4 The order of movement of the output link center (Xp, Yp, Zp); mm

Measuring order	Xp	Yp	Zp
1	0	0	−25
2	0	36	−25
3	−36	0	−25
4	0	−36	−25
5	36	0	−25
6	0	0	25
7	0	36	25
8	−36	0	25
9	0	−36	25
10	36	0	25

Table 5 Repeatability; mm

	(0,0,0)		
	δ_X	δ_Y	δ_Z
HC (1)	±0.0729	±0.0294	±0.0624
HC (2)	±0.0695	±0.0107	±0.0578
HC (3)	±0.0627	±0.0098	±0.0236
HC (4)	±0.0279	±0.0363	±0.0481
	(36,0,−25)		
	δ_X	δ_Y	δ_Z
HC (1)	±0.0290	±0.0061	±0.0107
HC (2)	±0.0121	±0.0041	±0.0239
HC (3)	±0.0191	±0.0055	±0.0097
HC (4)	±0.0158	±0.0084	±0.0210
	(36,0,25)		
	δ_X	δ_Y	δ_Z
HC (1)	±0.05325	±0.02805	±0.0078
HC (2)	±0.0641	±0.0081	±0.025
HC (3)	±0.0498	±0.0019	±0.03555
HC (4)	±0.04605	±0.00825	±0.08195

5 Conclusion

In the present study, polymer-made planar hinge and polymer-made spherical/S hinge are discussed. Especially, shape and dimensions of polymer-made S hinges made by a "Thermoplastic polyester elastomer" called Hytrel were discussed. The Hytrel was found to be a good material for the S hinge. Then, polymer-made three degrees-of-freedom (3DOF) spatial parallel manipulator, which obtained a mechanism synthesis were designed and manufactured by use of the Hytrel. The positioning accuracy and repeatability positioning accuracy of this parallel manipulator was revealed.

Acknowledgments I would like to thank Mr. Misao Kobayashi, Mr. Takehiko Saito and Mr. Naoto Mochizuki of NISCA CORPORATION which is a partner of the joint research.

References

Arai T et al (2012) Design of delta arm driven by linear actuator based on the kinetics. The 30-th annual conference of the RSJ, 2O1-6 (in Japanese)

DUPONT Co., Ltd. (2016) http://www.td-net.co.jp/hytrel/about/index.html

Horie M et al (2002) 3-DOF planar positioning-orientation mechanisms with links and large-deflective hinges. Romansy 14, theory and practice of robots and manipulators (pp 351–358)

Horie M et al (2006) Durability of large-deflective hinges made of blend polypropylene used for molded pantograph mechanisms. Romansy 16, robot design, dynamics, and control (pp 355–362)

Laribi1 MA et al (2008) Advanced synthesis of the DELTA parallel robot for a specified workspace. Parallel manipulators, towards new applications, I-Tech Education and Publishing, pp 207–224. ISBN: 978-3-902613-40-0

Lo′pez M et al (2006) Delta robot: inverse, direct, and intermediate Jacobians. Proc IMechE Part C: J Mech Eng Sci, vol 220

Miniaturization of Check Valves

M. Pendzialek, J. Schneider, K. Höhe and L. Zentner

Abstract This contribution deals with principles of developing and constructing technologies to enable the optimization of hydraulic components. The focus is on conventional check valves with a goal of achieving a reduction in size and an accompanying optimization of the housing. For this purpose, check valves, which up to now have had to be installed as additional equipment, are first analyzed, the requirements determined and then redesigned. Here, a conventional rigid-body valve is replaced by a compliant model, creating a monolithic component and uniting the functions of multiple elements, such as springs and closing members (steel balls). The model based development of the novel valve system was carried out by Finite Element Method (FEM). The new valve was improved, in relation to weight and design space, compared to the ball check valve.

Keywords Plate check valve · Miniaturization · Metallic layers · Compliant valve element · Hydraulic fluid

1 Introduction

The requirements for automobile suppliers and thus for their products have been increasing over the last few years. In essence, this trend stems from greater political emphasis on environment protection, where words such as "downsizing" and

M. Pendzialek (✉) · L. Zentner
Technische Universität Ilmenau, Ilmenau, Germany
e-mail: matthias.pendzialek@tu-ilmenau.de

L. Zentner
e-mail: lena.zentner@tu-ilmenau.de

J. Schneider · K. Höhe
Reinz-Dichtungs-GmbH, Neu-Ulm, Germany
e-mail: juergen.schneider@dana.com

K. Höhe
e-mail: kurt.hoehe@dana.com

L. Zentner et al. (eds.), *Microactuators and Micromechanisms*,
Mechanisms and Machine Science 45, DOI 10.1007/978-3-319-45387-3_2

"efficiency optimization" are used. What is meant in general is a reduction of pollutant emissions and a decrease in fuel consumption for the same or even improved performance. A very effective method to fulfill these requirements is a reduction in weight or rather the miniaturization of assemblies or individual components (Löwer 2010). The design discussed here primarily places an emphasis on weight reduction.

The analysis involves a check valve used to maintain oil flow in one direction and prevent back flow. Many different designs can be found, including those using either a ball or a cone as the closing member (Bauer 2011). The feasibility of a new type of compliant check valve for hydraulic fluids is to be investigated on the basis of an empiric investigation of a known ball check valve. The current state of the art contains a wide variety of solutions. As described in [DE 10 2009 056 135 A1 (27.11.2009)], a valve can be created by adding a compliant element connected on one side in order to achieve the one-way sealing function in combustion engines. Alternatively, this can also be done with a type of spring-loaded valve disk, which represent a one-way coolant flow valve for cylinder head gaskets, as shown in [WO 2006/002254 A3 (05.05.2007)]. Here, a solution is to be found for the rigid system of the ball check valve using a compliant valve element in order to achieve a reduction in weight and size.

2 Poperies of a Ball Check Valve

The ball check valve (BCV) is used in this analysis because it represents a good example of a common hydraulic element. The new design intends to replace the BCV with a compliant valve element. The corresponding requirements were compiled from the following analysis of the BCV. It contains a valve body made of synthetic material which functions as its housing. The metallic spherical seat defines the inlet cross-section and accordingly the narrowest point of the valve.

The polished steel ball (2) in Fig. 1 is held in position and pre-loaded by a helical spring (4). When the resulting force of hydraulic pressure exceeds the spring pre-tension, the valve begins to open and deformation of the spring increases until the ball reaches its maximum point of deflection.

An important parameter with respect to weight reduction is the mounting volume V_{BCV} required to mount the BCV and for it to fulfill its function. For simplification, the shape is assumed to be a cylinder, whose diameter is the largest outer diameter D_O of the ball and whose height is d_{vh}. This represents the actual space required by the valve. However, in order to calculate the total volume, a ring-shaped base (not shown) of surrounding material must be added to the volume to allow the element to be installed. For this, an estimated allowance of 5 mm for the assembly is assumed for both diameter D_O and height d_{vh} to generate D_a and d_{ah}. The mounting volume of the BCV was determined to be $V_{BCV} = 5026.5\,\text{mm}^3$.

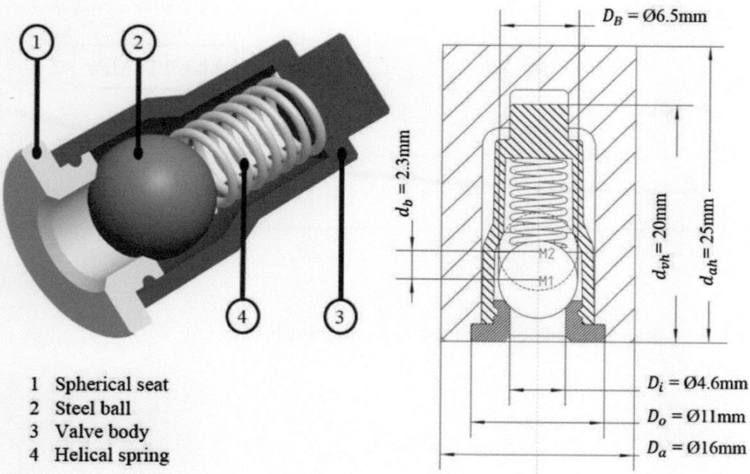

Fig. 1 *Left* Partial cutaway with component overview; *right* section drawing of the assembly with dimensions

Relevant factors for valve function include the spring preloading force F_p, and the maximum possible deflection d_{BCV}. The spring was measured while installed in the BCV. An analysis of the force-displacement curve of the helical spring yielded the following results using a material testing machine manufactured by Zwick GmbH & Co. KG [Fa. Zwick GmbH & Co. KG (10.2.2016)]. The measurement results are shown in Fig. 2.

Together with the inlet cross-section $D_i = 4.6$ mm and the spring preload force $F_p = 0.415$ N, it is now possible to calculate the opening pressure p_{open} of the BCV.

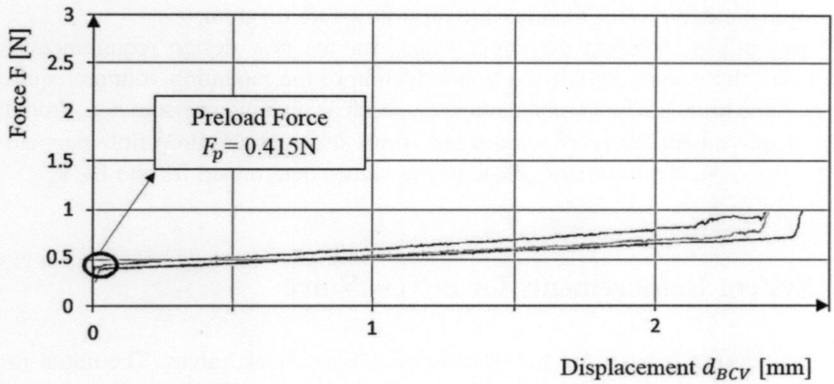

Fig. 2 Force-displacement curve of the BCV with spring preloading force F_p

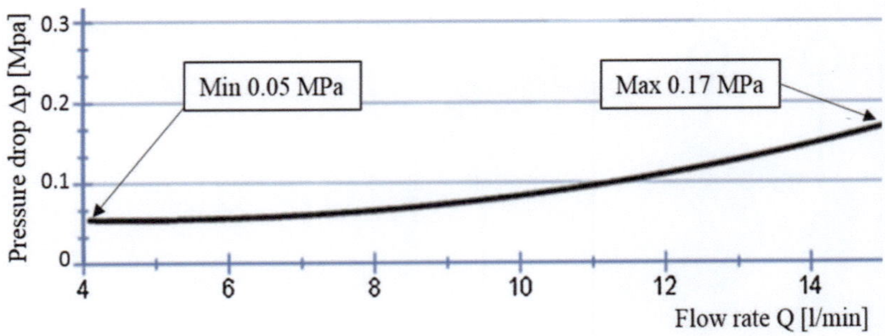

Fig. 3 Pressure-drop-flow-rate curve at the hydraulic test bench

$$\mathrm{p_{open}} = \frac{F_p}{A_{Di}} \tag{1}$$

Here, opening pressure $p_{open} = 0.025$ MPa refers to the minimum pressure that must act on the ball (2) in order to cause a displacement. Starting at that pressure p, oil flow Q is ensured. Another important parameter for a check valve is the pressure-drop-flow-rate curve ($\Delta p = \Delta p(Q)$—curve), which represents the hydraulic resistance of the open valve within the system.

The goal is to keep that resistance as low as possible in order to avoid losses and deterioration of the efficiency of the entire system. The curve shown in Fig. 3 was determined by installing the BCV in an experimental setup and exposing the valve to a hydraulic flow Q while maintaining a constant temperature. The system pressure p is then ramped up to 8 MPa. The volumetric flow rate Q was measured with an accuracy of 0.3 % of full scale using a VC 3 F1 P S volumetric flow sensor from Kracht GmbH [Fa. Kracht GmbH (15.3.2016)].

The basis of the new design is formed by three main pillars: The required mounting space, the spring parameters from the force-displacement curve and the characteristic curve from the pressure-drop-flow-rate diagram.

These pillars represent the target values for the new design requirements. In particular, the values should lead to a reduction of the mounting volume required. At the same time, the functional parameters, such as opening pressure p_{open} from the force-displacement curve or parameters from the pressure-drop-flow-rate curve (Figs. 2 and 3), should remain close to the values determined for the BCV.

3 System Requirements for a New Valve

The new design is based on the function of classic check valves. The initial form requirements originate from research into the state of the art as well as from the BCV parameters determined above. On the one hand, a high value is to be placed on the miniaturization of the mounting volume in the new design, through which the efficiency can be improved, too.

On the other hand, the parameters from the force-displacement curve and the pressure-drop-flow-rate curve needs to be reproduced as close to identical as possible. The following list serves both as a basis and as a guide for the optimization.

Requirements:

- required volume should be reduced by at least 50 %
- maximum outer diameter $d_V \leq$ 14 mm should not be exceeded
- inlet diameter of $D_i = 4.6$ mm as in the BCV should remain
- formation of the valve should be done with metallic layers
- the spring and sealing element should be generated using a single component
- the spring deflection should be limited by a mechanical stop
- the sealing surface should be continuous and line-shaped
- the valve geometry should not protrude beyond the metallic layers
- the opening pressure p_{open} must be 0.025 MPa
- the pressure drop should be smaller than 10 % of the system pressure p = 8 MPa.

4 Development of a New Compliant Valve

The method of achieving a significant reduction of the mounting volume is the development of a valve system formed from individual metallic layers with a compliant valve element (10) called plate check valve (PCV), while still fulfilling the basic functions of a check valve.

As shown in Fig. 4 symbolically with an arrow, the oil flow Q_{Oil} must be allowed in indicated direction and blocked in the other direction. By forming a fine-toleranced (±10 μm) wave element (11) in the bottom layer (8), it is possible to adjust the spring preload force F_{PCV} of the compliant element (10). It is formed with bridges out of the spring layer (7). The wave element (11) serves as the valve seal. The oil pressure supports the sealing function of the closed valve to prevent an oil flow Q_{Oil} in the opposite direction (Fig. 4). Here the parallelism between the wave element (11) and the compliant element (10) is given to reduce the hydraulic pressure drop of the valve.

In order for oil to be able to flow from one side of the valve to the other, the distance layer (6) is left open at the location of the compliant element (10), generating an opening valve path. This, in turn, is limited by the cover layer (5), while still offering the oil a flow path through the openings lateral.

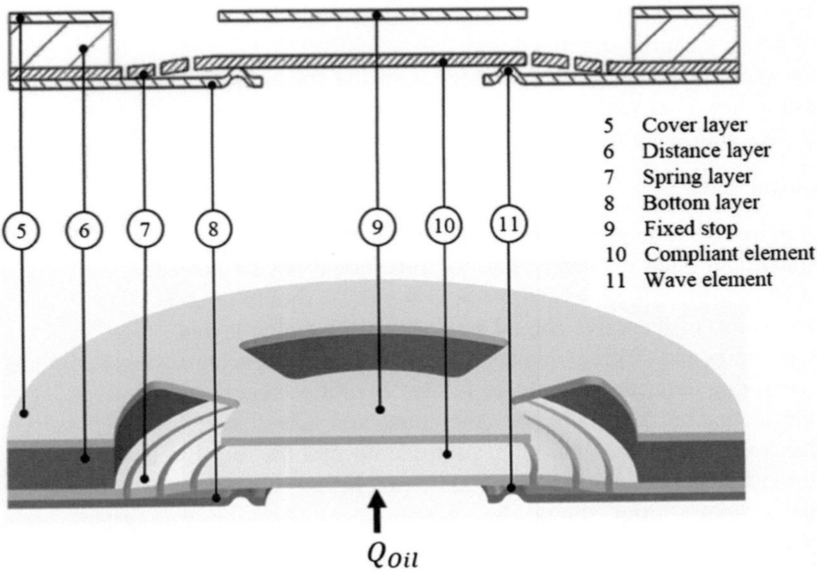

Fig. 4 *Top* section drawing of the valve with layer descriptions; *bottom* 3D section view

4.1 Design Variants of the Spring Layer

A compliant element (10) is provided for within the spring layer (7). The element combines the functions of the helical spring (4) and the steel ball (2) from the BCV (Fig. 1) and is made of spring steel in order to achieve the best possible properties. Two different variants were analyzed. These own the same compliant properties by adapted spring curves. Both variants were produced using the laser cutting process with a laser spot width of 150 μm.

The compliance of variant (a) is achieved by cutting three recesses around the center and offset by 120°. This spiral, known as an Archimedean spiral, arises from a circular motion which increases the radius in proportion to the rotation angle. In this variant, the flow surface affected by the pressure drop Δp has an area of $A_{(a)} = 10.8\,\text{mm}^2$.

Variant (b) possesses 10 circular segments cut out using a laser. They are arranged such that two segments are located at each radius, forming two bridges with a width of 0.3 mm. The radius values are set linearly and the segments at each radius are rotated 90° about the center with respect to the radii above and below. The flow surface here is $A_{(\mathbf{b})} = 27.7\text{mm}^2$ (Fig. 5).

Both variants (a) and (b) form a compliant element out of a flat metallic material. The outer diameter D_{OL} of both contours is 11 mm, fulfilling the requirements. The inner area with the contour diameter D_{iL} forms a spring plate, which provides for sealing of the wave element (11) (Fig. 6).

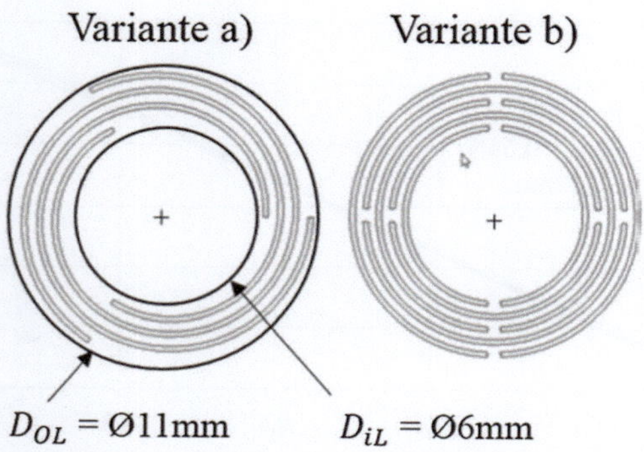

Fig. 5 Design variants for the valve compliant element; **a** spiral contour made of bridges rotated 120°; **b** 10 circular-segment contours with different radius

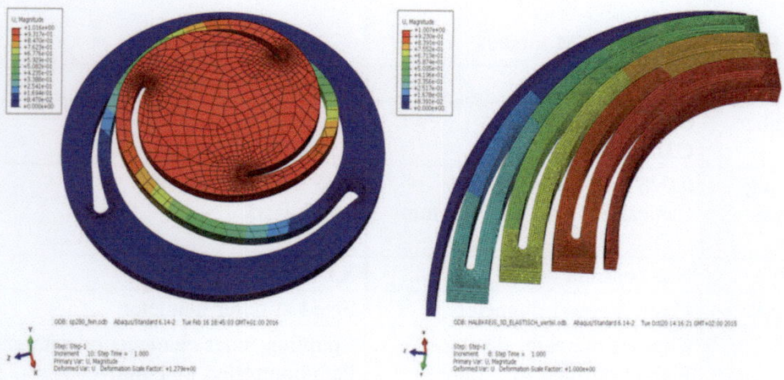

Fig. 6 Displacement simulation of variant (a) (*left*) and (b) (*right*)

In order to limit the number of variables, the protrusion of the wave element (11) above the bottom layer (8) will be set at 0.22 mm. Based on the spring parameters of the ball check valve from Sect. 2, the opening pressure p_{open} of the compliant valve is been determined at 0.025 MPa. Assuming a wave element (11) diameter of 5.7 mm, the spring preloading force F_{PCV} is 0.64 N. Variant (a) as well as variant (b) should have that value and they are shown in Fig. 7.

The characteristic curve of Fig. 7 is reached by variation the length of variant (a) and the number of variant (b) of the bridges in the compliant element (10) and thus the gradient of the spring curve is been influenced (Table 1).

With the help of FEM and the program Abaqus/CAE, the geometry is connected using primarily C3D8R elements and the calculations run. The resulting functionality corresponds to that of the ball check valve.

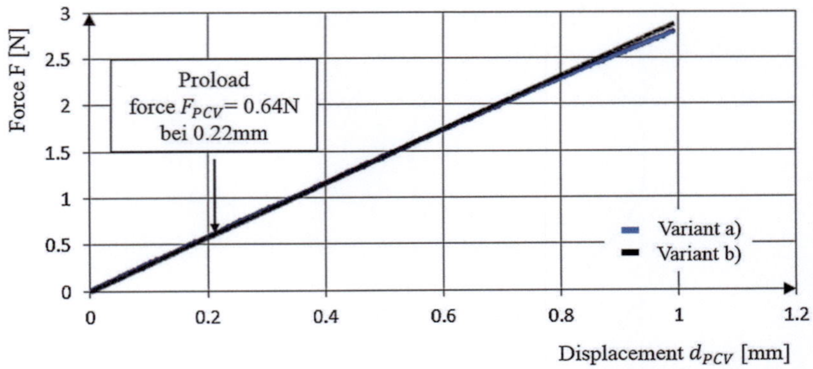

Fig. 7 force-displacement curve of variants (a) and (b) of the compliant element

Table 1 Parameter of the FEM models

	Variant (a)	Variant (b)
Type/number	C3D6/8070 C3D8R/83200	C3D6/3835 C3D8R/75985
Element size • Thickness • Surface	0.025 mm (10 elements) Critical Sects. 0.011–0.5 mm	0.05 mm (5 elements) 0.015–0.05
Symmetry	Not possible	XY- and ZY-plane
Boundary C	Fixed at outer diameter	Fixed at outer diameter
Deflection	Contour diameter displaced Path-controlled to 1 mm	Coupling inner elements on reference point Path-controlled to 1 mm

4.2 Comparison of Mounting Volumes

A comparison of the mounting volumes of the conventional BCV and the newly developed compliant valve PCV requires an explanation of the dimensions shown in Fig. 8, which are used to calculate the mounting volume V_{PCV}.

The diameter D_V arises due to trapezoid-shaped raised areas formed within the cover and bottom layers (5/8), which increase the diameter. These areas can be found around the valve recesses and are called “sealing beads”. The beads represent the valve system connection points with the surrounding parts and maintain an appropriate seal.

The total height d_V of 2.0 mm arises through the material thickness of the individual layers comprising the valve. The cover and bottom layers (5/8) contain the sealing beads, increasing the total height d_V by 0.4 mm. In contrast to the BCV, this setup does not require additional volume (base) to mount the valve. This yields

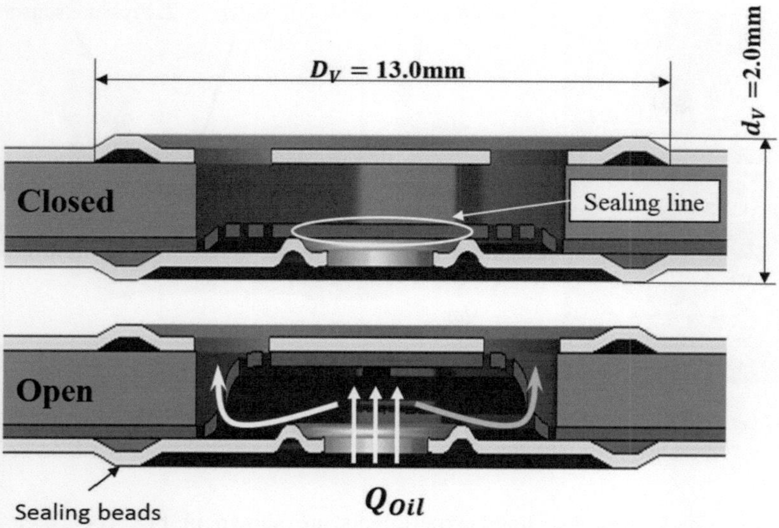

Fig. 8 *Top* closed compliant valve with dimensions; *bottom* open compliant valve with flow process Q_{Oil}

a mounting volume $V_{PCV} = 265.5\,\text{mm}^3$ for the compliant valve system. Here, the mounting volume of the BCV is $V_{BCV} = 5026.5\,\text{mm}^3$. This means that the necessary mounting volume has been reduced by 95 %.

4.3 Functional Comparison of the Valve Types

As previously described for the BCV in Sect. 2, the pressure-drop curve is a characteristic parameter for evaluating check valves. This curve is measured by hydraulic tests. These tests were carried out using a modified controllable hydraulic unit from the company NOLD Hydraulik + Pneumatik GmbH (Fig. 9 Left); the unit can generate a power output of 11 kW, a maximum pressure $p = 21$ MPa and a volume flow rate of $Q_{Oil} = 18\,\text{l/min}$. Measurement data are captured by two pressure sensors, in front of and behind the object under test. The sensors can record a pressure drop with an accuracy of 0.1 % at a maximum pressure of 10 MPa [Fa. Tecsis GmbH (14.3.2016)]. The volume flow Q_{Oil} rate is measured with the same gear flow meter as the BCV.

Compliant valve variants (a) and (b) were tested along with the BCV. The pressure sensors record the absolute pressure at a constant temperature. The data was digitally captured and processed with the help of LabVIEW measurement software and DIAdem data processing software, both from National Instruments.

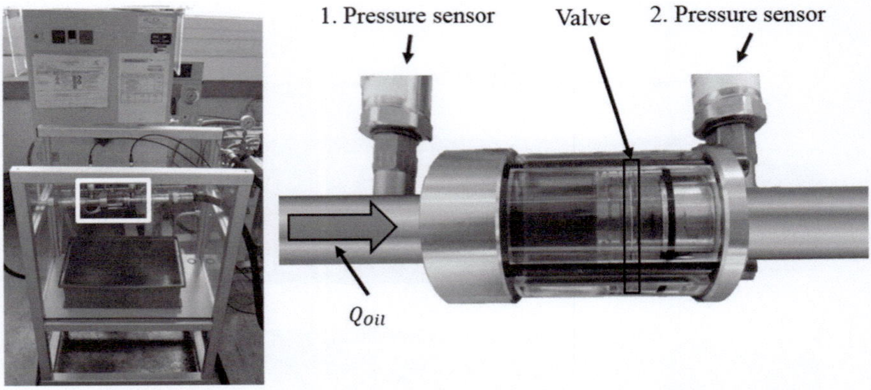

Fig. 9 *Left* setup of the hydraulic unit; *right* test setup with measurement connectors and symbolic volume flow Q_{Oil} direction

The results of the test stand experiments are shown in Fig. 10. Three measurement series were taken of each valve, the curves were averaged and smoothed. The pressure drop Δp can be calculated because the pressure p was measured both before and after the valves. The flow rate Q_{Oil} remains the same at all locations along a closed conduit, so it only needed to be measured once.

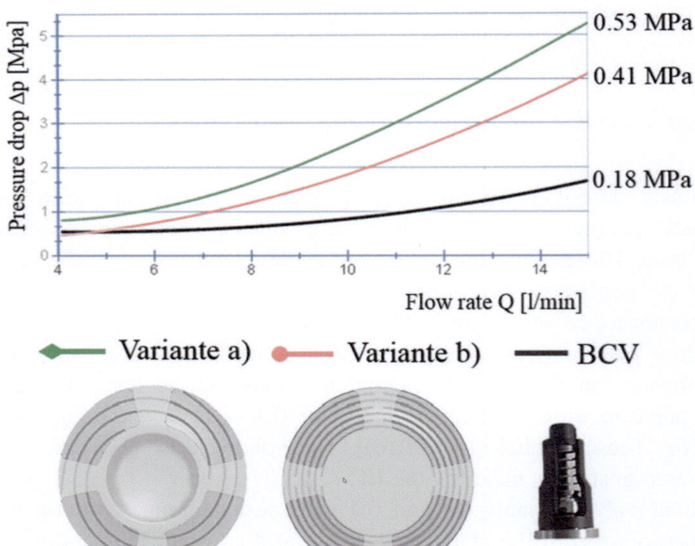

Fig. 10 Pressure-drop-flow-rate curve for variants (a)/(b) and the BCV, the images of the valve elements are not true to scale

Comparing the results of the individual measurement series with each other (see Fig. 10), it becomes evident that the geometry of the compliant element (10)—in particular the type of bridge connection used for the spring plate in variants (a) and (b)—has a significant influence on the pressure drop Δp of the check valve. Variant (a) causes a pressure drop Δp of 0.53 MPa at a flow rate Q_{Oil} of 15 l/min, whereas variant (b) generated a 23 % smaller drop. This can be traced back to the 256 % larger flow surface A in variant (a) compared to variant (b) as well as to the different overall forms.

Variant (b) also exhibits an advantageously smaller pressure drop Δp across the compliant valve at small flow rates Q_{Oil} between 0 l/min and 4 l/min even compared to the ball check valve.

However, both variants generate a larger pressure drop Δp in the range of flow above 4 l/min compared to the ball check valve tested, which generated a measured maximum value of 0.18 MPa at 15 l/min. This result was to be expected because the BCV uses a 2.3 times larger deflection d_{BCV} compared to the deflection d_{PCV} of the compliant valve. Nevertheless, the pressure drop Δp limit of 10 % of the system pressure (p = 8 MPa) was not exceeded.

5 Conclusion and Outlook

The analysis of a ball check valve provided the basis for the development of a miniaturized check valve. The new development of the compliant valve system is based on the reduction of the mounting volume. This could be reduced by approximately 95 % Geometric requirements have been met as well as the compliance with the material specifications. The spring characteristics of the compliant element could be set both in case of variant (a) and of variant (b) so that the hydraulic opening pressure corresponds to the default. It was possible to reduce the hydraulic pressure drop of the compliant valve system to approximately 49 % of the permissible limit. In partial areas of the pressure drop curve even an improvement compared to the ball check valve could be achieved.

The next step should be the optimization of the pressure drop at higher flow rates, so that the hydraulic resistance can be further reduced. In order to guarantee its long term function, the durability of the compliant element has to be analyzed. This includes both practice tests in the form of fatigue tests as well as analytical experiments with the finite element method. In addition, the opening movement of each of the variants to be investigated as a possible tipping behavior to prevent. In order to get from a development stage to a series product, it is not only required that the functional parameters fit, but that the most appropriate production method is established. Further investigation in this respect is required.

References

Bauer G (2011) Ölhydraulik; Grundlagen, Bauelemente, Anwendungen; Vieweg & Teubner, S.144

DE 10 2009 056 135 A1 (27.11.2009) Sealing with valve element

Fa. Zwick GmbH & Co. KG (10.2.2016) Datenblatt Statische Materialprüfmaschinen 5-250 KN

Fa. Kracht GmbH (15.3.2016) Datenblatt Zahnraddurchflussmesser VC

Fa. Tecsis GmbH (14.3.2016) Datenblatt Drucksensor P3290

Löwer C (2010) Autohersteller trimmen Motoren auf Effizienz

WO 2006/002254 A3 (05.05.2007) Cylinder head gasket with one-way coolant flow valve

A Biologically Inspired Sensor Mechanism for Amplification of Tactile Signals Based on Parametric Resonance

T. Volkova, I. Zeidis and K. Zimmermann

Abstract In this paper, the vibrational motion of an elastic beam under the parametric excitation is investigated theoretically and numerically. The problem is motivated by biological tactile sensors, called vibrissae or whiskers. Mammals use these thin long hairs for exploration of the surrounding area, object localization and texture discrimination. We propose a mechanical model of the vibrissa sweeping across a rough surface as a straight truncated beam stimulated by a periodic following force. The equation of transverse motion of the beam is studied using the Euler–Bernoulli beam theory and asymptotic methods of mechanics. The numerical analysis is performed by means of the finite element method. It is shown that the parametric resonance of the beam occurs at the specific ranges of the excitation frequency, which depend on the parameters of the beam and the amplitude of the applied force. For these frequency values, the vibrations of the beam are unstable with exponentially increasing amplitude. The comparison of the resonance ranges obtained theoretically and numerically is made. Thus, together with the realisation of the viscoelastic support of an artificial tactile sensor, the parametric resonance may be a potentially useful method for amplifying small signals arising from the contact with an object.

Keywords Parametric resonance · Vibration · Beam theory · Method of averaging · Vibrissa

---

T. Volkova · I. Zeidis (✉) · K. Zimmermann
Technical Mechanics Group, Department of Mechanical Engineering,
Technische Universität Ilmenau, Ilmenau, Germany
e-mail: igor.zeidis@tu-ilmenau.de

T. Volkova
e-mail: tatiana.volkova@tu-ilmenau.de

K. Zimmermann
e-mail: klaus.zimmermann@tu-ilmenau.de

L. Zentner et al. (eds.), *Microactuators and Micromechanisms*,
Mechanisms and Machine Science 45, DOI 10.1007/978-3-319-45387-3_3

1 Introduction

In nature, all mammals except humans possess specialised tactile hairs, called *vibrissae* or *whiskers*. These thin long hairs grow usually in groups in different locations on an animal's body. Among the most familiar to us are *mystacial* vibrissae orderly distributed over the whiskerpad on a muzzle, *supraorbital* vibrissae above the eyes, and *carpal* vibrissae that are located on the downside aspect of the forelimbs (Fig. 1). For terrestrial and marine mammals, vibrissae are significantly important tactile sensors used for the exploration of the surrounding area, texture discrimination, or even detection of water vibrations (Dehnhardt and Kaminski 1995; Niederschuh et al. 2014; Vincent 1912).

The structure and characteristics of vibrissae differ considerably from normal body hairs of the same mammal individual. They are thicker and stiffer than other types of hairs and have a conical shape with the linearly decreasing diameter (Voges et al. 2012). Another significant difference lies in the more complicated base of the vibrissa. The vibrissal hair grows from a special follicle, called the *follicle-sinus complex* (FSC). It incorporates lots of sensory nerve endings and a capsule of blood (Ebara et al. 2002). For some groups of vibrissae, the FSC is surrounded by two types of muscles that allow an animal to move its vibrissae back and forth (Dörfl 1982). These sniffing active movements of vibrissae, called *whisking*, play an important role during the exploratory behaviour of mammals.

The fascinating biological paradigm of the vibrissal tactile sensing has inspired a large number of research investigations. Scientists, biologists and engineers would like to understand the functional principles of the vibrissal sensing system from different aspects. Neuroscientists are trying to figure out how nerve impulses are processed from vibrissae to the brain and how an animal encodes the information. They perform laboratory experiments with living creatures, e.g. (Jadhav and Feldman 2010; Wolfe et al. 2008). Biologists usually describe the anatomy and morphology of vibrissae from species to species. They believe that the vibrissal sensing system played an important role during evolutionary development of mammals (Mitchinson et al. 2011).

For the engineers, the biological vibrissa gives an inspiration to design various artificial tactile sensors and find possible applications for them, as it promises to be competitive with artificial vision. Most often, the research focus lies on the

(a) (b)

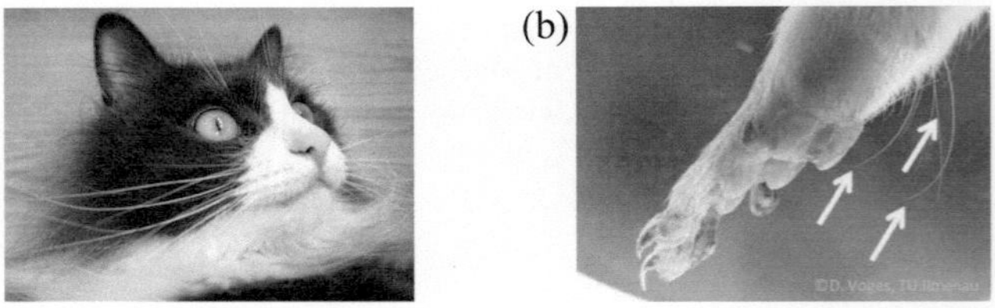

Fig. 1 **a** Mystacial and supraorbital vibrissae of a house cat; **b** Triple of carpal vibrissae at a forelimb of *Rattus norvegicus*, photo taken by Voges D (TU Ilmenau)

Fig. 2 The experimental setup at the Department of Mechanical Engineering (TU Ilmenau) for the quasi-static profile scanning

development of single sensors. Some interesting examples are the rotary actuated whisker for obstacle avoidance and the whisker-like sensor for underwater applications (Scholz and Rahn 2008; Valdivia y Alvarado et al. 2012). Furthermore, a variety of developed prototypes of mobile robots with a vibrissa-like sensory array can be found, e.g., in (Fend et al. 2004; Solomon and Hartmann 2006; Zimmer 1995). One of the latest platforms is the SCRATCHbot—a rat-like robot, which can move and position its whiskers toward specific objects with three degrees of freedom (Pearson et al. 2011).

Recent experimental investigations performed at the Department of Mechanical Engineering (TU Ilmenau) show that it is possible to reconstruct a profile of an object by one single quasi-static sweep of the straight technical vibrissa made of steel (Fig. 2). By measuring the clamping forces and the bending moment at the fixed base of the beam, the contact point with an object may be quantified. The theoretical analysis of the problem is based on the non-linear Euler–Bernoulli theory of the static deformable beam (Will et al. 2016).

Summarising the research on mathematical-mechanical modelling of vibrissae, two approaches may be marked out: rigid rod models, e.g. (Behn 2013; Berg and Kleinfeld 2003; Volkova et al. 2016a), and continuum elastic models, e.g. (Neimark et al. 2003; Scholz and Rahn 2008; Volkova et al. 2016b; Yan et al. 2013). The continuum models are in turn closer to the biological paradigm, as they take into account the inherent dynamical behaviour and the bending stiffness of the vibrissal hair. As for the modelling of the viscoelastic support of the hair inside the FSC, several spring and damping elements may be used, e.g. (Behn 2013). In (Volkova et al. 2016a), the FSC incorporating the blood capsule is modelled as a continuum volume filled with a viscous magnetic fluid. There, an approach to realise and control three-dimensional oscillations of the rigid rod is presented using an applied uniform magnetic field.

The present paper focuses on the non-linear process of the vibrissa sweeping across a surface of an object during texture discrimination. Frictional interactions between the tip of the vibrissal hair and a rough surface generate vibrations of the hair (Fig. 3). Since it has no receptors along the length, the tactile signals are transmitted by these mechanical vibrations to sensory receptors inside the FSC.

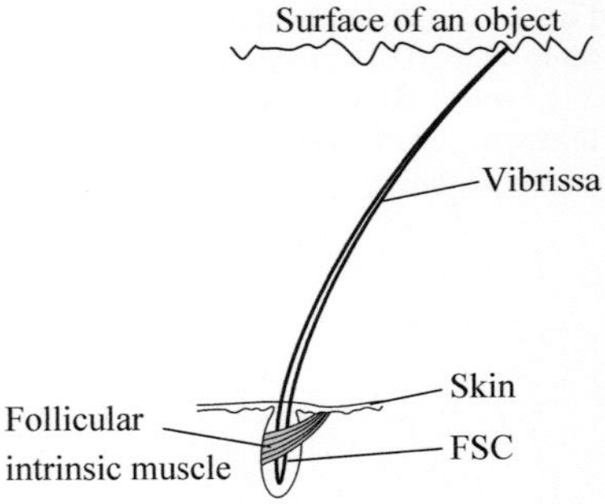

Fig. 3 Schematic drawing of a vibrissa sweeping past a rough surface

In this paper, the vibrissal hair is modelled as a straight elastic beam with linearly decreasing diameter along its length. The model describing in-plane vibrations of the beam stimulated by a periodic following force is developed. Based on asymptotic methods of mechanics, the *parametric resonance* of the beam is analysed theoretically. According to the so-called *resonance hypothesis* proposed in (Andermann and Moore 2008; Neimark et al. 2003), the vibrissa resonance may be a potentially useful mechanism during texture discrimination. It could amplify small tactile signals arising from the contact with an object and enchase the sensitivity of these biological tactile sensors. In the second part of the work, the numerical analysis of the problem using a finite element model is presented. The simulations are performed for various values of the excitation frequency of the applied force. The comparison of the theoretical and numerical results is made.

2 Theoretical Analysis

2.1 Geometry and Assumptions of the Model

Consider a truncated Euler–Bernoulli beam whose undeformed neutral axis is a straight line of a length L (Fig. 4). The radius $r(x)$ of the beam's circular cross-section evolves linearly along the axial direction:

$$r(x) = r_b - \frac{r_b - r_t}{L} x, \tag{1}$$

where r_b and r_t are, respectively, the radii at the base and the tip of the beam. The Cartesian coordinate system (x, y, z) is placed such that the x axis lies along the neutral axis of the beam, and the origin is in the middle of the base's cross-section. The Cartesian basis vectors are $\mathrm{e}_x, \mathrm{e}_y, \mathrm{e}_z$. The material properties of the beam are constant, i.e., the density $\rho = \mathrm{const}$ and the Young's modulus $E = \mathrm{const}$. The support condition of the beam at the base is assumed to be pinned. It allows the beam to rotate around the z axis, but not to translate in any direction. At the end of the beam, the roller support is considered, which provides free rotation around the

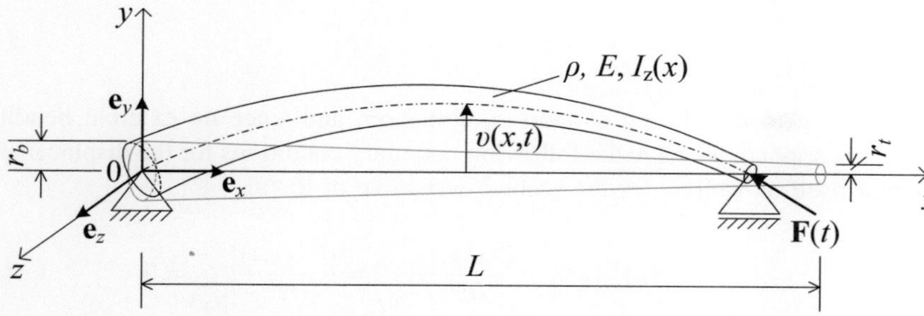

Fig. 4 Deflection of the straight truncated beam under the following force applied toward the cross-section at the tip

z axis and the horizontal deflection (Fig. 4). The beam is stimulated by a periodic force applied directly toward the cross-section of the beam at the tip:

$$\mathrm{F}(t) = -F_0 \cos(2\pi\,\Omega\, t)\mathrm{n}_t, \tag{2}$$

where F_0 is the amplitude, Ω is the excitation frequency, t is the time, and n_t is the normal to the cross-section at the tip. The assumed force $\mathrm{F}(t)$ corresponds to the frictional force, which arises from interactions between the vibrissal tip and the complex roughness profile of a surface (Fig. 3). Thus, it may be considered as a first harmonic component of the Fourier transform spectrum of the friction force.

It is supposed throughout that the transverse vibrations of the beam caused by the force $\mathrm{F}(t)$ about its straight natural configurations are in the x-y plane and *small*. The displacement vector of the axis points has the transverse component: $\mathrm{u}(x,t) = v(x,t)\mathrm{e}_y$.

2.2 *Equation of Motion and Its Approximation*

The small in-plain vibrational motion of the straight beam can be described with a single partial differential equation of the fourth order. With regard to the rotary inertia and in the absence of the force of viscous damping, the equation of motion in terms of $v(x,t)$ has the following form (Svetlitsky 2005; Zentner 2014):

$$m_0(x)\frac{\partial^2 v(x,t)}{\partial t^2} + \frac{\partial^2}{\partial x^2}\left(E\,I_z(x)\frac{\partial^2 v(x,t)}{\partial x^2}\right) + F_0\cos(2\pi\,\Omega\, t)\frac{\partial^2 v(x,t)}{\partial x^2} = 0. \tag{3}$$

Here, $m_0(x) = \pi\rho r^2(x)$ is the mass of the beam's element, and $I_z(x) = \pi\, r^4(x)/4$ is the moment of inertial of the cross-section.

The bending moment $\mathrm{M}(x,t) = M(x,t)\mathrm{e}_z$ at any cross-section of the beam can be expressed by the transverse displacement as:

$$M(x,t) = E\,I_z(x)\frac{\partial^2 v(x,t)}{\partial x^2}. \tag{4}$$

According to the chosen support combination, and since no external bending moment is applied, we have the following boundary conditions for the displacement function itself as well as for the second derivative of it:

$$v(0,t) = v(L,t) = 0, \quad \left.\frac{\partial^2 v(x,t)}{\partial x^2}\right|_{(0,t)} = \left.\frac{\partial^2 v(x,t)}{\partial x^2}\right|_{(L,t)} = 0. \tag{5}$$

In order to find the solution of the partial differential Eq. 3, let us first reduce it to an ordinary one using the Galerkin method (Kantorovich and Krylov 1958). We shall seek one-term approximate solution of Eq. 3 in the form:

$$v(x,t) = \sin\left(\frac{\pi x}{L}\right) f(t), \tag{6}$$

which satisfies the boundary conditions of the beam (Eq. 5). Substituting this expression in Eq. 3 with the requirement of the orthogonality of Eq. 3 to the function $\sin(\pi x/L)$, we obtain an ordinary second-order differential equation for the function $f(t)$. It can be written in the dimensionless form as:

$$\ddot{f}(\tau) + (1 - \varepsilon\cos(\gamma\tau))f(\tau) = 0. \tag{7}$$

Here, the dot notation is used to represent a time derivative of a function, and the dimensionless variables are introduced as follows:

$$\varepsilon = \frac{2\pi L^2 F_0}{b_1 E r_b^4},\ \gamma = \frac{\Omega}{\omega_1},\ \tau = 2\pi\omega_1 t,\ b_0 = \frac{1}{6}\left(1+\delta+\delta^2\right) - \frac{1}{4\pi^2}(1-\delta)^2 > 0, \tag{8}$$

$$b_1 = \frac{\pi^4}{10}\left(1+\delta+\delta^2+\delta^3+\delta^4\right) - \frac{\pi^2}{2}(1-\delta)\left(1-\delta^3\right) + \frac{3}{4}(1-\delta)^4 > 0,\ \delta = \frac{r_t}{r_b}\in[0, 1].$$

The natural frequencies of transverse vibrations of the beam are (in Hz):

$$\omega_n = \frac{n^2 r_b}{4\pi L^2}\sqrt{\frac{b_1 E}{b_0\rho}}, \quad n = 1, 2, 3, \ldots \tag{9}$$

In particular, for the cylindrical beam with constant radius $r(x) = r_b$, the assumed solution (Eq. 6) is an exact one. In this case, Eq. 3 reduces to the same Eq. 7 with the parameters $b_0 = 1/2$ and $b_1 = \pi^4/2$ as $\delta = 1$.

3 Parametric Resonance

Equation 7, known as the Mathieu equation, describes periodically excited vibrations of a system (McLachlan 1947). It has a cosinusoidal time-dependent coefficient of the term $f(\tau)$. The dimensionless parameters ε and γ, defined by Eq. 8, depend on parameters of the beam and the periodic applied force. It appears that for different values of ε and γ solutions of Eq. 7 may be periodic and bounded for all time, or the amplitude of the vibrations in solutions may grow progressively in time (*parametrical resonance*). Let us determine analytically and numerically ranges of the parameters, when the parametric resonance of the beam takes place.

3.1 Procedure of Averaging

For the approximate analysis of the oscillating process described by Eq. 7, the method of averaging is used (Bogolyubov and Mitropoliskii 1961). First, in order to reduce Eq. 7 to the standard form, it is assumed that the function $f(\tau)$ and its derivative have the form: $f(\tau) = a(\tau)\cos(\psi(\tau))$ and $\dot{f}(\tau) = -a(\tau)\sin(\psi(\tau))$.

Then, the differential equation of the second order (Eq. 7) converts to the system of two equations of the first order for the new variables $a(\tau)$ and $\psi(\tau)$:

$$\dot{a}(\tau) = -\frac{\varepsilon}{2}a(\tau)\sin(2\psi(\tau))\cos(\gamma\tau), \quad \dot{\psi}(\tau) = 1 - \varepsilon\cos^2(\psi(\tau))\cos(\gamma\tau). \tag{10}$$

As it will be shown, the most intense parametric resonance occurs for values of the excitation frequency Ω close to $2\omega_1$ of free vibrations of the beam. Therefore, we can set $\gamma = 2 + \varepsilon\Delta$, where Δ corresponds to the amplification factor of the parametric excitation. Consider the amplitude F_0 to be small in comparison with the elastic forces of the beam element, so that ε can be treated as a small positive parameter: $0 < \varepsilon \ll 1$. If $a(\tau)$ and $\xi(\tau) = \gamma\tau - 2\psi(\tau)$ are smooth functions of the time such that their derivatives are small terms of order ε, the values of these functions can be seen as the superposition of slowly varying part and small rapidly oscillating terms. Thus, they can be averaged on ψ over one period:

$$\dot{a}_0(\tau) = \frac{1}{2\pi}\int_0^{2\pi} \dot{a}(\tau)d\psi = \frac{\varepsilon}{4}a_0(\tau)\sin(\xi_0(\tau)), \tag{11}$$

$$\dot{\xi}_0(\tau) = \frac{1}{2\pi}\int_0^{2\pi} \dot{\xi}(\tau)\,d\psi = \varepsilon\Delta + \frac{\varepsilon}{2}\cos(\xi_0(\tau)).$$

3.2 Region of the Principal Parametric Resonance

The non-linear system (Eq. 11) may be simplified to a linear one with constant coefficients by defining new variables $\eta(\tau) = a_0(\tau)\cos(\xi_0(\tau)/2 + \pi/4)$ and $\varsigma(\tau) = -a_0(\tau)\sin(\xi_0(\tau)/2 + \pi/4)$:

$$\dot{\eta}(\tau) = -\frac{\varepsilon}{4}\eta(\tau) + \frac{\varepsilon\Delta}{2}\varsigma(\tau). \quad \dot{\varsigma}(\tau) = -\frac{\varepsilon\Delta}{2}\eta(\tau) + \frac{\varepsilon}{4}\varsigma(\tau). \tag{12}$$

The matrix corresponding to the system (Eq. 12) has the following eigenvalues:

$$\lambda^2 = \frac{\varepsilon^2}{4}\left(\frac{1}{4} - \Delta^2\right). \tag{13}$$

Thus, the solution of Eq. 12 is aperiodic and unstable, when there is an eigenvalue with positive real part, i.e., for $|\Delta| < 1/2$ around the dimensionless frequency value $\Omega/\omega_1 = 2$. This means that the resonance takes place within the interval

$$0 < \varepsilon \ll 1 \quad \text{and} \quad 2 - \frac{\varepsilon}{2} < \frac{\Omega}{\omega_1} < 2 + \frac{\varepsilon}{2}. \tag{14}$$

It is called the *region of the principal parametric resonance*. The width of this range is proportional to the parameter ε.

Other frequency ranges of the parametric resonance are close to the values $\Omega = 2\omega_1/n$ for any natural number n (Landau and Lifshitz 1969). However, the width of them gets narrow proportionally to the value ε^n as n increases. In practice, the cases for n = 1, 2, and 3 (rarely) are usually observed.

4 Numerical Analysis and Results

The small transverse vibrations of the parametrically excited beam are described by the partial differential Eq. 3. The solution of it with respect to the boundary conditions (Eq. 5) may be obtained by means of the finite element method. The numerical simulations are performed using software ANSYS Workbench 16.2.

Two particular models of the beam are considered:

1. A straight cylindrical beam (CB) with a constant radius of the cross-section.
2. A straight truncated beam (TB), when the radius at the tip is half of that at the base: $r_t = r_b/2$.

Table 1 Values of the parameters used in numerical simulations

Parameters	Cylindrical beam (CB)		Truncated beam (TB)
	Line body	Solid cylinder	Solid body
Length L (mm)	50	1	50
Radius r_b (mm)	1	–	1
Radius r_t (mm)	1	1	0.5
Density ρ (kg m^{-3})	1000	10^{-6}	1000
Young's modulus E (Pa)	10,313	2×10^{13}	10,313

4.1 Remarks on the Simulations

For both beam models, the vibrational characteristics (natural frequencies and mode shapes) are evaluated using an analysis system ANSYS "Modal". The dynamic response of a structure under the action of a periodic following load is calculated within the ANSYS module "Transient structural".

The CB is modelled as a line body with a thin circular solid cylinder attached at the right end. The latter one is needed to apply the periodic following force perpendicular to the cross-section. The cylinder may be considered massless and rigid. The transition between beam elements (beam 188, 50 elements) of the line body and solid elements (solid 186, 16 elements) of the cylinder is defined with the multi-point constraint bonded contact. Both translational and rotational degrees of freedom are accounted for. The TB is modelled as a whole using solid elements (solid 186, 504 elements). The values of the parameters for both models are given in Table 1. The boundary conditions are applied according to the theoretical formulation of the problem. Within the simulations, no damping is considered. The numerical damping is defined manually to be zero.

For the chosen parameter sets, the natural frequencies are listed in Table 2. It is shown that the values of the CB obtained numerically for three different meshing elements coincide with the theoretical eigenfrequencies given by Eq. 9. It has to be noted that the values of the parameters assigned to the beam (Table 1) do not correspond to vibrissae values reported in literature, e.g. (Neimark et al. 2003; Voges et al. 2012). They are chosen in such a way that $\omega_1 \approx 1$ Hz of the CB. This in turn further allows using a reasonable adequate time step size (10^{-2}–10^{-3} s) in transient simulations. The theoretical natural frequencies of the TB give upper boundary values, i.e., they are slightly higher than the numerical results (Table 2).

The transient analysis of parametric vibrations requires an initial deflection of the beam from the equilibrium position. Thus, the load process is developed in three steps: (1) [0, 0.25] s, (2) [0.25, 0.5] s, and (3) [0.5, 30] s. Within the first two steps, the impulse force is applied and discharged linearly at the midst of the beam in the transverse direction. Its maximum magnitude is 10–7 N. Then, within a time range [0.25, 30] s, the main load is applied as a pressure $P(t) = -P_0 \cos(2\pi\,\Omega t)$, $P_0 = 3$ Pa, normal to the cross-section at the right end.

Table 2 Natural frequencies of transverse vibrations of the CB and TB

Frequency(Hz)	Cylindrical beam (CB)				Truncated beam (TB)	
	Theory	Numerical simulations (ANSYS)			Theory	ANSYS
	Equation 9	Beam 3	Beam 188	Solid 186	Equation 9	Solid 186
ω_1	1.0089	1.007	1.0068	1.0053	0.7835	0.709
ω_2	4.0355	4.008	4.007	4.0016	3.1338	2.9593
ω_3	9.08	8.9526	8.9439	8.9324	7.0511	6.6026
Elements		50 + 16	50 + 16	512		504
Nodes		101 + 93	101 + 93	2201		2167

Table 3 Regions of the parametric resonance of the CB and TB (in Hz)

	Theory	Numerical simulations (ANSYS)
Region of the principal parametric resonance: $\Omega \approx 2\omega_1$		
CB	$1.8691 < \Omega < 2.1665$	$1.87 \leq \Omega \leq 2.15$
TB	$1.483 < \Omega < 1.6508$	$1.32 \leq \Omega \leq 1.50$
Second range of the parametric resonance: $\Omega \approx x_1$		
CB	$0.9907 < \Omega < 1.0125$	$0.99 \leq \Omega \leq 1.01$

4.2 Regions of the Parametric Resonance

The dynamic response of the system under the parametric excitation is simulated for various frequency values. For the CB, the solution is found to be unstable for the frequencies close to the doubled eigenfrequency of the beam (region of the principal parametric resonance), as well as for the frequencies around this value (second range of the parametric resonance). For the TB, the region of the principal parametric resonance is obtained. Within these frequency intervals, the vibrations of the beam are unstable, i.e., their amplitude increases exponentially with time.

The comparison of the theoretical (Eq. 14) and numerical regions of the parametric resonance is given in Table 3. It may be seen that the results of the CB correspond to each other. Moreover, it is shown that the response of the system on the parametric excitation is qualitatively stronger within the region of the principal parametric resonance than within the second range. For the TB, a slight discrepancy between the theoretical and numerical results is observed. It is associated with the accounting of the variable radius of the cross-section.

In Fig. 5, some simulation results of the CB and TB are presented. The averaged amplitude of the theoretical solution is plotted for comparison as a dash line.

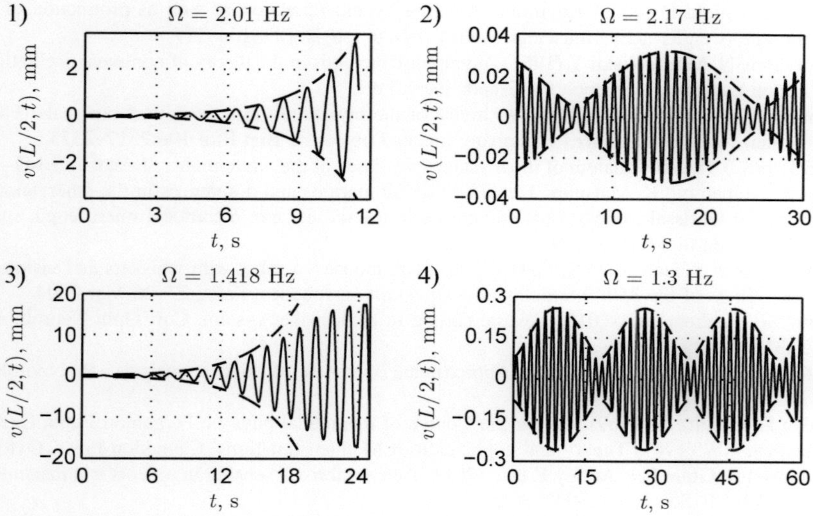

Fig. 5 Transverse displacement of the CB (**a**, **b**) and the TB (**c**, **d**): **a**, **c** unstable solution of the principle parametric resonance; **b**, **d** steady-state solution; — numerical simulations; – – the theoretical averaged amplitude $a_0(t)$ for: **a** $a_0(0) = 0.024$ mm, $\psi(0) = 0$; **b** $a_0(0) = 0.0265$ mm and $\psi(0) = 1.4573$; **c** $a_0(0) = 0.18$ mm, $\psi(0) = 0$, $\Omega = 2\omega_1 = 1.5669$ Hz; **d** $a_0(0) = 0.078$ mm, $\psi(0) = 1.7453$, $\Omega = 1.461$ Hz

5 Conclusions

In this work, a model of a straight truncated beam is developed to study the vibrational motion of mammal's vibrissae. The transverse oscillations of the beam stimulated by a periodic following force are investigated theoretically and numerically. It is shown that the parametric resonance of the beam occurs at the specific ranges of the excitation frequency, within which the vibrations of the beam are unstable with exponentially increasing amplitude. Together with the realisation of the viscoelastic support of an artificial tactile sensor, it may be a potentially useful method for amplifying small signals arising from the contact with an object. Future investigations will also focus on the design of sensor arrays consisting of several vibrissa elements.

Acknowledgments The work was supported by the Deutsche Forschungsgemeinschaft (DFG) within the Grant ZI 540-16/2.

References

Andermann M, Moore C (2008) Mechanical resonance enhances the sensitivity of the vibrissa sensory system to near-threshold stimuli. Brain Res 1235:74–81

Behn C (2013) Mathematical modeling and control of biologically inspired uncertain motion systems with adaptive features. Habilitation thesis, TU Ilmenau

Berg R, Kleinfeld D (2003) Rhythmic whisking by rat: retraction as well as protraction of the vibrissae is under active muscular control. J Neurophysiol 89:104–117
Bogolyubov N, Mitropoliskii Y (1961) Asymptotic methods in the theory of nonlinear oscillations. Gordon and Breach Science Publishers, New York
Dehnhardt G, Kaminski A (1995) Sensitivity of the mystacial vibrissae of harbour seals (Phoca vitulina) for size differences of actively touched objects. J Exp Biol 198:2317–2323
Dörfl J (1982) The musculature of the mystacial vibrissae of the white mouse. J Anat 135:147–154
Ebara S, Kumamoto K, Matsuura T et al (2002) Similarities and differences in the innervation of mystacial vibrissal follicle-sinus complexes in the rat and cat: a confocal microscopic study. J Comp Neurol 449:103–119
Fend M, Bovet S, Hafner VV (2004) The artificial mouse—a robot with whiskers and vision. In: Proceedings of the 35th international symposium on robotics, Paris, 23–26 Mar 2004
Jadhav SP, Feldman DE (2010) Texture coding in the whisker system. Curr Opin Neurobiol 20 (3):313–318
Kantorovich L, Krylov V (1958) Approximate methods of higher analysis. P. Noordhoff, Groningen
Landau L, Lifshitz E (1969) Mechanics. Course of theoretical physics. Pergamon Press, Oxford
McLachlan NW (1947) Theory and application of Mathieu functions. Clarendon Press, Oxford
Mitchinson B, Grant RA, Arkley K et al (2011) Active vibrissal sensing in rodents and marsupials. Phil Trans R Soc B 366:3037–3048
Neimark M, Andermann M, Hopfeld J et al (2003) Vibrissa resonance as a transduction mechanism for tactile encoding. J Neurosci 23:6499–6509
Niederschuh S, Witte H, Schmidt M (2014) The role of vibrissal sensing in forelimb position control during travelling locomotion in the rat (Rattus norvegicus, Rodentia). J Zool 118 (1):51–62
Pearson MJ, Mitchinson B, Sullivan JC et al (2011) Biomimetic vibrissal sensing for robots. Phil Trans R Soc B 366:3085–3096
Scholz G, Rahn C (2008) Profile sensing with an actuated whisker. IEEE T Robot Autom 20:124–127
Solomon JH, Hartmann MJ (2006) Biomechanics: robotic whiskers used to sense features. Nature 443(7111):525
Svetlitsky V (2005) Dynamics of rods. Springer, Berlin
Valdivia y Alvarado P, Subramaniam V, Triantafyllou M (2012) Design of a bio-inspired whisker sensor for underwater applications. In: IEEE sensors 2012, Taipei, 28–31 Oct 2012, pp 92
Vincent S (1912) The function of vibrissae in the behavior of the white rat. Behav Monogr 1:1–81
Voges D, Carl K, Klauer G et al (2012) Structural characterization of the whisker system of the rat. IEEE Sens 12(2):332–339
Volkova T, Zeidis I, Naletova VA et al (2016a) The dynamical behavior of a spherical pendulum in a ferrofluid volume influenced by a magnetic force. Arch Appl Mech (First online)
Volkova T, Zeidis I, Witte H et al (2016b) Analysis of the vibrissa parametric resonance causing a signal amplification during whisking behaviour. J Bionic Eng 13:312–323
Will C, Steigenberger J, Behn C (2016) Bio-inspired technical vibrissae for quasi-static profile scanning. Lect Notes Electr Eng (LNEE) 370:277–295
Wolfe J, Hill DN, Pahlavan S et al (2008) Texture coding in the rat whisker system: slip-stick versus differential resonance. PLoS Biol 6:e215
Yan W, Kan Q, Kergrene K et al (2013) A truncated conical beam model for analysis of the vibration of rat whiskers. J Biomech 46:1987–1995
Zentner L (2014) Nachgiebige Mechanismen. De Gruyter Oldenbourg, München
Zimmer U (1995) Self-localization in dynamic environments. In: IEEE/SOFT international workshop BIES'95, Tokio, 30–31 May 1995, p 8

Towards the Development of Tactile Sensors for Determination of Static Friction Coefficient to Surfaces

M. Scharff, M. Darnieder, J. Steigenberger and C. Behn

Abstract Natural vibrissae fulfill a lot of functions. Next to object distance detection and object shape recognition, the surface texture can be determined. Inspired by the natural process of surface texture detection, the goal is to adapt this feature by technical concepts. Modeling the vibrissa as an Euler–Bernoulli bending beam with a quasi-statically moving support and the vibrissa–surface contact with respect to Coulomb's Law of Friction, a first approach was formed by the group of Behn and Steigenberger. Due to the motion of the support (pushing the vibrissa) and the surface contact, the vibrissa gets deformed. Firstly, the beam tip is sticking to the surface. The acting friction force prevents a movement of the beam tip until the maximal stiction is reached. The displacement of the support corresponds to changes in the acting forces and moments. Out of these changes the coefficient of static friction can be determined. The analytical results of Steigenberger and Behn are verified and validated by numerical simulations and an experiment.

Keywords Surface detection · Vibrissae · Friction · Mechanical contact · Beam

1 Introduction

Rodents scan the surrounding environment with different sensory organs. Besides of eyes and ears they use tactile hairs, named *vibrissae*. Vibrissae exist, e.g., at the paws (*carpal vibrissae*) and in the snout region (*mystacial vibrissae*) (Helbig et al. 2014).

In order to simplify the natural sophisticated example of a vibrissa and related periphery from a mechanical point of view, usually the following assumptions are

M. Scharff (✉) · M. Darnieder · C. Behn
Technical Mechanics Group, Department of Mechanical Engineering,
Technische Universität Ilmenau, Ilmenau, Germany
e-mail: moritz.scharff@tu-ilmenau.de

J. Steigenberger
Faculty of Mathematics and Natural Sciences, Institute of Mathematics,
Technische Universität Ilmenau, Ilmenau, Germany

L. Zentner et al. (eds.), *Microactuators and Micromechanisms*,
Mechanisms and Machine Science 45, DOI 10.1007/978-3-319-45387-3_4

done. Each vibrissa is held by a viscoelastic support, which is maybe controllable. A vibrissa has a tapered shape, whereby the diameter of the cross-section is decreasing from base to tip. The diameter of the cross section is much smaller than the length of a vibrissa. Along their whole length, vibrissae are naturally pre-curved (stress free).

From the inside to the outside of the vibrissa there are three layers made out of different materials. The thickness of each layer changes along the length of the vibrissa (Voges et al. 2012).

The complex morphology offers a wide functionality of the vibrissa. They are used for distance evaluation, object shape recognition and surface texture detection (Carl et al. 2011).

2 State of the Art

The natural process of surface texture detection is not completely understood yet. It is observed that, e.g., a rat follows a special control strategy. If the vibrissae get into contact with a surface the rat will position its head in the way that a maximum quantity of vibrissae are in touch while they are minimally deformed, (Mitchinson et al. 2007; Grant et al. 2009). Out of this position, the rat starts whisking. Three to five whisk-cycles are needed to collect the information (Carvell and Simons 1990).

For the procedure of the transduction of the mechanical stimuli into a signal and finally into a property of the surface, there are several hypotheses. The authors of Moore and Andermann (2005) formulate the *resonance hypothesis*. They assume that there is a relation between the frequency of the vibrissa and the surface roughness inducing vibration. Alike to the resonance hypothesis are the works (Arabzadeh et al. 2005; Hipp et al. 2006). They are based on the recording of pattern of vibrissa motion events under different velocities, which represent the surface texture. The authors of Wolfe et al. (2008) analyze another concept. Based on the *Stick-Slip-Effect*, the frequency of stick-slip events and the distance between the sticking periods are related to the surface roughness. A high frequency of stick-slip events and a large distance between the sticking periods indicate a high roughness of the surface.

Under aspects of applied mechanics, these three theories need further analytical treatment. A first approach that gives a complete, analytical description is done in Steigenberger et al. (2015), which forms the basis of this work.

3 Aim and Scope

The surface texture detection is a critical point in literature. Known hypotheses were studied and proofed empirically, but there is still a lack of knowledge as well. Adding new aspects to the present works, a discussion of the mechanical theory behind the hypotheses—described above—has to be done. Using the theory of Steigenberger et al. (2015), a *first* approach to a sensor concept is discussed, where we try to determine the static coefficient of friction μ_0 at first.

4 Modeling

Using the *Euler–Bernoulli beam theory* with respect to large deflections, the single vibrissa is modeled as a straight beam with a cylindrical cross-section. In a very first step, the clamping replaces the viscoelastic support. The resulting frictional force acts at the tip of the beam. The load acts only in the x-y-plane and is considering *Coulomb's Law of Friction* with respect to stiction. With a view to a technical sensor, the beam consists of *Hooke material.*

To find generally valid statements about the discussed theory, a nondimensionalization takes place with respect to the following basic parameters:

$$\text{units:}\quad [\text{length}] = L\ ;\ [\text{force}] = EI_zL^{-2}\ ;\ [\text{moment}] = EI_zL^{-1}$$

whereby L is the length, E is the Young's Modulus, I_z is the second moment of area. For example: if we consider a steel beam with the basic parameters $E = 2.10 \cdot 10^5\,\text{MPa}$, $I_z = 250\,\text{mm}^4$, $L = 100\,\text{mm}$, than a value $f = 1.33$ during dimensionless calculation results in the following real force:

$$F = f \cdot [\text{force}] = f \cdot \frac{EI_z}{L^2} = 7000\,\text{N}$$

The system also gets normalized by assuming the arc length coordinate s in the interval:

$$s \in [0\,,\,1]$$

The idea behind the model out of Steigenberger et al. (2015) is sketched in Fig. 1. The beam is deformed and the tip is touching a surface at the point (x_1, y_1), see Fig. 1a. The parameter η describes the distance between the clamping and the assumed horizontal surface. Initializing a displacement (quasi-static) of the foot point (clamping) in x-direction, the beam gets further deformed. If the static friction coefficient μ_0 (respectively the angle of static friction α_0) of the surface combination is large enough, there is an *inflection point*, see Fig. 1b. An inflection point marks a sign change of the curvature of the deformed shape of the beam. The sign change is located at the point (x_b, y_b).

The dash-dotted line means the following:

- Passing right to the clamping there exist no inflection point;
- Passing through the clamping means a zero clamping moment;

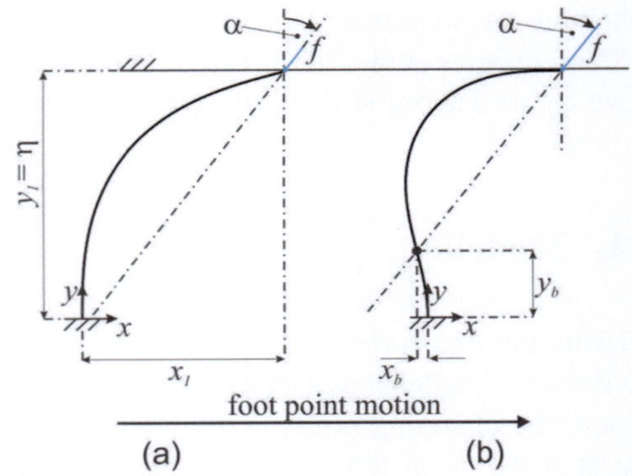

Fig. 1 Relations and definitions of the system states, adapted from Steigenberger et al. (2015)

- Passing left to the clamping results in an inflection point, whereby we have a zero bending moment at the intersection point to the beam.

To study the relations between the present μ_0 and the resulting reaction forces and moments in the clamping, the equilibrium state conditions are formulated in the following way:

$$\mathbf{M}_{bz}(s) = \mathbf{r} \times \mathbf{f} = f\,(\cos(\alpha)\,(x(s) - x_1) + \sin(\alpha)\,(y(s) - y_1))\,\vec{e}_z \tag{1}$$

According to the parametric representation the system of equations yields to:

$$\left.\begin{aligned} x'(s) &= \cos(\varphi(s)) \\ y'(s) &= \sin(\varphi(s)) \\ \varphi'(s) &= f\,(\cos(\alpha)\,(x(s) - x_1) + \sin(\alpha)\,(y(s) - y_1)) \end{aligned}\right\} \tag{2}$$

with boundary conditions:

$$\left.\begin{aligned} x(0) &= 0 \quad ; \quad x(1) = x_1 \\ y(0) &= 0 \quad ; \quad y(1) = y_1 \\ \varphi(0) &= \tfrac{\pi}{2} \quad ; \quad \varphi(1) = \phi_1 \end{aligned}\right\} \tag{3}$$

The function φ is the slope of the deformed shape of the beam. The system of Eq. (2) and the corresponding boundary conditions (3) form a free boundary value problem (BVP) with two unknown quantities.

Remark Using Matlab R2016a, the system of Eqs. (2) and (3) is solved in applying a *Shooting-Method*: starting from a guess for the unknown quantities, the values of the unknown quantities are optimized by performing the 2D-optimizaion function *fminsearch()*. In every step of the optimization, the system of equations is solved by the *Runge-Kutta-Method of 4th order*.

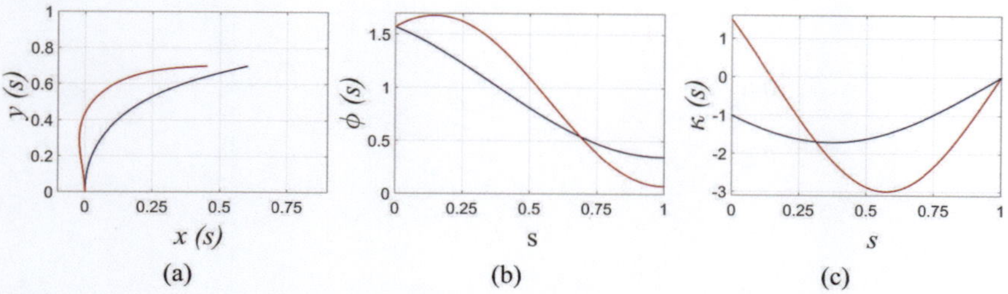

Fig. 2 Relations and definitions of the two system states, adapted from Steigenberger et al. (2015). **a** The deformed shapes of both states in dependence on cartesian coordinates. The related slope and curvature trends are plotted in (**b**) and (**c**)

As already mentioned, the inflection point occurs when there is a sign change in the curvature, see Fig. 2. The blue line corresponds to a solution without an inflection point and the red line to one with one.

The trends of the functions for the curvature show the assumed behavior. There is no sign change of $\kappa(s)$ for the state without a inflection point. On the contrary, a sign change of $\kappa(s)$, from positive to negative, belongs to a state with an inflection point.

5 Simulation and Experiment

The system of Eq. (2) with its boundary conditions (3) is solved for a sequence of $x_1 = -0.2\ (0.05)\ 0.6$ and a fixed $\eta = 0.75$, see Fig. 3.

Remark From now on, the representation of the results reverses reality: fixed support and moving contact point (tip of the beam).

There are two shapes without and fifteen with an inflection point. The appearance of an inflection point corresponds to smaller x_1. From the left to right, α is negative and decreasing until the first shape without inflection point occurs, see Fig. 4a. For this shape α takes larger values. This trend also goes on for the last shape. In this configuration α becomes positive. That means the complete force starts pulling the beam, with view to real scenario this case is out of interest.

Figure 4b indicates that there are two solutions for the same φ_1. So, φ_1 cannot be used to identify any surface properties because the result is not unique. Figure 5 shows the deformed shapes for a sequence of $y_1 = 0.4\ (0.025)\ 0.875$ and a fixed $x_1 = 0.45$. From top to bottom, there are three solutions without and seventeen with an inflection point. At $y_1 = 0.875$, α and the acting forces are positive again, see Fig. 6a, c. This is equal to the previous finding; in this configuration the beam gets pulled.

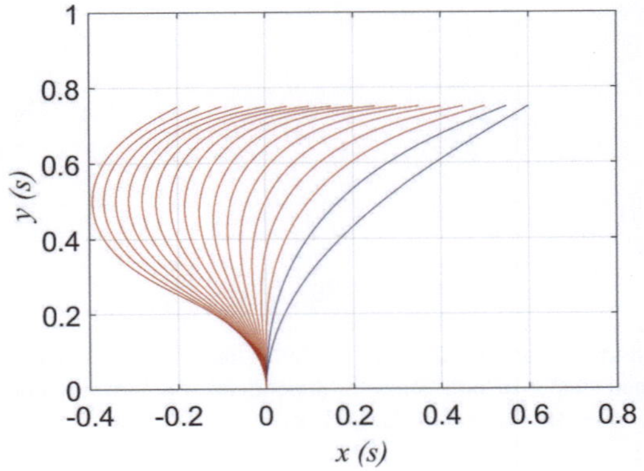

Fig. 3 In contrary to the two *blue* shapes, the fifteen *red* ones show an inflection point. The inflection point occurs in dependence on the tip coordinate x_1

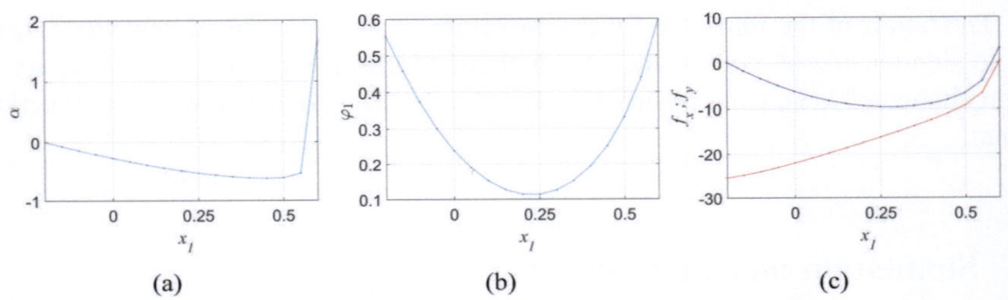

Fig. 4 **a** The static friction angle α versus x_1; **b** Relation between the slope in the beam tip φ_1 and x_1; **c** The trend of the acting forces f_x (*blue*) and f_y (*red*) in dependence on x_1

Fig. 5 For a sequence of y_1 and a fixed x_1 twenty solutions are calculated. There are three solutions without and seventeen ones with inflection point. There are negative values for φ_1 when y_1 is small

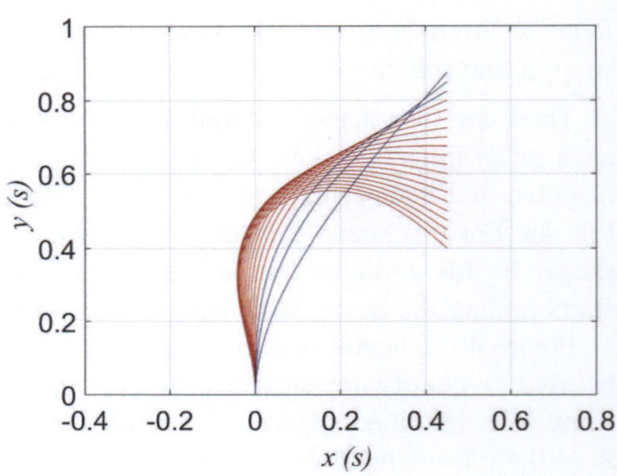

There are solutions that belong to negative values of φ_1. These solutions show values for $y(s) \geq y_1$. In reality, this is impossible. If $\eta \leq y(s)$, the beam has to penetrate the surface. These solutions are correctly computed but should be omitted, because $\varphi_1 < 0$.

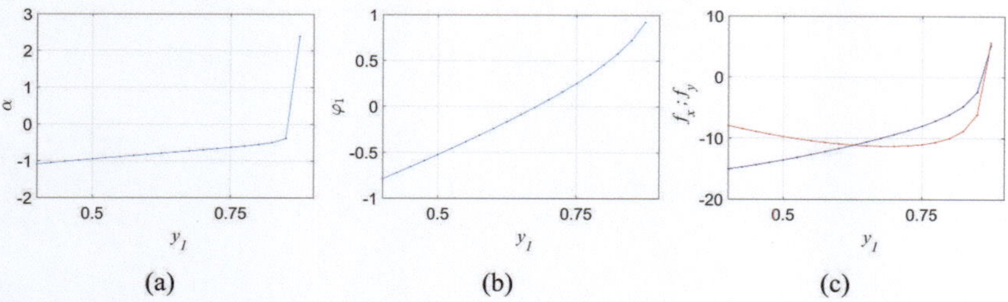

Fig. 6 **a** The static friction angle α versus y_1; **b** Relation between the slope in the beam tip φ_1 and y_1; **c** The trend of the acting forces f_x (*blue*) and f_y (*red*) in dependence on y_1

While y_1 is increasing, so φ_1 does. The slope $\varphi_1 = \frac{\pi}{2}$ when $y_1 = L$ and $x_1 = 0$. Combing the findings of the Figs. 3, 4, 5 and 6, it follows that the appearance of an inflection point depends on, e.g., the coordinate of the beam tip (respectively the value of α). This means, for every η, there is a μ that belongs to a shape with an inflection point. The relation is shown in Fig. 7. For each value of a sequence of $\eta = 0.05\,(0.05)\,0.85$ solutions for a sequence of $x_1 = -0.02\,(0.01)\,0.6$ are computed. Out of these solutions for each value of η, the last one without and the first one with an inflection point are collected. The exact limit for the appearance of an inflection point is located between these two solutions. Figure 7 indicates that for large values of η smaller values of μ are required to get an inflection point.

To validate the numerical simulations an experiment is performed. Using a stepping motor as a drive and a steel band as a vibrissa/sensor, the deformations of the structure are analyzed while it is touching a surface. Using an image tracking system, the collected information are compared to the simulation, see Fig. 8. There

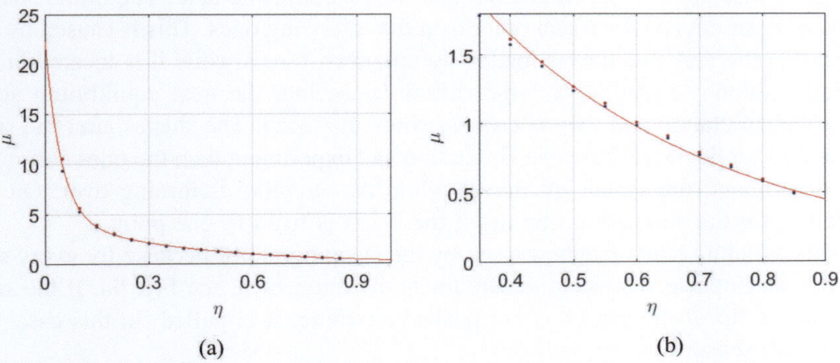

Fig. 7 Trend of the appearance of an inflection point. The *blue points* are the last solution without and the *red points* the first solution with inflection point. The *red, solid line* illustrates a fitted exponential function due to the mean values of the point pairs. The figure part (**b**) shows a zoom on the area of interest

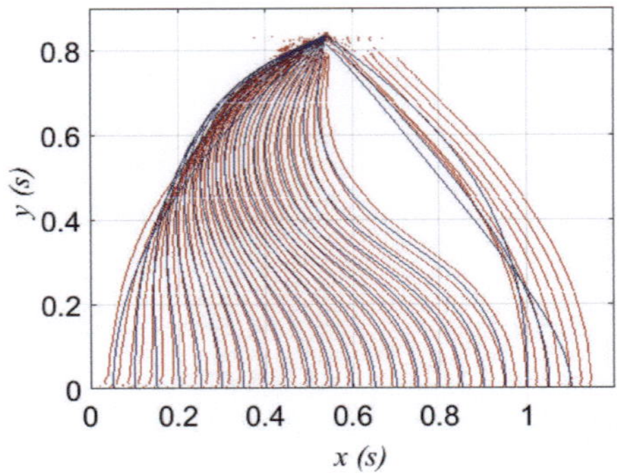

Fig. 8 The simulated shapes (*blue*) are compared to the shapes out of measured data (*red*)

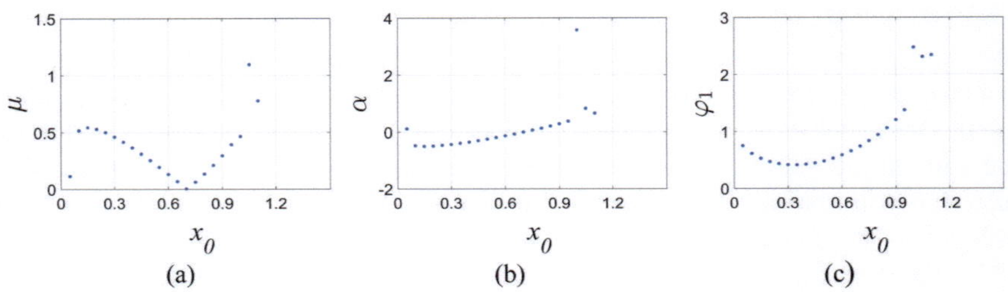

Fig. 9 **a** Static friction coefficient μ versus foot point coordinate x_0. **b** In addition to (**a**), instead of μ, α is plotted. **c** Trend of φ_1 over x_0

is a good matching of the simulation and the experimental data. The initial shapes show a larger deviation to each other than the following ones. This is caused by the effect that the real structure is loaded by unknown forces while it is located in the initial position. Around $x_1 \approx 1$ the sensor *jumps* into the next equilibrium state. Both the simulation and the experiment show this *snap*. The shapes after the snap show a larger deviation between simulation and experiment than the ones before. In the experiment, the sensor tip moves while the structure is turning over. On the contrary, for the simulation, the tip of the beam is fixed to one point.

This dynamic effect maybe caused by the stepping motor because for every step there is an impulse. The snap occurs for high values of μ, see Fig. 9a. If the snap occurs, the tip of the sensor is not pushed anymore, it is pulled. In this case, the mentioned sensor concept will fail.

In this experiment, the snap probably occurs when $\varphi_1 = \frac{\pi}{2}$, see Figs. 8 and 9c. The trend of φ_1 in Fig. 9c indicates a similar behavior than the one in Fig. 4b. In both cases there is a parabola-like trend.

6 Conclusions

The analytical work in Steigenberger et al. (2015) was verified by numerical simulations and validated by an experiment. It is possible to determine the coefficient of static friction μ_0 out of the deformation of the sensor.

$$\tan(\alpha) = \frac{F_T}{F_N} = \mu \leq \mu_0 = \tan(\alpha_0) \tag{4}$$

Measuring the reaction forces to the tangential and normal force F_T, F_N at the sensor support, μ can be determined by using (4). In the course of further moving of the support μ increases until μ_0 is reached. For $\mu_0 \leq \mu$ the beam tip snaps off. So, the last measured values of the reaction forces before the snap off corresponds to μ_0.

The simulations show, if the distance η between vibrissa base and contact surface is small, there are no solutions of practical interest. This leads immediately to the claim in formulating a stringent restriction of the deformations $(y(s) \leq \eta,\ \forall s \in (0,1))$. Both, simulation and experiment show that inflection points and a snap occur.

While the snap leads to a fail of the sensor, the inflection point can be used as an input signal for the sensor. When the inflection point occurs, the curvature $\kappa(0)$ at the vibrissa base respectively the bending moment $M_{bz}(0)$ are equal to zero. This configuration can be easily detected by a torque sensor. Out of the information that there is an inflection point the belonging μ_0 can be determined. The inflection point configuration for any surface can be found by controlling η and sweeping the sensor tip over the surface until $M_{bz}(0) = 0$. Then the value of μ_0 follows out of the relation of Fig. 7 in combination with the known η.

This work is restricted due to various simplifications and is continued under various aspects. The influence of the mentioned real morphology of a vibrissa for a technical sensor has to be analyzed. To study a more realistic scenario the boundary conditions have to be modified, in the way that they represent a contact plane and not only a contact point. Inspired by the findings out of literature, the analytical considerations are enhanced to dynamics, to consider effects like stick-slip. In the course of this, the used mechanical model will be optimized under aspects of tribology.

References

Arabzadeh E, Zorzin E, Diamond ME (2005) Neuronal encoding of texture in the whisker sensory pathway. PLoS Biol 3(1):1–11

Carl K, Hild W, Mämpel J, Schilling C, Uhlig R, Witte H (2011) Characterization of statical properties of rat's whisker system. IEEE Sens J 12(2):340–349

Carvell GE, Simons DJ (1990) Biometric analyses of vibrissal tactile discrimination in the rat. J Neurosci 10(8):2638–2648

Grant R, Mitchinson B, Fox C, Prescott T (2009) Active touch sensing in the rat: anticipatory and regulatory control of whisker movements during surface exploration. J Neurophysiol 101 (2):862–874

Helbig T, Voges D, Niederschuh S, Schmidt M, Witte H (2014) Characterizing the substrate contact of carpal vibrissae of rats during locomotion. In: Proceedings of the 3th international conference, living machines, Milan, Italy, 30 July–1 Aug 2014, 3 pages

Hipp JE, Arabzadeh E, Zorzin E, Conradt J, Kayser C, Diamond ME, König P (2006) Texture signals in whisker vibrations. J Neurophysiol 95(3):1792–1799

Mitchinson B, Martin C, Grant R, Prescott T (2007) Feedback control in active sensing: rat exploratory whisking is modulated by environmental contact. Proc R Soc B 274:1035–1041

Moore C, Andermann M (2005) The vibrissa resonance hypothesis. In: Somatosensory plasticity. CRC Press, Boca Raton, pp 21–60

Steigenberger J, Behn C, Will C (2015) Mathematical model of vibrissae for surface texture detection. Preprint No. M 13/03. Technische Universität Ilmenau. 16 pages

Voges D, Carl K, Klauer G, Uhlig R, Schilling C, Behn C, Witte H (2012) Structural characterization of the whisker system. IEEE Sens J 12(2):332–339

Wolfe J, Hill DN, Pahlavan S, Drew PJ, Kleinfeld D, Feldman DE (2008) Texture coding in the rat whisker system: slip-stick versus differential resonance. PLoS Biol 6(8):1–17

Development and Investigation of Photoelastic Sensor for Torque Measurement

A. Bojtos, N. Szakály and A. Huba

Abstract This article describes the development and the calibration of the photoelastic contactless torque sensor and it also presents the measurement of the starting torque of a disk micro motor, as a function of the rotor angle. Finite element analysis was performed to prove the functionality of the sensor and to determine its required sizes. In order to define the torque—rotor angle characteristics of the micro motor, electromagnetic simulation was also performed and the results were compared to each other.

Keywords Photoelastic · Optical · Torque sensor · FEM analysis · Silicone elastomer

1 Introduction

The aim of our investigation was to develop an optical torque sensor for the measurement of starting torque depending on the start angle position of low power disc-shape micro motor. Because of the very low torque level of the motor (<10 mNm) the search for suitable measurement principle—which ensures the demanded accuracy at low cost—was not easy. The aim of our research additionally to that was to analyse the characteristic and the parameter of the developed optical torque sensor for further special applications.

A. Bojtos (✉) · N. Szakály · A. Huba
Department of Mechatronics Optics and Mechanical Engineering Informatics,
Budapest University of Technology and Economics, Budapest, Hungary
e-mail: bojtos@mogi.bme.hu

N. Szakály
e-mail: szakaly@mogi.bme.hu

A. Huba
e-mail: huba@mogi.bme.hu

L. Zentner et al. (eds.), *Microactuators and Micromechanisms*,
Mechanisms and Machine Science 45, DOI 10.1007/978-3-319-45387-3_5

There are several technical solutions for contactless optical measurement of low level torque based on the changing of light intensity or the phase changing of light wave occurred on a special deformation body (Zi et al. 2015; Chung et al. 1998; Horváth and Nagy 2009). The solution presented in this paper uses the changing of the transmitted light intensity through a photoelastic deformation body.

1.1 Working Principle of Photoelastic Torque Sensors

The measurement principle based on the birefringence of light occurred by the deformation. Photoelastic sensors published in the literature, can be operate with the reflection or transmission of the light. The measurement arrangement can be axial or radial (Hajdu and Horváth 2012). The deformation body of the radially arranged reflection torque sensor was a tube made from photoelastic material. The shaft surface inside of the tube reflected the polarized light similar to optical stress analysis using layer coating (Fig. 1a) (Hajdu and Horváth 2012). The photoelastic strain gauge fixed on the surface of the shaft in 45° angle to the shaft axis worked with similar principle. The arrangement was similar to the conventional torque transducers with strain gauges (Sakai 1993; Puente and Jessop 2002). In the axially arranged transmission torque sensor the ends of the shafts were coupled by a photoelastic disc and by a transparent disc without birefringence (Fig. 1c), (Hajdu and Horváth 2012). The sensor implemented a hybrid arrangement of radial reflection and axial transmission effects using the total reflection. The light reflects on the inner surface of the material with birefringence(Fig. 1b), (Chung et al. 1998).

Based on the special construction of the proposed silicone rubber sensor, the transmitting arrangement could be applied in radial direction (Fig. 4a–b). This allowed dispersions free light transmission and the application of optimal optical length. A further advantage of the proposed silicone rubber sensor material is that it can to function as an integrated sensor in a flexible structure (Chaykina et al. 2012;

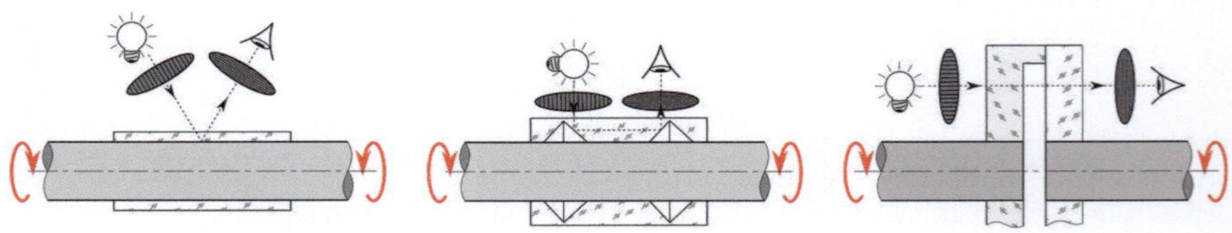

a.) b.) c.)

Fig. 1 Photoelastic measurement principles of torque. **a** Reflection. **b** Total reflection. **c** Transmission

Issa and Zentner 2012). For example, the proposed hyperelastic torque sensor can function as a flexible coupling when using the appropriate material and geometry.

1.2 Theoretical Background

The basis of the phenomena is the birefringence that occurs by deformation or stress of transparent materials with amorphous molecule structure. The equation above describes the intensity of the light which is transmitted though the material with birefringence (situated between parallel polarizers), (Thamm et al. 1968):

$$I \approx I_0 \cdot \left(1 - \sin^2(2\alpha) \cdot \sin^2(\pi \cdot m)\right) \quad (1)$$

where: I_0 is the light intensity leaving the polarizer, α is the angle between polarizer position and the principal axes of the strain (45°), m—fringe order (as degree of the retardation) (Fig. 2).

The Eq. (1) shows that the light intensity depends on the fringe order m and on the angle α between the principal stress axes and the polarizer position (Ábrahám 1998; Thamm et al. 1968). For silicone elastomers in view of their wide deformation range and their nonlinear characteristic it is reasonable to give the value of the m by the true or logarithmic strain (Thamm et al. 1968).

$$m = \frac{l \cdot D}{\lambda_0} \cdot (\varepsilon_{t1} - \varepsilon_{t2}) = \frac{l \cdot D}{\lambda_0} \cdot \ln\left(\frac{1 + \varepsilon_1}{1 + \varepsilon_2}\right) \quad (2)$$

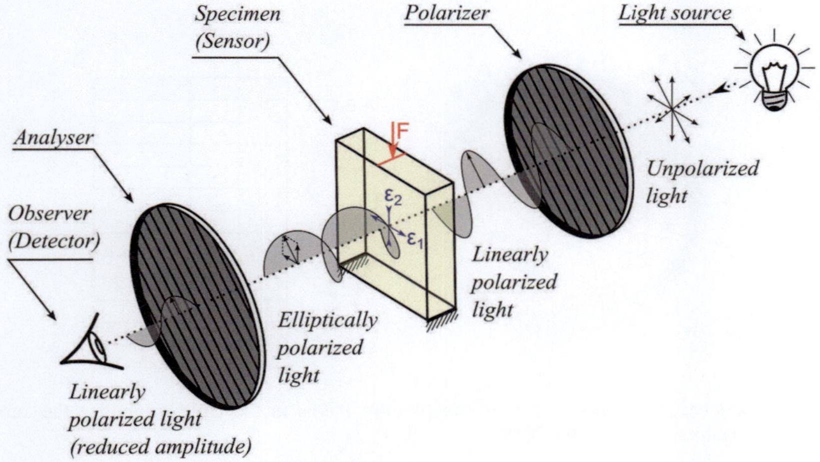

Fig. 2 Theoretical arrangement of the optical stress analysis

where: l—thickness of the body with birefringence (12 mm), λ_0—wavelength of the applied light in vacuum (635 nm), D—strain-optic coefficient, ε_i—engineering strain in the principal axes, $\varepsilon_{ti} = \ln\lambda_i = \ln(1 + \varepsilon_i)$ true or logarithmic strain in principal axes. Using the Eqs. (1) and (2) the intensity of light transmitted through the photoelastic materials with birefringence (by parallel positioned polarizers) can be determined by the following equation (Bojtos and Huba 2011):

$$I \approx 1 - \sin^2(2\alpha) \cdot \sin^2\left(\pi \cdot \frac{l \cdot D}{\lambda_0} \cdot \ln\left(\frac{1+\varepsilon_1}{1+\varepsilon_2}\right)\right) \tag{3}$$

2 FEM Analysis of the Motor

Subject of the investigations was the disc shaped DC micromotor type *T2607*. The stator consists of a four-segment permanent magnet besides the iron parts of the magnetic circuit. The rotor consists of three kidney shaped coils and the commutator. The coils are in star-connection with the commutator which allows the operation of two coils at the same time.

For a full turning it needs 12 commutation (per 30° angle). The motor torque depends on the angle position of the rotor since the magnetic flux changes during the rotation of the coils. Maximum torque can be detected in commutation angles.

The numeric analysis of this complex electromagnetic system was performed using Ansys FEM program. During the modelling the relative angle position of the rotor was changed (1° steps in the angle range of 0°–120°) and the simulations resulted the theoretical torque changing characteristic for the commutation process (Fig. 3), (Halmai and Szakály 2010).

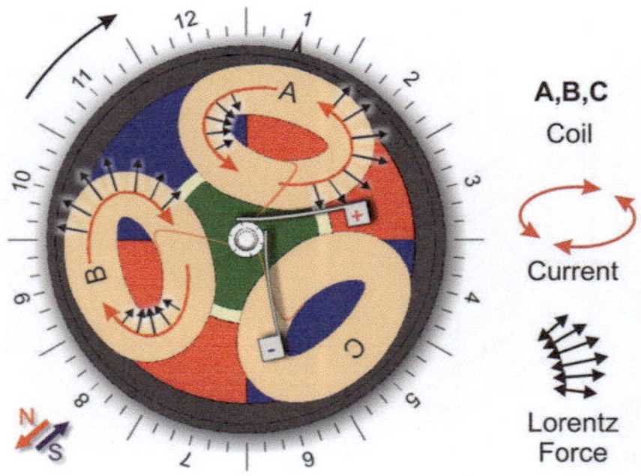

Nr.	A	B	C
1	+	-	
2		-	+
3	-		+
4	-	+	
5		+	-
6	+		-
7	+	-	
8		-	+
9	-		+
10	-	+	
11		+	-
12	+		-

Fig. 3 Commutation process of the DC micromotor (signs in the table symbolize the current directions), (Halmai and Szakály 2010)

3 FEM Analysis of the Sensor

The aim of the finite element analysis was to prove the functionality of the sensor and to determine the required sizes. The finite element analysis was performed using the ANSYS Workbench 14. Mooney-Rivlin material model was used for modelling the structure because of the non-linear behavior of silicone rubber sensor body. The material model parameters were determined by uniaxial and biaxial tensile and uniaxial compression tests.

In the FEM analysis, supports and torque were applied to the model, similar to the real situation during the operation (Fig. 4a–b). Using the above equations the characteristic of the optical transmittance of the sensor can be calculated. The sensitivity depends on the principal strains (ε_{t1}, ε_{t2}), perpendicular to the LASER beam. Therefore the relationship between light intensity and torque, and the desired measurement range of the sensor can be calculated in advance. The torque maximum will be 50 % higher than the nominal torque, namely 15 mNm. The output quantity of the sensor was represented by the difference of its principal strains ($\Delta\varepsilon_t = \varepsilon_{t1} - \varepsilon_{t2}$) which were approximately perpendicular to the light beam. As results of the simulations the deformations along the length axis of the sensor were determined by steps *dl*.

The Eq. (3) was usually applied in optical stress analysis as due to the applied load affect plain strain state occurs which does not change along the thickness of the specimen. In our case the strain state ($\Delta\varepsilon_t$) changed along the length *l* of the sensor (Fig. 4c). As a result Eq. (3) is only valid for sufficiently small segment *dl* of the

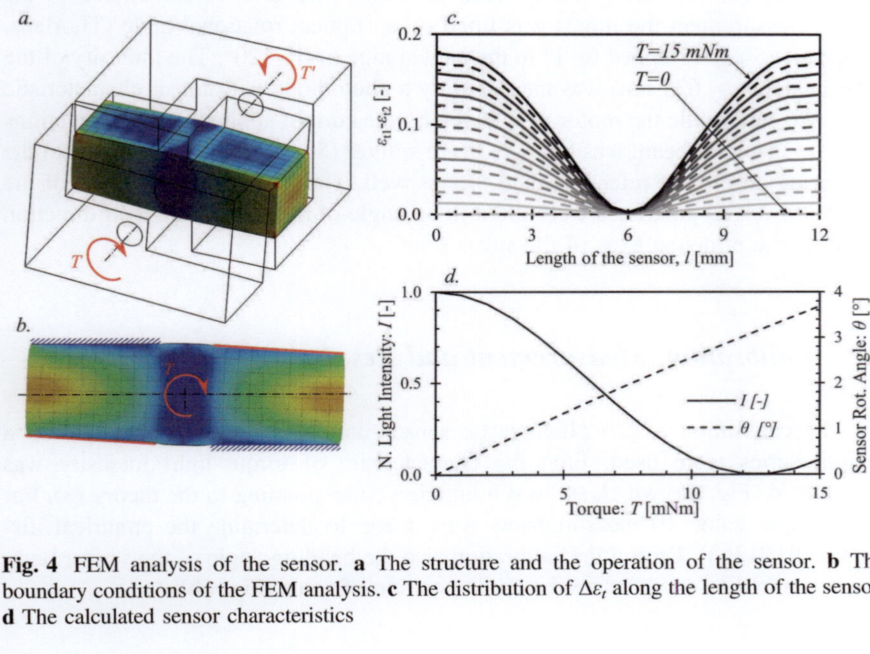

Fig. 4 FEM analysis of the sensor. **a** The structure and the operation of the sensor. **b** The boundary conditions of the FEM analysis. **c** The distribution of $\Delta\varepsilon_t$ along the length of the sensor. **d** The calculated sensor characteristics

full length l. Therefore, the intensity of light transmission through the whole sensor can be calculated according to the following equation:

$$I \approx 1 - \sin^2(2\alpha) \cdot \sin^2\left(\frac{\pi \cdot D}{\lambda_0} \cdot \int_0^l \Delta\varepsilon_t dl\right) \quad (4)$$

The value of integral was calculated numerically with the value $\Delta\varepsilon_t$ resulting from FEM analysis along the light axis.

$$\int_0^l \Delta\varepsilon_t dl \approx \sum_{i=1}^{N} (b_i - a_i)\frac{\Delta\varepsilon_t(b_i) - \Delta\varepsilon_t(a_i)}{2} \quad (5)$$

where: b_i—the upper limit of interval i, a_i—the lower limit of interval i.

Theoretically for each input value of $\Delta\varepsilon_t$ the following characteristic of light intensity and torque could be obtained (Fig. 4d).

4 Measurement of Starting Torque

4.1 *Measurement Arrangement*

The conceptual measuring arrangement is shown in Fig. 5. In the course of the torque measurement the motor was fitted on an optical rotational table (Thorlabs, resolution: 5′) and stepped by 1° in the angle range of 0°–120°. The intensity of the laser beam (λ_0 = 635 nm) was measured by a photodiode with linear characteristic (350–820 nm) while the motor was driven by the current applied in the simulations (0.8 A). The laser beam was led by a beam splitter (50 %) in both directions to the measuring and to the reference detectors as well. The plane of polarization of the parallel arranged polarizers were fitted in an angle of α = 45° to the load direction meaning the principal axes of the stress state.

4.2 *Calibration, Measurement and Results*

For the calibration of the photoelastic sensor dark and bright references and a weight series were used. First the characteristic of torque-light intensity was determined (Fig. 6a) which showed a function corresponding to the theory (3). For each torque value 10 measurements were made to determine the empirical dispersion (0.013[-]). During the calibration also the bending angle of the sensor body was measured to compensate the angle position of rotor (Fig. 6b).

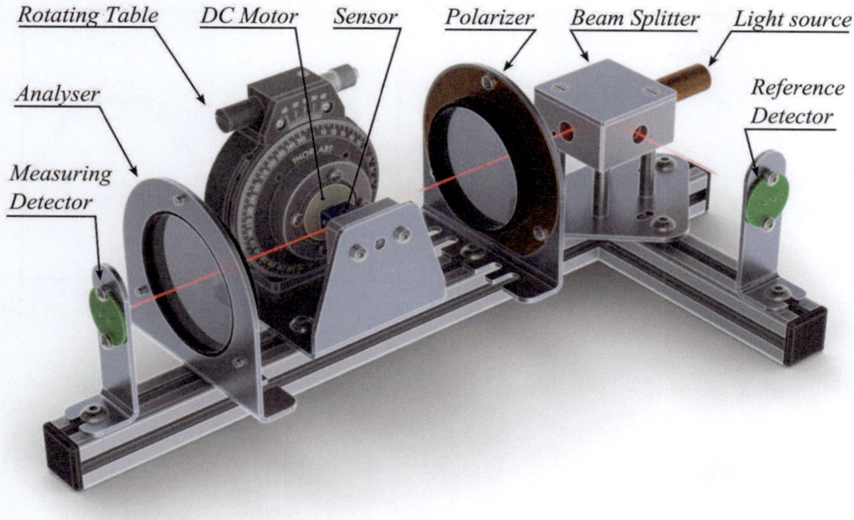

Fig. 5 Measurement arrangement

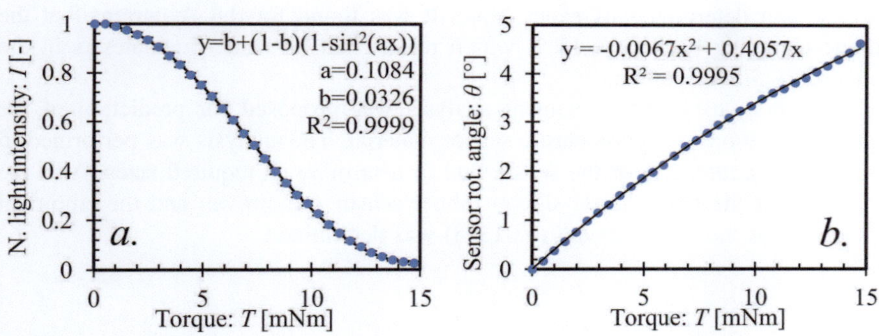

Fig. 6 Calibration characteristics of the photoelastic torque sensor. **a** Normalised light intensity—torque characteristic. **b** Sensor rotation angle—torque characteristic. The coefficients of determination (R^2) indicate the goodness of fit

The results of the measurement and the FEM simulations are shown in Fig. 7 where the starting motor torque can be followed depending on the relative angle position between stator and rotor.

The correspondence of the FEM simulations and the results of the measurement shows the fluctuation of the starting torque between ~5.5 and 6.5 mNm (15 %) and squarely the separate commutation cycles per 30° rotational angle.

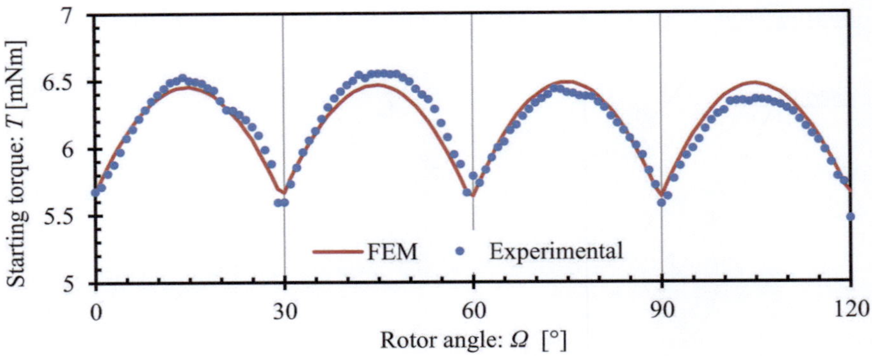

Fig. 7 Starting torque of disc shape micro motor versus angle position of the rotor

5 Conclusions

Photoelastic torque sensor was developed and calibrated for measurement of the starting torque of a disk micro motor, as a function of the rotor angle. Measurement showed the torque pulsation of a disc shape micro motor during the commutation process and determined its exact value. It was found for 0.8 A current that the torque changing reaches $\sim 15\ \%$ which means that its value fluctuates is in the range of ~ 5.5–6.5 mNm.

A method, using finite element analysis was proposed for prediction of the optical operation of a photoelastic sensor material. The analysis was performed to prove the functionality of the sensor and to determine its required sizes. Also the eligibility of the photoelastic silicon rubber sensor was proven and the empirical dispersion of the measurement (0.013[-]) was determined.

References

Ábrahám G (1998) Optics (in Hungarian: Optika). Panem Press, Budapest

Bojtos A, Huba A (2011) Application of transparent silicone rubber for force and strain sensor development. 28th Danubia-Adria-symposium on advances in experimental mechanics. Siófok, Hungary, pp 219–220

Chaykina A, Griebel S, Zentner L (2012) Richtungsabhäniger Berührungssensor zur Sensorisierung von Nachgiebigen Mechanismen. In: Mechanismentechnik in Ilmenau, Budapest und Niš. Deutschland, Ilmenau, pp 79–90

Chung D, Merat FL, Discenzo FM, Harris JS (1998) Neural net based torque sensor using birefringent materials. Sens Actuators A Phys 70(3):243–249

Hajdu F, Horváth P (2012) Contactless torque sensor development. Acta Technica Jaurinensis 5(2):115–120

Halmai A, Szakály N (2010) Electromagnetic finite element simulation of the disk micromotor. In: Seventh conference on mechanical engineering. Budapest, pp 835–840

Horváth P, Nagy A (2009) Optical torque sensor development. In: Brno University of Technology, Faculty of Mechanical Engineering, Luhacovice, Czech Republic, Brno, pp 91–96

Issa M, Zentner L (2012) Sensorelemente aus Leitfähigem Silikon für einen Nachgiebigen Greifer. In: Mechanismentechnik in Ilmenau, Budapest und Niš. Deutschland, Ilmenau, pp 65–78

Puente GDL, Jessop PE (2002) Temperature insensitive fiber-optic torque and strain sensor. CA, Patent No.: US006647800B2

Sakai I (1993) Apparatus for measuring torque. EU, Patent No.: EP0578422A2

Thamm F, Huszár I, Ludvig G, Szántó I (1968) Experimental methods of the mechanics (in Hungarian: A szilárdságtan kísérleti módszerei), Budapest: Műszaki Könyvkiadó, pp 218–227, 265

Zi XY, Geng S, Wu H, Zhao SF (2015) The application of torque-sensitive materials in shaft dynamic torque measurement. In: Manufacturing and engineering technology, pp 111–113

Flexural Body for a Wireless Force/Displacement Sensor

J. Hricko and S. Havlik

Abstract This paper deals with the design of the flexural body for a two axis force/displacement sensor based on electro-magnetic sensing principle. The compact elastic structure consists of two independent parallelograms that provide decoupled flexural motions of sensing elements in two rectangular directions. The procedure proceeds by experimental verification of this new sensing principle for building sensors that exhibit specific functional features.

Keywords Force/displacement sensor · Elastic structure · Flexural analysis · Stiffness

1 Introduction

For sensing the components of external force/torque several sensing principles and many different constructions of sensors have been developed for past years (Stefanescu 2011; Wu and Cai 2013). Such multi-component force and torque (F/T) sensors are frequently used to provide the adaptation ability for robots performing contact tasks.

In design procedure of any sensor there are two main decisive steps: choosing the sensing principle and design of an appropriate mechanical structure. It is obvious that both problems are closely related. As to the sensing principle it depends on transducers and further processing output signals. On the other hand; the design of mechanical structure directly corresponds to correct function and

J. Hricko (✉) · S. Havlik
Institute of Informatics, Slovak Academy of Sciences, Banská Bystrica, Slovakia
e-mail: hricko@savbb.sk

S. Havlik
e-mail: havlik@savbb.sk

L. Zentner et al. (eds.), *Microactuators and Micromechanisms*,
Mechanisms and Machine Science 45, DOI 10.1007/978-3-319-45387-3_6

quality of sensor, as whole. For this reason it is very important to pay attention to design the of sensor mechanisms.

There are several works dealing with problem of multi-component sensor design (Uchiyama et al. 1988). The most general approach is based on building the linear mathematical model represented by the sensitivity/compliance matrix that relate the force components and measured elastic deflections—strains, as sensor readings. As the optimality criterion the ratio of maximal and minimal sensing sensitivity of the particular components is defined. By using the singular value decomposition the condition number as the measure of structural isotropy is defined (Uchiyama et al. 1987).

This paper presents the design procedure and building the force/displacement sensor using the new sensing principle working on measurement of the electromagnetic field around the flexural sensor body (Maršalka and Harťanský 2013). The first task is the design of the sensor elastic structure that exhibits displacements of two pairs of capacitor plates in two orthogonal directions.

In order to verify functional assumptions the first experimental tests were performed. Main parts of proposed test bed are in Fig. 1:

- experimental force/displacement sensor
- micro-positioning screw
- calibration force sensor
- E-M cover
- evaluation electronics

As can be seen in Fig. 1, the measured parameter of E-M field—the peak frequency depends on deflection of the elastic body, or, acting force. This dependence represents output reading from the sensor.

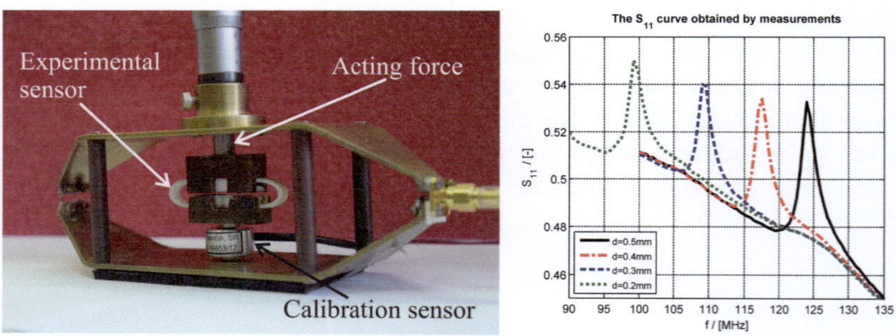

Fig. 1 Arrangement of experimental measurement (*left*); the S11 curve obtained by measurements (Maršalka and Harťanský 2013)

2 Design of Sensor Mechanisms

The two stage design procedure was accepted. The first stage includes building the functional—bigger scale model for verification of the assumed flexural performance and sensing characteristics. The second step includes design that corresponds to real dimensions and using appropriate manufacturing technology too.

2.1 Functional Requirements

Designing the sensor the input requirements related to expected performance should be considered. There are:

- The output—flexural displacements in particular directions should correspond to parameters of LC circuit. For the case of building the first stage model the capacitor plates should move strongly in parallel, within 2 mm.
- Accuracy. For many sensory control applications the sensor activity in quasi-static and dynamic processes is assumed. The sensor should have a prescribed accuracy in a supposed frequency range of force/displacement changes. When consider the dynamical range of sensor use a satisfactory high limit of natural frequency of mechanical oscillation is required. At the stage of the conceptual design the requirement for the sensor dynamics can be formulated in the following way:

Let ω be the supposed maximal frequency of the measured force/displacement oscillations. To have the proposed sensing accuracy within the whole frequency range 0–ω, the minimal frequency of free sensor oscillations should be (Havlik 1995).

$$\omega_0 \geq \frac{\omega}{\sqrt{\xi}} \tag{1}$$

where ξ is the error due to sensor dynamics.

- Compactness of the whole elastic mechanisms. In an ideal case the best quality of sensing will be reached when each component is measured with the same sensitivity. This practically means that the stiffness in all flexural directions should be equal.
- Decoupled motions of capacitor plates for sensing in *x* and *y* directions. This requirement results in minimal cross-sensitivity what means that particular force–displacement components can be directly evaluated.

In principle, for manufacturing complex forms there are two technologies: precise machining or 3D printing. In building models for first—experiments both technologies were used. Standard 3D printers support only specific types of materials like PLA or ABS; but both these materials exhibit small flexibility. As a suitable material can be processed by 3D printers seems the polyamide (PA2200). First samples were manufactured by precise machining the Teflon (PTFE). But when using this technology the restriction on the minimal thickness of flexural hinges/links should be considered.

2.2 *Proposed Compliant Structures*

Principal requirement related to the proposed sensing principle is that capacitor plates must be in mutual parallel positions. Then, the deformable structure should satisfy decoupled parallel motion of two pairs plates both in x, and y directions. Some examples of such flexural structures are discussed in (Hricko 2014; Beroz et al. 2011; Tian et al. 2009). In principle, the design of such compliant mechanisms proceeds from similarity with classic mechanisms.

The first design_1 of deformable part ideologically follows the design presented in (Peirs et al. 2004). The output characteristics in Fig. 2 are non-linear due to high number of flexure hinges. On the other side, such design provides displacements of plates bigger then the input deflections.

The designs in Fig. 3, use parallelograms as flexural element/prismatic joint. These compliant structures are designed for two axis sensor measuring forces in x and z axes.

Main disadvantage of both designs (design_2 and design_3) are mutually bounded motions of two parallelograms; i.e., if the only load in x axis is acting, the plate for z axis moves too. This fact naturally results in cross sensitivity between sensor readings. For more details see Maršalka and Harťanský (2013).

Our final design in Fig. 4 (left) consists of two independent parallelograms that deflect independently in two directions. Calculated force–displacement characteristics are shown in Fig. 4 (right). Both characteristics are practically linear and mutually decoupled.

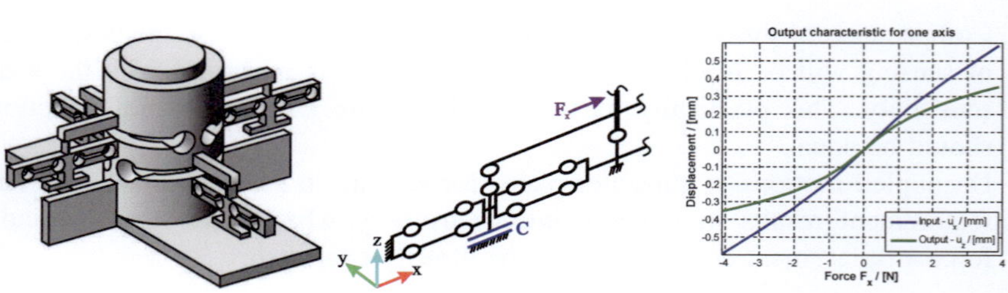

Fig. 2 Compliant structure (design_1), output characteristic for one axis

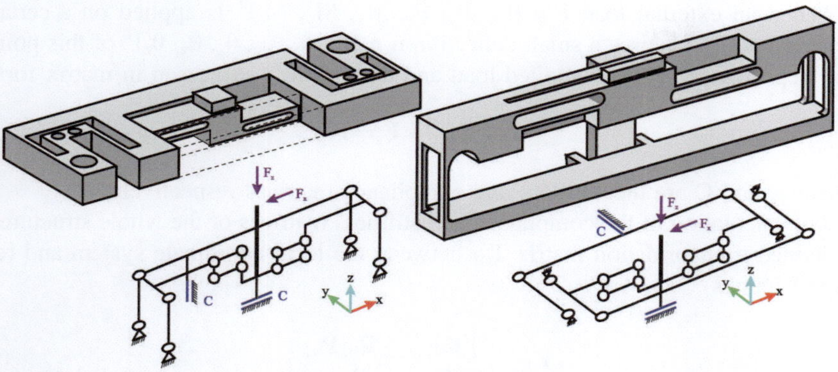

Fig. 3 Proposed compliant structures of deformable part of force/displacement sensor design_2 (*left*) and design_3 (*right*)

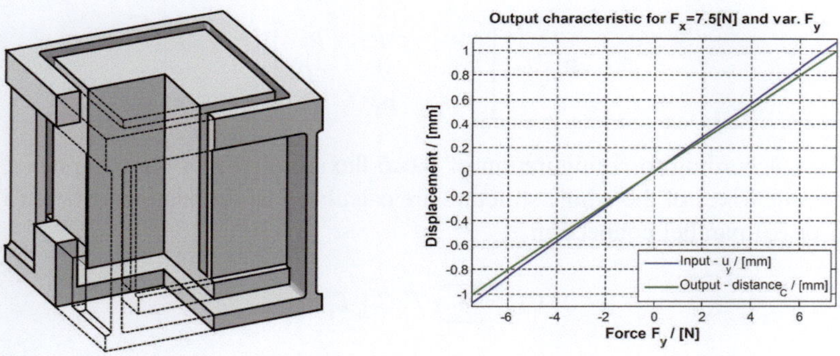

Fig. 4 Proposed compliant structure of sensor (design_4); Force–displacement characteristics of the proposed structure

2.3 *Stiffness Analysis*

In order to satisfy design criteria the careful compliance analysis of this flexure was made. The procedure how deflections of the structure in given points and directions were calculated is outlined below.

To build the full stiffness model of the flexure the approach the deflections in terms of bending, torsion, and tensile/compression of each part of the mechanism are calculated. Such model can be derived effectively by the matrix method under the assumption of Hooke's law for the material (Sciavicco and Siciliano 1996; Xu and Li 2006; Li and Xu 2009).

When an external load $F = [F_x, F_y, F_z, M_x, M_y, M_z]^T$ is applied on a certain (contact) point it causes a small deflection $u = [u_x, u_y, u_z, \theta_x, \theta_y, \theta_z]^T$ of this point. The dependence between applied load and deflection is expressed in matrix form

$$\mathbf{u} = \mathbf{CF} \rightarrow \mathbf{F} = \mathbf{Ku} \quad (2)$$

where K and C are the stiffness and compliance matrices respectively.

For calculation of the compliance and stiffness matrices of the whole structure it is defined transformation matrix T_{01} between the local coordinate system and reference frames

$$\mathbf{T}_{01} = \begin{bmatrix} \mathbf{R}_{01} & -\mathbf{R}_{01}\mathbf{P}_{01} \\ 0 & \mathbf{R}_{01} \end{bmatrix} \quad (3)$$

where R_{01} is rotation matrix between coordinate systems and P_{01} is position matrix of point O_1 expressed in reference coordinates O_0 (see Fig. 5)

$$\mathbf{P}_{01} = \begin{bmatrix} 0 & -p_z & p_y \\ p_z & 0 & -p_x \\ -p_y & p_x & 0 \end{bmatrix} \quad (4)$$

Then, according to configuration of those flexure elements the compliance, or, stiffness matrixes of the whole structure are calculated using relations where (5) for serial or (6) parallel connection.

$$C_1 = \sum_n {}^iT_1^* C_i \left({}^iT_1^*\right)^T \quad (5)$$

$$K_1 = \sum_n {}^iT_1^* K_i \left({}^iT_1^*\right)^T \quad (6)$$

The relative motion of the capacitor plate in x direction is given by relevant components of compliance matrix related to point O_6. Similarly, the motion of

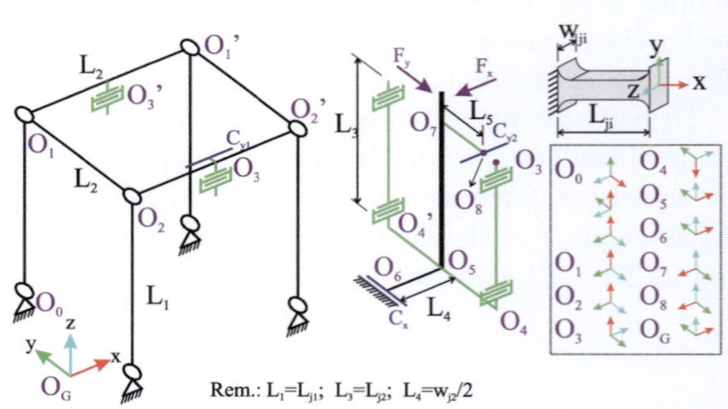

Fig. 5 Geometry of the final design, orientation of local coordinate system in observed points

capacitor plates for sensing y direction is calculated as displacement of point O_8 with respect to O_3.

Transformation matrices between particular local frames to that the compliance matrices were calculated are as follows:

$$\begin{gathered}
\mathbf{T}_{01} = \begin{bmatrix} \mathbf{R}_z\left(\frac{\pi}{2}\right)\mathbf{R}_x\left(\frac{\pi}{2}\right) & 0 \\ 0 & \mathbf{R}_z\left(\frac{\pi}{2}\right)\mathbf{R}_x\left(\frac{\pi}{2}\right) \end{bmatrix}; \quad \mathbf{T}_{12} = \begin{bmatrix} \mathbf{I} & -\mathbf{P}_z(L_2) \\ 0 & \mathbf{I} \end{bmatrix} \\
\mathbf{T}_{23} = \begin{bmatrix} \mathbf{R}_x\left(\frac{\pi}{2}\right) & -\mathbf{R}_x\left(\frac{\pi}{2}\right)\mathbf{P}_y\left(\frac{L_2}{2}\right) \\ 0 & \mathbf{R}_x\left(\frac{\pi}{2}\right) \end{bmatrix}; \quad \mathbf{T}_{34} = \begin{bmatrix} \mathbf{R}_z(\pi) & 0 \\ 0 & \mathbf{R}_z(\pi) \end{bmatrix} \\
\mathbf{T}_{45} = \begin{bmatrix} \mathbf{R}_y\left(\frac{\pi}{2}\right) & -\mathbf{R}_y\left(\frac{\pi}{2}\right)\mathbf{P}_y\left(\frac{L_2}{2}\right) \\ 0 & \mathbf{R}_y\left(\frac{\pi}{2}\right) \end{bmatrix}; \quad \mathbf{T}_{56} = \begin{bmatrix} \mathbf{I} & -\mathbf{P}_x(-L_4) \\ 0 & \mathbf{I} \end{bmatrix} \\
\mathbf{T}_{57} = \begin{bmatrix} \mathbf{R}_z(\pi) & -\mathbf{R}_z(\pi)\mathbf{P}_z(L_3) \\ 0 & \mathbf{R}_y(\pi) \end{bmatrix}; \quad \mathbf{T}_{78} = \begin{bmatrix} \mathbf{I} & -\mathbf{P}_y(L_5) \\ 0 & \mathbf{I} \end{bmatrix}; \quad \mathbf{T}_{j2} \\
= \begin{bmatrix} \mathbf{I} & -\mathbf{P}_y(-L_2) \\ 0 & \mathbf{I} \end{bmatrix}
\end{gathered} \tag{7}$$

Rem.: Orientation of local frame in point O_0 is equal with orientation of coordinate system to which is calculated compliance matrix of flexure hinge.

By substitution of Eq. (7) into Eqs. (5) and (6) it is possible to calculate compliance/stiffness matrix of whole structure as

$$\begin{aligned}
\mathbf{C}_7 = \mathbf{T}_{57}\mathbf{T}_{45}\Bigg(\mathbf{T}_{34}\Big(2\mathbf{T}_{23}\Big(\big(\mathbf{T}_{01}\mathbf{C}_{j1}\mathbf{T}_{01}^T\big)^{-1} + \big(\mathbf{T}_{12}\mathbf{T}_{01}\mathbf{C}_{j1}\mathbf{T}_{01}^T\mathbf{T}_{12}^T\big)^{-1}\Big)\Big)^{-1}\mathbf{T}_{34}^T \\
+ \Big(\mathbf{C}_{j2}^{-1} + \mathbf{T}_{j2}\mathbf{C}_{j2}^{-1}\mathbf{T}_{j2}^T\Big)^{-1}\Bigg)\mathbf{T}_{45}^T\mathbf{T}_{57}^T
\end{aligned} \tag{8}$$

3 Conclusions

Paper presents an innovative solution of the two component force and displacement sensor working on the electro-magnetic sensing principle. The development of sensors working on this principle is actually in initial phase. There are shown first results from measurements have been made on experimental one-component force/displacement sensor. Such sensor consists of two principal functional parts: elastic–mechanics and transducers. Main attention in this paper is devoted to design of flexural mechanics for building the two-component sensor.

Acknowledgments This work was supported by the Slovak Research and Development Agency under the contract No.: APVV-14-0076—"MEMS structures based on load cell" and by the national scientific grant agency VEGA under project No.: 2/0154/16—"Network management of heterogeneous multi-agent systems"

References

Beroz J, Awtar S, Bedewy M, Sameh T, Hart AJ (2011) Compliant microgripper with parallel straight-line jaw trajectory for nanostructure manipulation In: Proceedings of 26th American society of precision engineering annual meeting, Denver, CO

Havlik S (1995) Modeling, analysis and optimal design of multicomponent force/displacement sensors, Proceedings of ECPD conference, Athens, 1995

Hricko J (2014) Straight-line mechanisms as one building element of small precise robotic devices. In: Hajduk M, Koukolova L (eds) Applied mechanics and materials, vol 613, pp 96–101. ISSN 1660-9336

Li Y, Xu Q (2009) Design and analysis of a totally decoupled flexure-based XY parallel micromanipulator. In: IEEE transactions on robotics, vol 25, No 3, June 2009

Maršalka L, Harťanský R (2013) Proposal of novel sensor applicable to contactless displacement measurement. In: Measurement 2013: Proceedings of the 9th international conference on measurement. Smolenice, Slovakia, May 27–30, 2013. Bratislava: Slovak Academy of Sciences, pp 287–290. ISBN 978-80-969672-5-4

Peirs J, Clijnen J, Reynaerts D, Brussel H-V, Herijgersm P, Corteville B, Boonea S (2004) A micro optical force sensor for force feedback during minimally invasive robotic surgery. Sens Actuators A 115(2004):447–455

Sciavicco L, Siciliano B (1996) Modeling and control of robot manipulators, Electrical Engineering Series. McGraw-Hill International Editions. ISBN 0-07-114726-8

Stefanescu DM (2011) Handbook of force transducers: principles and components. Springer Science & Business Media 16(3):2011

Tian Y, Shirinzadeh B, Zhang D, Alici G (2009) Development and dynamic modelling of a flexure-based Scott-Russell mechanism for nano-manipulation. Mech Syst Sig Process 23 (2009):957–978

Uchiyama M, Nakamura T, Hakomori K (1987) Evaluation of robot force sensor structure using singular value decomposition. J Robot Soc Jpn 5(1):4–10

Uchiyama M, Bayo E, Palme-Villalon E (1988) A mathematical approach to the optimal structural design of a robot force sensor. Paper ASME, reprint from Crossing bridges: advances in flexible automation and robotics, vol I, pp. 539–546. (Book No. 10271A)

Wu B, Cai P (2013) Decoupling analysis of a sliding structure six-axis force/torque sensor. Measur Sci Rev 13(4):2013

Xu Q, Li Y (2006) Stiffness modeling for an orthogonal 3-PUU compliant parallel micromanipulator. In: Proceedings of the 2006 IEEE international conference on mechatronics and automation, Luoyang, China, 25–28 June 2006

Capsule Micromechanism Driven by Impulse—Wireless Implementation

T. Ito and S. Murakami

Abstract We have developed a traveling small capsule, which has a smooth outer surface and is driven by vibration. Measuring only 11 mm in diameter and 25 mm in length, it is sufficiently small to be placed in the human gullet or intestines. Although the capsule does not have wheels nor legs, it can travel utilizing vibration. For these reasons, there is nothing protruding to harm the soft tissue of intestine. The capsule contains a small magnet, a coil, batteries and a circuit, and an electric pulse drives the coil to move the capsule. We did the experimental investigation that our capsule can travel in a plastic pipe and it can also travel on pig intestine surface. We also have developed liquid medicine dispenser that can be put inside the capsule. Our capsule may be useful for medical treatments such as inspection and drug delivery.

Keywords Capsule · Micromechanism · Magnetic force · Friction force

1 Introduction

Most part of the human body is composed of soft tubes with diameter ranging from μm order to cm order. Therefore, capsules which can travel inside these tubes are useful for medical treatments such as inspection, drug delivery and operation. Various kinds of machines for this purpose have been proposed (Taylor 2006) but most of them had hands in order to crawl in tubes (Adachi 1992; Masuda et al. 1994; Nakazato et al. 2001; Oura and Hayashi 1997; Takahashi 1992). However, the outer surface of the machine should not have any projections so as not to injure the surrounding site. Machines that have a smooth outer surface and crawl have been developed. Among them, some utilized a fluid actuator to move (Yokota 2006; Yoshida et al. 2001), but it was more than 30 mm long and needed a high-power

T. Ito (✉) · S. Murakami
Faculty of Computer Science and Systems Engineering,
Kyushu Institute of Technology, Iizuka, Fukuoka, Japan
e-mail: ito@mse.kyutech.ac.jp

L. Zentner et al. (eds.), *Microactuators and Micromechanisms*,
Mechanisms and Machine Science 45, DOI 10.1007/978-3-319-45387-3_7

fluid pump. To avoid using legs or hands to crawl, eel-like mechanisms were theoretically analyzed (Boyer et al. 2006; Wakimoto et al. 2006), and inchworm-type robots were developed (Hirano et al. 2006), but they had relatively complicated mechanisms and it was difficult to make them as small as medicine tablets. This paper describes a traveling small capsule, which has a smooth surface and is driven by vibration. Measuring only 11 mm in diameter and 25 mm in length, our capsule is sufficiently small to be placed in the human gullet or intestines. In this paper, we present how the capsule travels with a simple mechanism, and the wireless implementation of the capsule to achieve its traveling capability.

2 Structure and Principle of the Traveling Capsule

Our capsule travels by a to-and-fro motion of an inner mass. The capsule consists of the body (M) with magnet and the moving mass (m) with coil. The moving coil is made by 200 turns of φ0.05 mm copper. The moving mass coil is driven by magnetic force with an Nd–Fe–B permanent magnet by applying a step-shaped current to the coil. The outside appearance of the capsule is shown in Fig. 1.

2.1 Structure of the Capsule

The inside of the capsule is shown in Fig. 2. The movement of the coil as a moving mass is guided by a permanent magnet in the center. For the theoretical analysis, the motion of the capsule is divided into the following four steps.

Step 1: As shown in Fig. 2, M and m pull each other by magnetic force fm. The mass m that starts from the left end is accelerated by the magnetic force f_m and also the body M is accelerated by the force $(f_m - f_r)$, when the friction force between the capsule body and the base is f_r, until m

Fig. 1 The outside appearance of the capsule

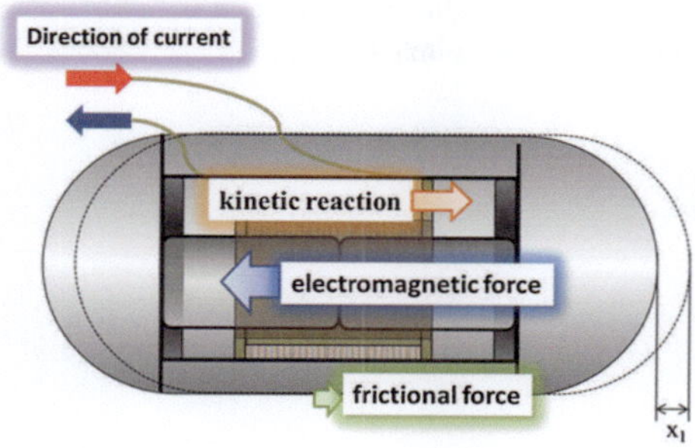

Fig. 2 Step 1: Moving mass (m) is driven by electro-magnetic force

collides with M at the right end. Defining that M moves by distance x_1 to left side and m to right side in time t_0 by that motion, we will be able to determine these values.

Step 2: As shown in Fig. 3, we assume that after collision, M and m move together with initial velocity v0, are decelerated by friction force f_r and finally come to a stop. Defining that M and m move by x_2 to the right within time t_2 by the motion shown in Fig. 3, we can determine these values.

Step 3: Like Step 1, upon applying magnetic force $-f'_m$ by adding a (−) step-shaped current to the coil, M and m push each other. The mass m that starts from the right end is accelerated by magnetic force f'_m and also the body M is accelerated by the force $(f'_m - f_r)$ until m collides with M at left end. Defining that M moves by distance x_3 to the right within time t_3 in that motion, we can determine these values (Fig. 4).

Step 4: Like Step 2, M and m move together with initial velocity v'_0, are decelerated by friction force f_r and then come to a stop. Defining that M and m move by x_4 to the left within t_4 in that motion, we can determine these values (Fig. 5).

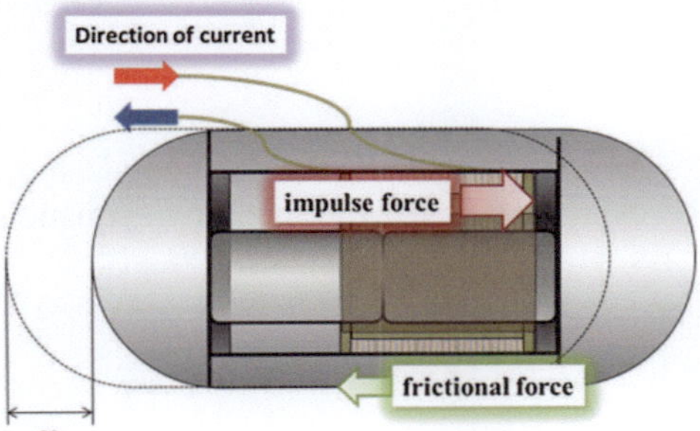

Fig. 3 Step 2: Moving mass (m) collides with the body (M)

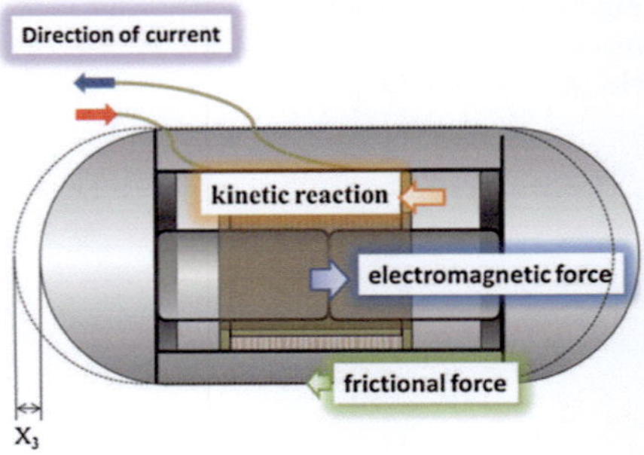

Fig. 4 Step 3: Moving mass (m) goes back by magnetic force

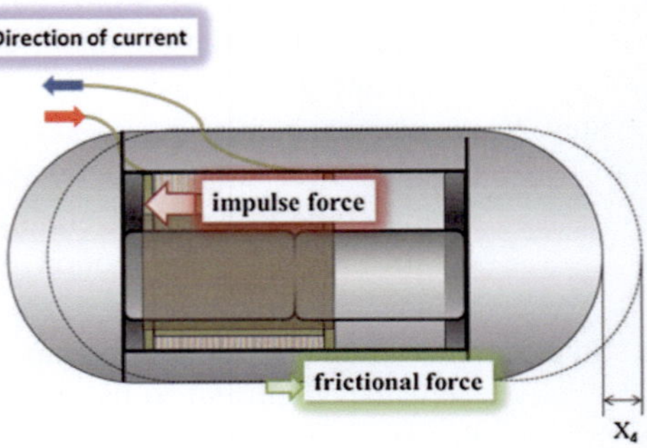

Fig. 5 Step 4: Moving mass (m) collides with the other side

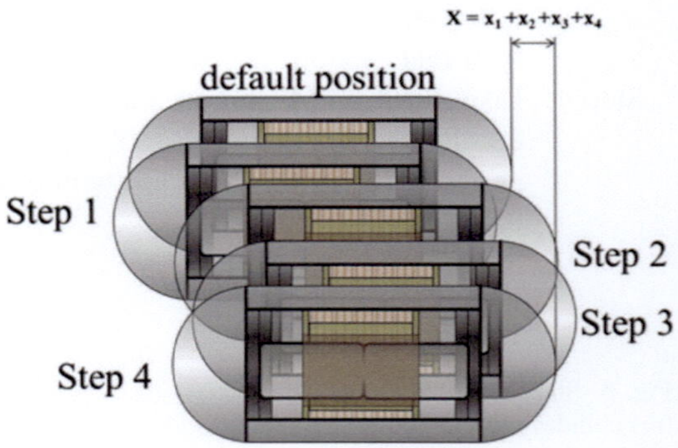

Fig. 6 Total displacement x with the step 1–4

As a result, defining that the body M moves x to the right within time t, they will be expressed by the following two equations:

$$x = x_1 + x_2 + x_3 + x_4 \tag{1}$$

$$t = t_1 + t_2 + t_3 + t_4 \tag{2}$$

In the design, we should aim to increase the value of x and decrease the value of t (Fig. 6).

2.2 Theoretical Analysis

Our capsule travels by a to-and-fro motion of an inner mass. The capsule consists of the body (M) with coil and the moving mass (m). We can obtain equations of motion in each step as shown below. For M and m, equations of motion are expressed by,

$$M \cdot \ddot{x} + f_r = f_m \tag{3}$$

$$m \cdot \ddot{u} = f_m \tag{4}$$

In the equation of motion above, the displacement of the body M is x, and the displacement of the inner mass m is u. The simplified physical model of the capsule is shown in Fig. 7. The movement of mass m is restricted between two stoppers A and B. The stroke of m is expressed as a in the figure. The resistance force f_r acts against the movement of the body M from the base. For simplicity, we neglected the friction force between m and M, which is relatively very small and does not play a major role in propelling the capsule. For the theoretical analysis, the motion of the capsule is divided into the four steps already shown above.

The traveling speed is the most important factor when designing the traveling capsule. From the above Eqs. (1)–(4), the total speed S of one stroke motion is derived as follows.

$$S = \frac{x}{t} = C_s \cdot K_s, \quad K_s = \sqrt{\frac{a \cdot f_r}{2M}} \tag{5}$$

$$C_s = \frac{C_x}{C_t} = \frac{(2n+1) - (n+1)f}{2(n+1) - \sqrt{(n+1)f - 1}} C_x = \frac{x}{a}, C_t = \frac{t}{K_t}, K_t = \sqrt{\frac{2aM}{f_r}}, n = \frac{M}{m} \tag{6}$$

The effect of n and f on the coefficient of traveling speed C_s is summarized in Fig. 8. In the figure, force ratio f (= f_m/f_r) should be greater than 1, because friction force between the capsule and the base cannot be greater than magnetic force f_m

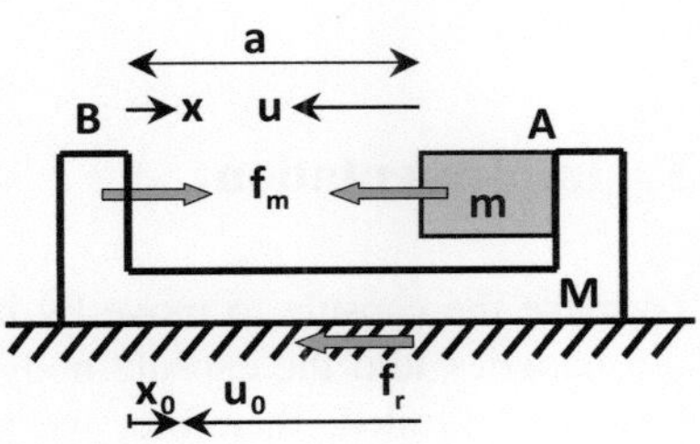

Fig. 7 The simplified physical model of the capsule

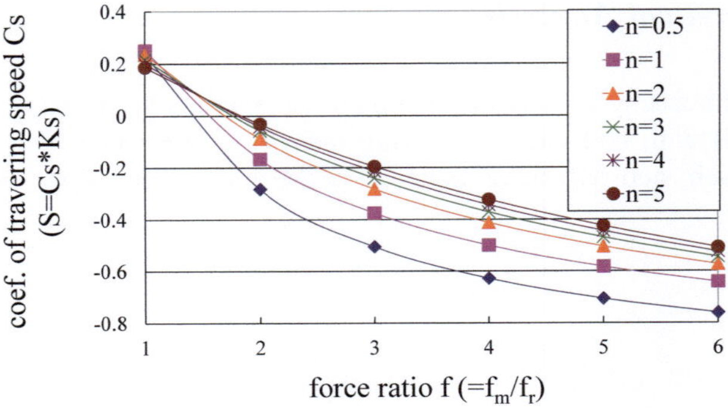

Fig. 8 Speed-force characteristics (calculation)

Fig. 9 Speed characteristics (experiments)

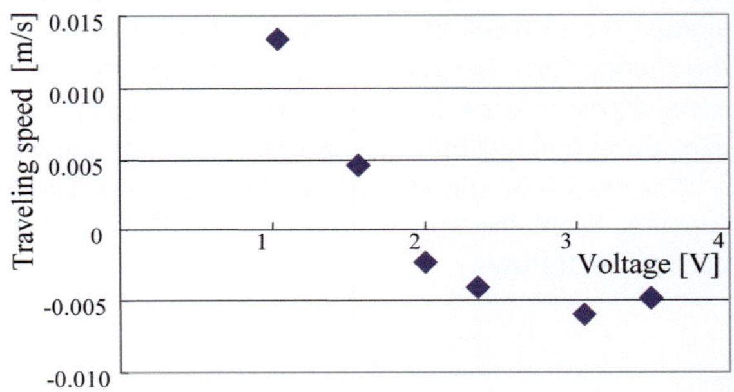

while the capsule is moving. From the figure, it is found that we can choose the parameters as follows to design the high-speed capsule:

(a) Mass ratio n should be small.
(b) Force ratio f should be large.

The result of theoretical analysis calculation matches well with an experimental result by a capsule shown in Fig. 9. As the voltage at the coil increased, which means higher magnetic force, the direction of travel changed.

3 Implementation

To make the capsule to move by itself, we implemented actuator, control circuit, and batteries into the capsule body as shown in Fig. 10. The inside space of the capsule is limited, therefore, we utilized the space around the actuator for the control circuit, which is implemented on a flexible film.

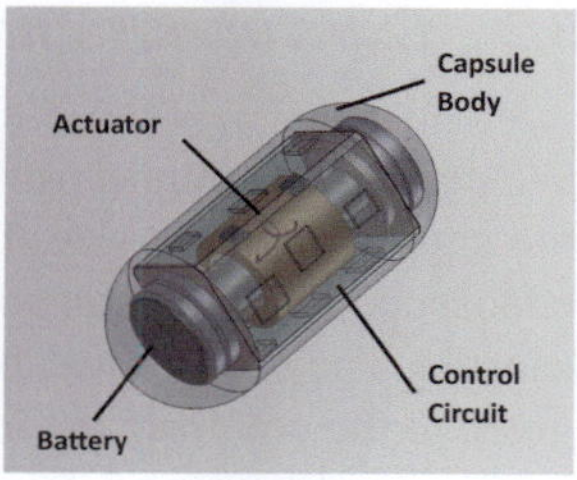

Fig. 10 Schematic of the inside of the capsule

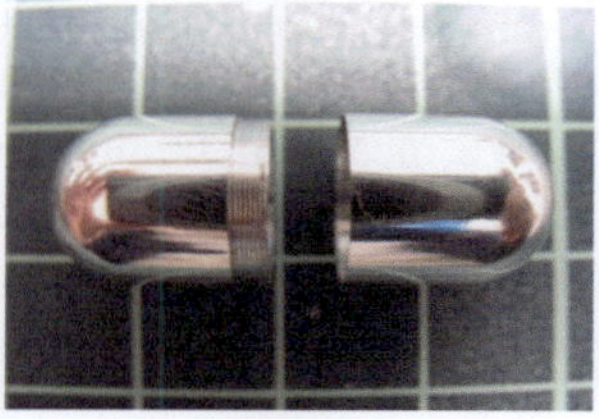

Fig. 11 Capsule body

All the parts are stored in the capsule body, shown in Fig. 11. As shown in the figure, the capsule body is divided into two pieces and are screwed to be assembled. This construction helps the capsule batteries to contact electrodes firmly and, seal off the capsule firmly too.

3.1 Control Circuit

To drive the capsule, step-shaped signal is generated from the control circuit including microcomputer (PIC) and the signal is amplified by driver circuit. These circuits are implemented on a flexible film and the film is folded to wrap the actuator in the capsule as shown in Fig. 12.

As shown in Fig. 12, batteries are also connected by the circuit film. Control command to move the capsule forward or backward is sent from outside capsule by wireless. The coil of the actuator also serves as the antenna to catch the signal. This configuration contributed to make the capsule volume smaller.

Fig. 12 The folded circuit with batteries and the actuator in the capsule

3.2 Experiments and Results

We made a traveling capsule as shown in Figs. 1 and 10. The major requirement for the capsule was the size. Diameter of 11 mm is the maximum for the capsule or endoscope to go in human throat without hardship and that was achieved by our capsule. The capsule was put on the flat plastic plate and a step-shaped current was given to the coil from the inside circuit. The capsule traveled forward and backward as the wireless signal from the outside controller was sent. The traveling speed measured in the experiment was 8.3 mm/s. We set the target speed more than 5 mm/s on a plate, and that was achieved. The power consumption was 250 mW (2.5 V, 100 mA) and was small enough for the battery of the capsule.

We measured friction force f_r between the capsule and a plate. For the plate material, plastic, rubber, and phantom, which was a plastic material having similar elasticity characteristics to the living body, were chosen. The measurement results are shown in Table 1.

We also measured friction force between the capsule and a pig's intestine. The measurement setup is shown in Fig. 13 and the result was approximately 0.05 N. Therefore, it is necessary to increase capsule driving force up to more than twice that of a conventional capsule traveling on a plate. Our capsule has a driving force of 0.2 and 0.4 N at its maximum, which is estimated to be enough to move on an intestine.

Table 1 Summary of friction force measurements

Capsule	Aluminum			
Sample	Rubber plate	Phantom		Plastic plate
Sample condition	Dry	Dry	Wet	Dry
Friction force f_r [N] (Average of 10 measured values)	0.020	0.015	0.018	0.005
Standard deviation	0.0034	0.0021	0.0016	0.0004

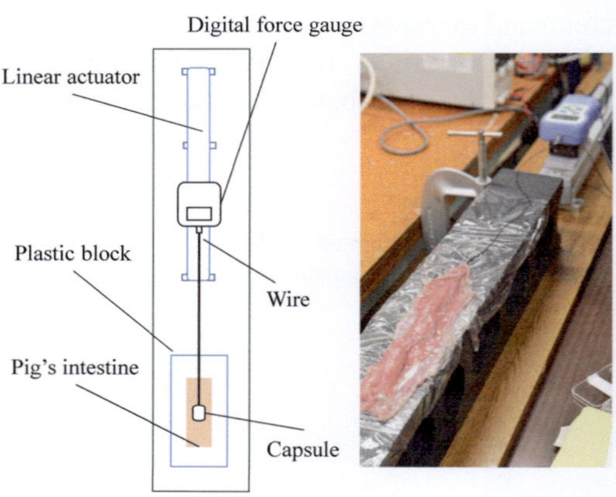

Fig. 13 Setup for measuring friction force between a capsule and a pig's intestine

4 Medicine Releasing Mechanism for the Capsule

Our capsule can be utilized for drug delivery inside the human body, when it contains medicine dispenser. Therefore, we have developed a medicine releasing mechanism that can be implemented inside the capsule.

The schematic of the medicine releasing device is shown in Fig. 14. The device is 2 mm square size and can be set in the capsule.

The fabrication process flow of the device is shown in Fig. 15. The device consists of a medicine container and a cover and is fabricated using Micro Electro Mechanical Systems (MEMS) process.

In the figure, Si substrate is etched using deep Reactive Ion Etching (RIE) to make a medicine container. Then, liquid medicine is filled in the container. At last, the container exit is covered with hydrogel. The hydrogel cover is removed when the current is put at the electrode by the container exit to release the liquid medicine.

To test the medicine releasing device, we used an experimental setup as shown in Fig. 16. The device filled with colored liquid was dipped in the water under a

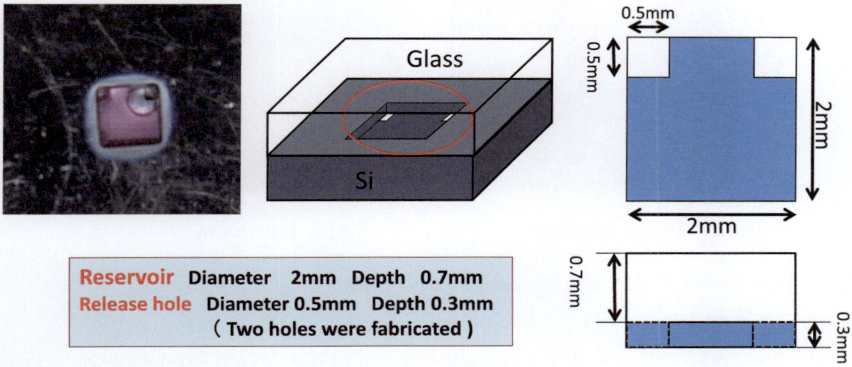

Fig. 14 The schematic of the medicine releasing device

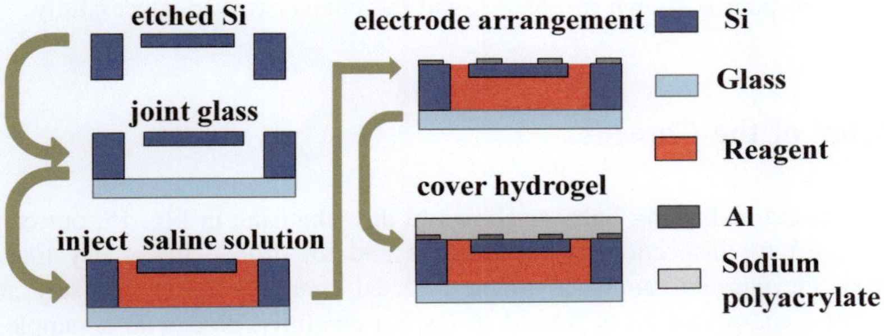

Fig. 15 The flow of the process to fabricate the medicine releasing device. (sectional view)

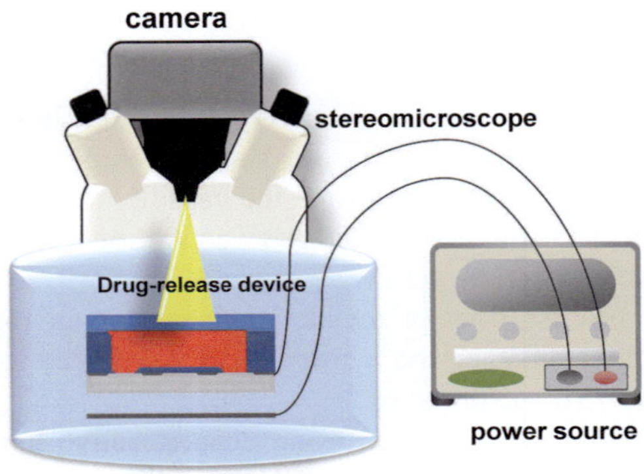

Fig. 16 The experimental setup to test medicine releasing device

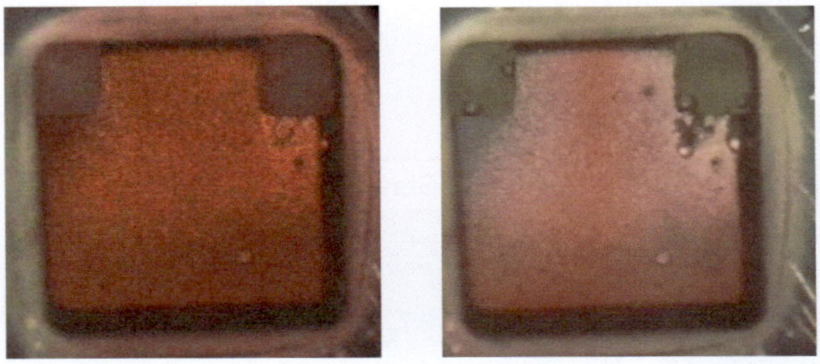

(a) Before experiment (b) After experiment

Fig. 17 The results of the medicine releasing device test

microscope and a camera. When current was put at the device from a power source, the hydrogel cover was removed and the colored liquid was dispensed out of the device. At the same time, water outside came into the device instead of the colored liquid. Then the color of the inside of the device became thinner. The result of above experiment is shown in Fig. 17, and the device worked successfully.

5 Goal of the Capsule

With a camera and a ring light, as shown in the schematic in Fig. 18, our capsule can be used for inspection in the intestine and for drug delivery. By traveling function, the capsule can reach to the affected area quickly or can stay at the important area, or can go backward to inspect carefully. And with a sample collecting mechanism, the capsule can be used for the check-up or surgery.

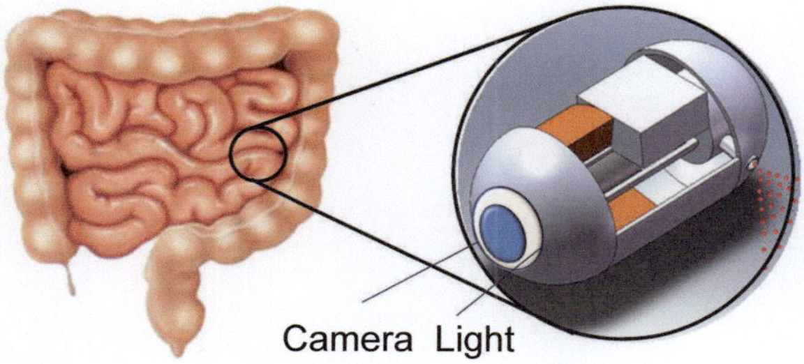

Fig. 18 Goal of the capsule

6 Summary

Even though the capsule has no wheels or legs, it can travel utilizing vibration. We conducted analyses for the capsule movement and showed how the capsule mechanism worked. We implemented actuator, control circuit, and batteries, all inside the capsule and achieved wireless traveling of the capsule. With the wireless command, the capsule can proceed forward and backward. Our capsule may be useful for medical treatments such as inspection, drug delivery and operation.

Acknowledgments The research work reported here was supported by Grant-in-Aid for Scientific Research from Japan Society for the Promotion of Science (#26350538).

References

Adachi H (1992) Japan Patent #H06-154191

Boyer F, Porez M, Khalil W (2006) Macro-continuous computed torque algorithm for a three-dimensional eel-like robot. IEEE Trans Robot 22(4):763–775

Hirano S, Luo Zhi-Wei, Kato A (2006) Development of an inchworm-type searching robot. J Robot Soc Jpn 24(7):838–844

Masuda N, Hayashi T, Cai Y (1994) Developments of guidance machine for colonoscope. Int J JSPE 28(1)

Nakazato Y, Ishii T, Sato H, Ariga Y (2001) Development of in-pipe mobile robot driven by hydraulic pressure. In: Proceedings of the 3rd IFToMM international micromechanism symposium, pp 13–16

Oura M, Hayashi T (1997) Traveling machine with six legs for inside of tube. In: Proceedings of JSPE 1997 spring conference

Takahashi K (1992) Japan Patent #H05-212093

Taylor RH (2006) A perspective on medical robotics. Proc IEEE 94(9):1652–1664

Wakimoto S, Suzumori K, Kanda T (2006) A bio-mimetic amphibious soft cord robot. Trans Jpn Soc Mech Eng Ser C 72(714):471–477

Yokota S (2006) Functional fluids from point of view of actuators. J Robot Soc Jpn 24(4):25–31

Yoshida K, Park J-H, Shimizu T, Yokota S (2001) A micropump-mounted in-pipe mobile micromachine using homogeneous electro-rheological fluid. In: Proceedings of the 3rd IFToMM international micromechanism symposium, pp 2–7

Development of Peristaltically Propelled Active Catheter Used in Radial Artery

Y. Nakazato, K. Kawanaka, K. Takita and M. Higuchi

Abstract In our previous studies, many experiments using pseudoblood vessels with an inner diameter of ≤6 mm have been carried out assuming that the catheter is inserted in a femoral artery. In recent years in Japan, however, along with the trends of reducing the burden on patients and the downsizing of catheters, in more than half of the cases, catheters are inserted in radial arteries with inner diameters of 2.0–3.5 mm for the treatment of dysfunctional coronary arteries. The purpose of this study was to develop a small active catheter so that the catheter can navigate in a radial artery. The effectiveness of the catheter was confirmed by the driving experiment using silicon tubes with inner diameters of 2.5–3.0 mm and an endovascular surgery simulator.

Keywords Micro-robot · Travelling wave · Peristaltic motion · Active catheter

1 Introduction

In clinical settings of the current aging society, less invasive medical methods have been demanded to reduce the burden on patients. In 2012, the second and fourth most common causes of death of Japanese were cardiac and cerebrovascular

Y. Nakazato (✉) · K. Takita · M. Higuchi
Department of Innovative Systems Engineering, Nippon Institute of Technology, Saitama, Japan
e-mail: nakazato@nit.ac.jp

K. Takita
e-mail: takita@nit.ac.jp

M. Higuchi
e-mail: mhiguchi@nit.ac.jp

K. Kawanaka
Department of Mechanical Systems Engineering Major, Nippon Institute of Technology, Saitama, Japan
e-mail: aase42battaion@yahoo.co.jp

L. Zentner et al. (eds.), *Microactuators and Micromechanisms*,
Mechanisms and Machine Science 45, DOI 10.1007/978-3-319-45387-3_8

diseases. The annual number of deaths due to these two diseases was approximately 310,000, which accounts for approximately 25 % of the total number of deaths (Annual Health 2013). Cardiovascular disease is caused by the narrowing of blood vessels and has been treated by endovascular surgery (Tsuchjkane et al. 1998; Kunio et al. 2004). The use of a less invasive catheter is one type of endovascular procedure. Catheterization is endovascular surgery using a polymer tube, called a catheter, with an external diameter of 2–3 mm (Miyazaki et al. 2007). During the surgery, a catheter is inserted into a blood vessel through a small cut made on the surface of the patient for vasodilatation. However, catheter insertion requires experienced skill and the success of the treatment depends on the competence of the practitioners and the facilities of medical institutions (Koga et al. 1992). In addition, the practitioner carries out catheterization as he/she observes X-ray images simultaneously taken, during which time the practitioner is exposed to a small dose of radiation. Therefore, the practitioner is required to wear a radiation protection suit and takes other protective measures such as keeping a large distance from the patient (Fukutomi et al. 2002).

The purpose of this study was to automate catheterization surgery using robotics to address the above problems. In particular, an automated catheter system is realized through the development of a micro-robot that can navigate in a blood vessel. The automated catheter system moves through two balloons that are alternately expanded and contracted by feeding fluid to emulate the peristaltic movement of an earthworm. During the expansion of the balloon, the area of contact between the balloon and the inner wall of the blood vessel is large, ensuring a sufficient frictional force to drive the movement and maintenance of the catheter. In addition, saline solution is used as the medium with which hydraulic pressure is transmitted. Therefore, there is no concern about electricity leakage in case the catheter breaks down in the human body, providing a higher safety than electrically operated catheters (Morise et al. 2000).

In our previous studies, many experiments using pseudoblood vessels with an inner diameter of ≤6 mm have been carried out assuming that the catheter is inserted in a femoral artery. In recent years in Japan, however, along with the trends of reducing the burden on patients and the downsizing of catheters, in more than half of the cases, catheters are inserted in radial arteries with inner diameters of 2.0–3.5 mm for the treatment of dysfunctional coronary arteries. The purpose of this study was to develop a small active catheter so that the catheter can navigate in small-diameter blood vessels.

2 Principle of Movement of Active Catheters

The following are limitations to the movement of active catheters.

(1) It is difficult to manufacture a catheter with complicated components, such as a turning mechanism with high precision, because of its minute size. The friction

between the rotation axis and a bearing caused by the small-scale effect significantly affects the turning mechanism.

(2) A catheter with components that may damage the inner wall of blood vessels should not be used. In addition, irregular morphological structures, such as folds and valves, and mucosal fluid (blood) existing in blood vessels may block its movement.

(3) The catheter should be designed considering the effect of electrical current on human bodies to ensure safety. In particular, when the catheter is used for the treatment of the heart, even a low current of 100 μA may induce ventricular fibrillation, that is, a microshock (Nogami 2011).

Focusing on the mechanism of movement of living organisms, some insects and microorganisms solve the above problems using the frictional force of their mucosal membrane and fine hairs. In this study, we applied the peristaltic movement of earthworms as the mechanism of movement of catheters and reproduced the mechanism mechanically by alternately expanding and contracting rubber balloons. The advantages of the peristaltic movement using balloons are as follows.

(1) Compared with a propulsion mechanism using wheels, this method is less likely to cause damage to the inner wall of blood vessels.

(2) The breakage of catheters has minimal effect because saline solution is used as the driving fluid.

(3) It is possible to use the catheter in treatments using balloons or stents without the need for an actuator.

From the above reasons, we have developed a catheter based on the mechanism of movement utilizing the frictional force between the surface of the active catheter and the inner wall of the blood vessel (Nakazato et al. 2010a, b). The mechanism utilizes the peristaltic movement generated by the alternating expansion and contraction of two balloons induced by supplying a driving fluid. During the expansion of the balloon, the area of contact between the balloon and the inner wall of the blood vessel is large, ensuring a sufficient frictional force for the movement and maintenance of the catheter.

In this study, we aimed to develop a catheter that functions like a micro-robot and can move in blood vessels. Therefore, the diameter of the catheter should be smaller than the inner diameter of the blood vessel. The diameter of the human coronary artery is very small (≤6 mm). Therefore, mounting all the necessary components, such as an energy source and an actuator, onto the main body of the active catheter is difficult. Therefore, a long self-propelled tubular micro-robot, to which driving power is provided from outside, was developed. With the adoption of this design, a self-propelled active catheter was realized without a guide wire.

Figure 1 shows the schematics of the operation principle. First, a driving fluid is pressure-fed to section B to expand its crosssection, and then, section B comes into contact with the inner wall of the blood vessel and expands axially. Next, the driving fluid is pressure-fed to section A to expand the section similarly to section B.

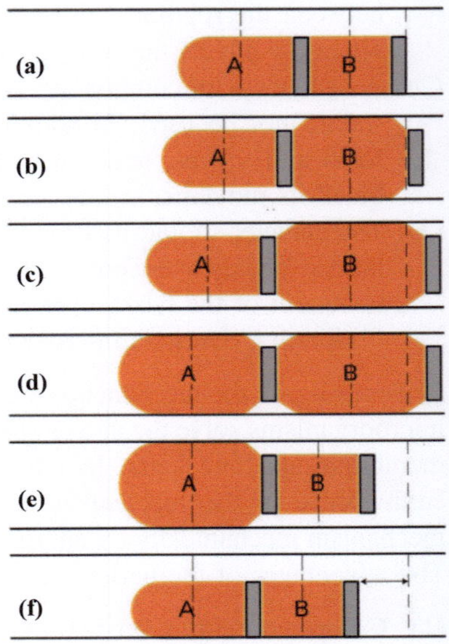

Fig. 1 Schematic of basic mechanism of catheter movement. The catheter moves forward through the generation of traveling waves by alternately expanding and contracting sections A and B, as shown in (**a**)–(**f**)

Here, the frictional force applied to section B is larger than the force of expansion applied to section A, preventing the backward movement of section B. Then, the saline solution is aspirated from section B first and then from section A to move the catheter forward. This two-step feeding method is the most basic mechanism developed in our laboratory and has already been applied to various prototype catheters (Nakazato et al. 2010a, b).

3 Development of Active Catheter System

3.1 Structure of Active Catheter

The design principles of the active catheter are as follows.

(1) It is not possible to prepare sections with a precise length because the catheter is manually manufactured. The length of one segment is designed in millimeter units.
(2) It was demonstrated in previous studies that the target movement is nearly obtained when the rear sections are made longer and the front sections are gradually shortened (Nakazato et al. 2010a, b). In this study, we designed the catheter on the basis of this finding.
(3) The pressure of the driving fluid decreases, owing to pressure loss, as the driving fluid passes through the feeding pipe and septa. Therefore, the number of sections per feeding pipe should be less than 4 Sects.

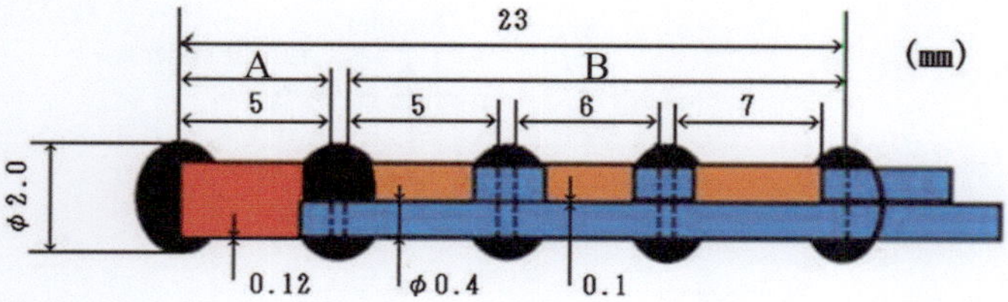

Fig. 2 Cross-sectional view of the proposed mechanism of movement. Section B in Fig. 1 consists of three subsections. The subsection in the rear is longer than that in the front (the part marked in *blue* is indicate feeding pipe, marked in *red* is section A and *orange* is section B)

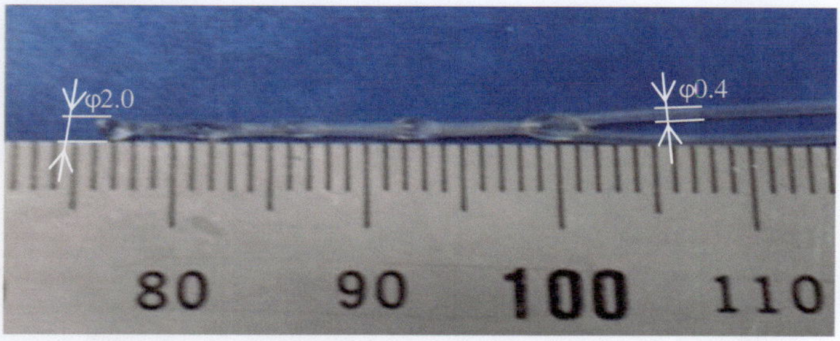

Fig. 3 External appearance of catheter developed in our laboratory [23.0 mm long by φ 2.0 (feeding pipe φ 0.4)]

(4) The treatment procedure using catheters is as short as 1–2 h. Therefore, the target movement speed is set at 5 mm/s so that the catheter can reach the heart within ~3 min.

On the basis of the above design principles, we fabricated an active catheter consisting of section A (5 mm) and section B (with three subsections 5, 6, and 7 mm long), as shown in Fig. 2. Figure 3 shows the external appearance of the fabricated active catheter. Sections A and B, which expand and contract alternately, are connected to two different silicon tubes and driven by different actuators. The catheter consists of four silicon rubber sections (outer diameter, 0.8 mm; thickness, 0.12 mm; Fuji Latex Co., Ltd.) and silicon tubes connecting the two sections (outer diameter, 0.4 mm; inner diameter 0.3 mm).

3.2 *Mechanisms of Pressure Feeding and Aspiration of Fluid*

Regarding the mechanisms of pressure feeding and aspiration of saline solution, we manufactured an actuator by combining an electric slider (XA-42-D, SUS

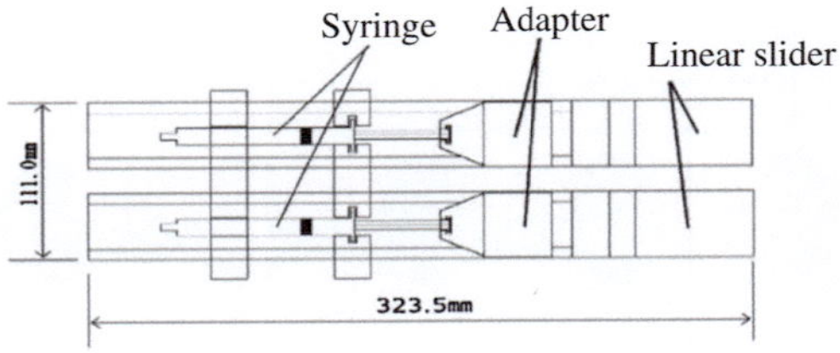

Fig. 4 Structures of syringe used to feed and aspirate driving fluid in the main body of the catheter and linear slider used for reciprocating motion of syringe piston. The linear slider is a directly operated actuator consisting of an electric motor and ball screw (323.5 mm long by 111.0 mm wide)

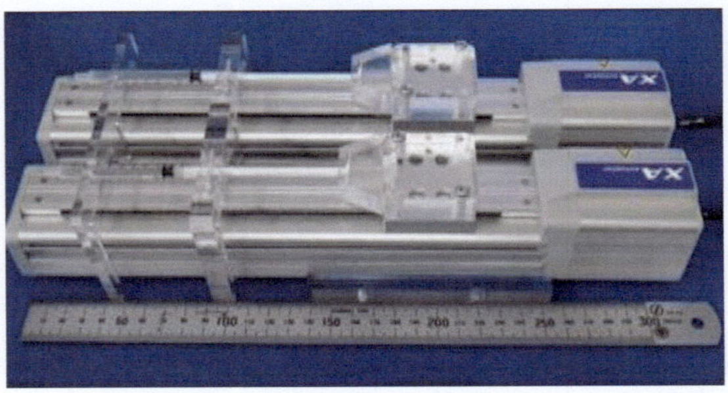

Fig. 5 External appearance of syringe used to feed and aspirate driving fluid in the main body of the catheter and linear slider used for reciprocating motion of syringe piston (323.5 mm long by 111.0 mm wide)

Corporation), consisting of a stepping motor and ball screws, with a syringe. Figures 4 and 5 show the structure of two actuator systems connected in parallel.

The mechanisms of pressure feeding and aspiration, as well as the micro-robot, are connected to the actuator via a silicon rubber tube (inner diameter, 0.5 mm; outer diameter, 1.0 mm; length, 1 m). When the saline solution is fed inside the balloon from the syringe via the tube, the catheter expands and contracts as shown in Fig. 1. Figure 6 shows the schematic of the entire experimental setup.

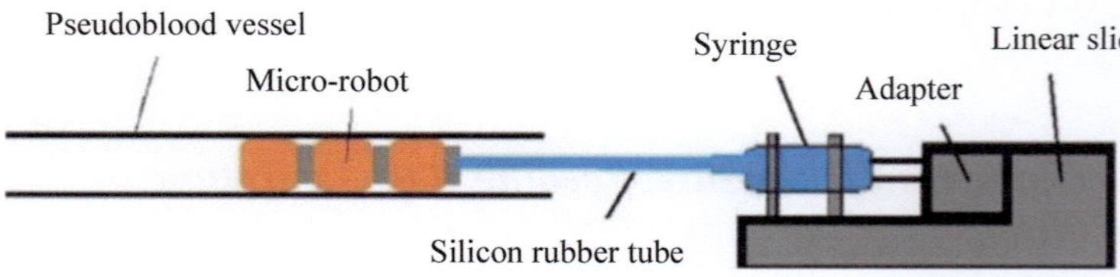

Fig. 6 Arrangement of syringe and linear slider used to feed and aspirate driving fluid in the feeding system and the catheter. Schematic of linear driving motor used to induce reciprocating motion of syringe piston to feed saline solution used as the driving fluid to the catheter through a feeding pipe

4 Driving Experiment Using Silicon Tubes

4.1 *Outline of Experiment*

We experimentally confirmed whether the active catheter developed in this study can perform the series of operations shown in Fig. 1 and move forward in silicon tubes that simulate blood vessels. The experiment was carried out using silicon tubes with inner diameters of 2.5 and 3.0 mm to measure the changes in the feeding rate (μl/s) of the driving fluid and movement speed. The active catheter shown in Figs. 2 and 3 was used in the experiment.

4.2 *Experimental Setup*

Table 1 shows a summary of the mechanical properties of blood vessels of adult men and of the silicon tube used in the experiment. The silicon tubes used showed mechanical properties (e.g., inner diameter, Young's modulus, Poisson's ratio, and coefficient of viscosity) similar to those of actual blood vessels. A mixture of water and glycerin was used as pseudoblood; its coefficient of viscosity was 5.0 mPa s, which is close to that of actual blood (4.7 mPa s). The actuator system shown in Figs. 4 and 5 was used to control the driving.

4.3 *Experimental Method*

The quantity of saline solution pressure-fed to the active catheter was 230–400 μl. The movement speed of the catheter was measured at various feeding rates in the range of 140–570 μl/s. In the experiment, the catheter was moved forward in the silicon tube for 30 mm and the time required for each operation was measured.

4.4 *Results*

Figure 7a, b show the results. The developed active catheter was confirmed to move forward in the silicon pseudoblood vessel with an inner diameter in the range of

Table 1 Mechanical properties of blood vessels of adult men and silicon tubes used in the experiment

	Adult men	Silicon tubes
Internal diameter (mm)	2.0–3.5	2.5/3.0
Young's modulus (MPa)	1–3	1.7
Poisson's ratio	0.5	0.5

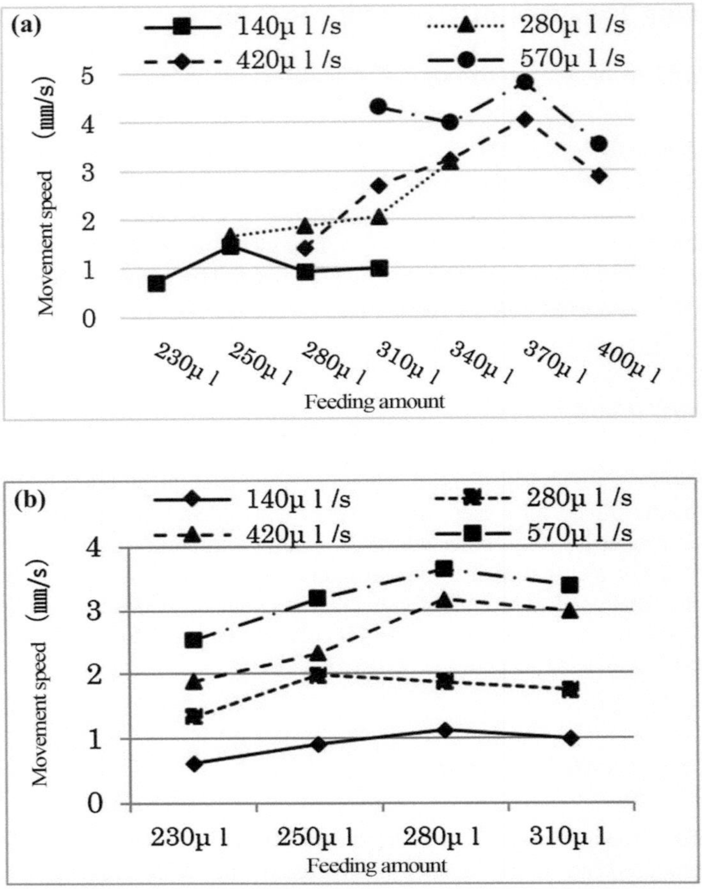

Fig. 7 **a** Relationship between movement speed and feeding amount of driving fluid using feeding rate as a parameter (inner diameter 2.5). **b** Relationship between movement speed and feeding amount of driving fluid using feeding rate as a parameter

2.5–3.0 mm. The movement speed of the active catheter increases with increasing feeding rate of the saline solution: the maximum movement speed of 4.79 mm/s was obtained at the maximum feeding rate of 570 μl/s and feeding amount of 370 μl. However, the movement speed does not necessarily increase with increasing feeding amount. The appropriate feeding amount differs depending on the inner diameter and feeding rate.

5 Driving Experiment Using an Endovascular Surgery Simulator

5.1 Outline of Experiment

As described in Sect. 4, a driving experiment using silicon tubes with diameters of 2.5–3.0 mm was carried out to confirm the effectiveness of the active catheter developed in this study. However, actual blood vessels are tortuous and their inner walls have irregular morphological structures. To address these problems, a driving experiment using the carotid artery of an endovascular surgery simulator for medical training and evaluation was carried out. Table 2 shows a summary of the mechanical properties of blood vessels of adult men and of the silicon tube used in the endovascular surgery simulator (FAIN-Biomedical Inc. 2015).

5.2 Experimental Method

Figure 8 shows a photograph of the endovascular surgery simulator. Two arteries, a right external carotid artery and a right internal carotid artery, which are relatively

Table 2 Mechanical properties of blood vessels of adult men and the silicon tubes used in endovascular surgery simulator

	Adult men	Endovascular surgery simulator
Young's modulus (MPa)	1–3	2.0
Poisson's ratio	0.5	0.5
Coefficient of viscosity	0.048	0.038

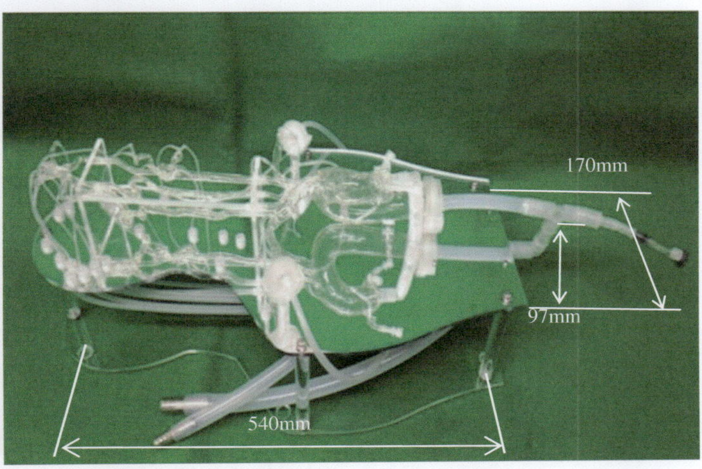

Fig. 8 External appearance of endovascular surgery simulator used in experiment (width: 540 mm, depth: 170 mm, height: 97 mm)

close to the radial artery, were used in the experiment. The inner diameters and the radius of curvature of the right external carotid artery were ~3–4 mm and ~1.2 or 0.8 cm, respectively. Those of the right internal carotid artery were ~4–5 mm and ~1.3 cm, respectively. The pseudoblood explained in Sect. 4 was also used in this experiment.

5.3 Results

Figure 9 shows photographs of the driving experiment. It was confirmed that the catheter moves forward in the right internal carotid artery and right external carotid artery with a radius of curvature of 1.2 cm. However, the catheter moves backward

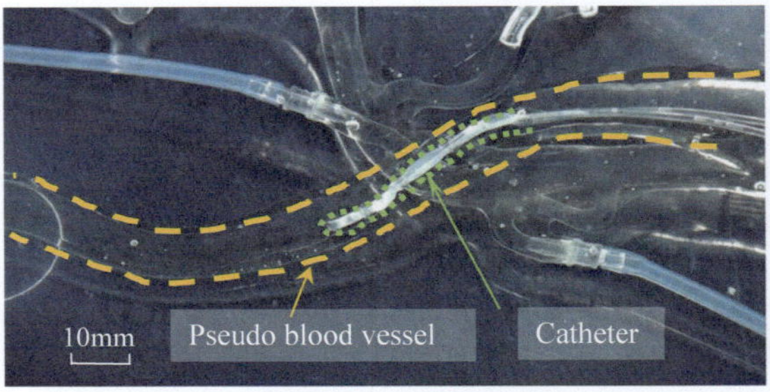

(a) Default position

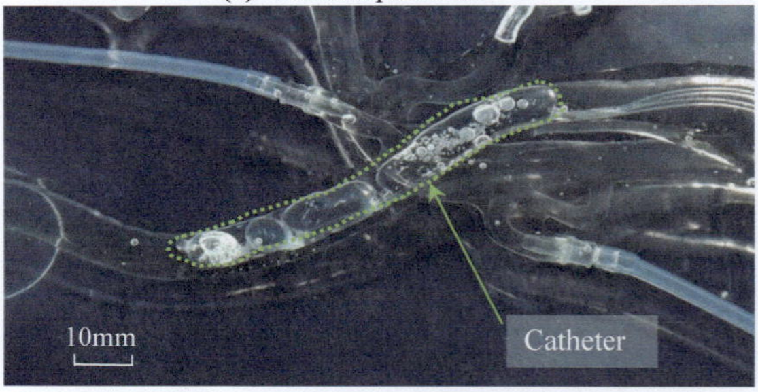

(b) After a lapse of 10 seconds

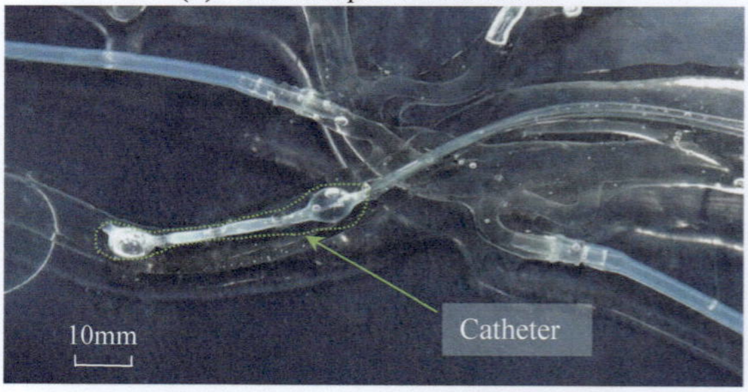

(c) After a lapse of 30 seconds

Fig. 9 Photographs of the catheter moving in endovascular surgery simulator (lines are added to clarify the boundaries). **a** Default position. **b** After a lapse of 10 s. **c** After a lapse of 30 s when the balloon contracts in the right external carotid artery with a radius of curvature of 0.8 cm.

6 Discussion

It is considered that the peristaltically propelled active catheter proposed in this study moves forward in a straight blood vessel without irregular morphological structures on the inner wall at a maximum movement speed of 4.79 mm/s, as determined from the results obtained in the experiment using silicon tubes. The developed active catheter can mostly satisfy the requirements because the movement speed of 5.0 mm/s is required in actual surgery. In the experiment using an endovascular surgery simulator, which simulates the shape of actual blood vessels, the developed catheter was confirmed to move forward in the right internal carotid artery and right external carotid artery, which are close to the radial artery, with a radius of curvature of up to 1.2 cm.

7 Conclusion

In this study, we developed a small active catheter for use in a radial artery. The effectiveness of the catheter was confirmed by the driving experiment using silicon tubes with inner diameters of 2.5–3.0 mm and an endovascular surgery simulator.

Acknowledgments A part of this study was supported by Grant-in-Aid for Scientific Research C (20500397).

References

Annual Health, Labour and Welfare Report 2013–2014 (2014) Overview of the system and the basic statistics, p 10

FAIN-Biomedical Inc. website (2015) Surgical Simulator from FAIN-Biomedical Training-Evaluation-Simulation. In: Challenging the Frontier of Surgical Simulation FAIN-Biomedical, Inc. http://fain-biomedical.co.jp/about_eve. Accessed 30 June 2015

Fukutomi T, Suzuki T et al (2002) Early and late clinical outcomes following coronary perforation in patients under-going percutaneous coronary intervention. Circ J 66(4):349–356

Koga H, Mukawa J et al (1992) Analysis of retained ventricular shunt catheters. Neurol Med Chir 32(11):824–828

Kunio Y, Naohiko T et al (2004) Improved hypertrophic obstructive cardiomyopathy by left ventricular apex epicardial pacing. Intern Med 43(4):295–299

Miyazaki S, Takahashi A et al (2007) Randomized comparison of the continuous vs point-by-point radiofrequency ablation of the cavotricuspid isthmus for atrial flutter. Circ J 71(12):1922–1926

Morise T, Horita M et al (2000) Noninvasive, continuous evaluation of peripheral vascular resistance in humans. Hypertens Res 23(1):15–19

Nakazato Y, Sonobe Y, Toyama S (2010a) Development of an in-pipe micro mobile robot using peristalsis motion. J Mech Sci Tech 24(1):51–54

Nakazato Y, Sonobe Y, Toyama S (2010b) Development of in-pipe micro mobile robot using peristalsis motion driven by hydraulic pressure. Micromechanics and microactuators. In: Proceedings of MAMM 2010, Aachen, Germany, 27–29 May 2010, pp 23–30

Nogami A (2011) Noninvasive, catheter ablation of ventricular fibrillation: state of the art. J Arrhythmia 27(Supplement), SDJ_3

Tsuchjkane E, Nakarnura T et al (1998) Transluminal percutaneous septal myocardial ablation in a patient with hypertrophic obstructive cardiomyopathy. Jpn Circ J 62(7):537–540

Locomotion Principles for Microrobots Based on Vibrations

F. Becker, V. Lysenko, V.T. Minchenya, O. Kunze and K. Zimmermann

Abstract Microrobotics is a growing field with great advances in recent years. New applications in the fields of medicine, biology, manufacturing and maintenance technologies are developed. They require mobile systems with enhanced motion abilities. The present paper concerns principles of terrestrial locomotion for vibration-driven microrobots. Such systems are characterized by an internal periodic excitation, which is transformed to a directed motion due to asymmetric system properties. An extensive overview on the state of the art shows the great potential of the vibration-driven locomotion for miniaturized applications in technics. To perform a controllable two-dimensional locomotion with only one actuator, it is needed to overcome limits of rigid body systems. The proposed approach uses the frequency-dependent vibration behavior of elastic systems, like beams and plates. Experimental investigations are supported by finite element method. It is shown that the two-dimensional locomotion on a flat and solid ground can be controlled by only one actuator using the resonance characteristics of elastic systems.

Keywords Locomotion · Mobile robotics · Microrobot · Vibration · Resonance

---

F. Becker (✉) · O. Kunze · K. Zimmermann
Department of Mechanical Engineering, Technical Mechanics Group,
Technische Universität Ilmenau, Ilmenau, Germany
e-mail: felix.becker@tu-ilmenau.de

O. Kunze
e-mail: oliver.kunze@tu-ilmenau.de

K. Zimmermann
e-mail: klaus.zimmermann@tu-ilmenau.de

V. Lysenko · V.T. Minchenya
Department of Construction and Production of Instruments,
Belarusian National Technical University, Minsk, Belarus
e-mail: victor_lysenko@mail.ru

V.T. Minchenya
e-mail: vlad_minch@mail.ru

L. Zentner et al. (eds.), *Microactuators and Micromechanisms*,
Mechanisms and Machine Science 45, DOI 10.1007/978-3-319-45387-3_9

1 Introduction

A mobile robot can perform locomotion and, due to it, has an increased working space. It is characterized amongst others by its size and mass. The term microrobot has been defined "to include the robotic manipulation of objects with characteristic dimensions in the millimeter to micrometer range (micromanipulation), as well as the design and fabrication of robotic agents in a similar size range (microrobots)" (Abbott et al. 2007). In the literature, the use of the term is not consistent. Robots with physical dimensions in the range of millimeters or centimeters are called mini, micro, or miniaturized. The complexities of fabrication and application of a mechatronic system strongly increase by decreasing the size. That is why the volume of a robot should be chosen as a characteristic physical parameter. Following the volume dependent classification proposed by Caprari (2003), the locomotion principles discussed in this work are applicable for microrobots ($\leq 1\ cm^3$) and minirobots ($\leq 1\ dm^3$).

Macroscopic robotic systems with volumes around cubic meters are often de-signed as rigid body systems with actuated joints (Zimmermann et al. 2009). Classical locomotion principles like rolling and walking are applied. In new fields of applications, especially in the vast fields of manufacturing engineering (handling and assembly of parts from micro fabrication), medicine (drug delivery, minimal invasive surgery, endoscopy), biology (handling of single cells, fluid injection into cells) and maintenance technology (inspection of hard-to-access areas, e.g. complex pipe systems), the application of classical robot designs is limited (Abbott et al. 2007; Sahu et al. 2010; Diller and Sitti 2013; Sitti et al. 2015; Bogue 2015). Systems are needed characterized by:

- small size and mass,
- minimal costs for production and application,
- great working space, and
- long operation time.

The adaption of established systems to the recent challenges by miniaturization is limited due to scaling properties of the physical effects that are used in classical actuators and joints. The production costs and energy consumption of classical systems rise substantially with decreasing physical dimensions. The coordinated control of several energy consuming actuators is needed to create complex multi-directional motion gaits.

Our aim is to show that locomotion principles based on vibrations can be used to create robots, which are suitable for recent requirements, with a minimal number of actuators. Driven by internal high-frequency vibrations, mobile robots can perform controllable locomotion with account of asymmetric system properties depending on frequency. From the methodological point of view it is focused on model-based design, which is not fully based on constructive know-how.

2 Locomotion Principles

The term *locomotion* is commonly used in the biological context and comprises processes of walking, crawling, climbing, digging, swimming, and flying (Zimmermann et al. 2009). That it is rather difficult to give a general definition from the mechanical point of view is discussed in Steigenberger (2011).

In general, actuating forces F_A are needed to be applied to realize locomotion. Resistive forces F_R impede the movement. It is focused on terrestrial locomotion which means that the actuation forces are created relative to a dry and solid ground. The motion of the system is not substantially deforming the environment.

2.1 Nature Inspiration

Nature has created organisms which can perform effective and energy efficient locomotion in any natural environment. Bionics researchers use biological examples for technical solutions (Lepora et al. 2013). Analyzing biological systems, the principles have been categorized, which can be applied to create terrestrial robots with a minimal number of actuators (Lysenko et al. 2007):

- reducing the number of active legs (kangaroo, basilisk),
- periodic change of the body shape (lizard, salamander),
- anisotropy of friction (earthworm, snake),
- multidimensional resonant vibrations of elastic extremities (mosquito).

2.2 Vibration Driven Locomotion

The microrobots considered in this paper are driven by internal actuators. As the deflection of the actuator in general is considerably smaller than the displacement of the robot, a periodic character of the internal excitation is needed. Systems with a periodic drive are called vibratory or undulatory locomotion systems (Ostrowski et al. 1995; Steigenberger 2011). Vibratory locomotion is characterized by "fast" actuation vibrations which overlay the "slow" translocation. Following Blekham (2000), it can be classified by means of a characteristic time interval. The maximal velocity of vibration driven robots can often be observed at an excitation close to the resonance of a system.

A great number of locomotion principles for microrobots are known from the literature. Using the Coulomb model of dry friction, a coherent classification for the motion on flat solid ground can be given (Driesen 2008). It is illustrated in Table 1.

Table 1 Classification of terrestrial locomotion principles based on dry (Coulomb) friction and vibratory actuation, following (Driesen 2008)

		Vibratory actuation	
		Relative motion	Inertia
Asymmetry in the system properties	Normal force $N_+ \neq N_-$	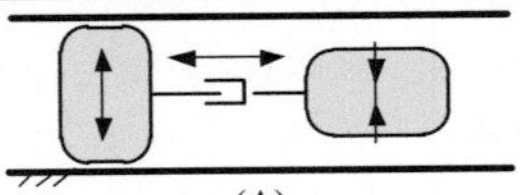 (A)	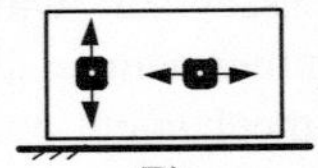 (B)
	Friction coefficient $\mu_+ \neq \mu_-$	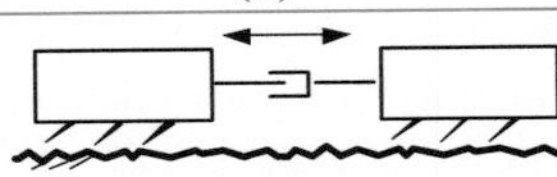 (C)	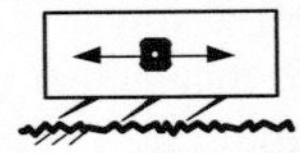 (D)
	Actuation force $F_{A+} \neq F_{A-}$	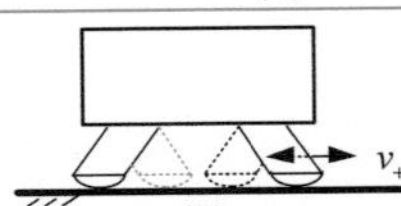 (E)	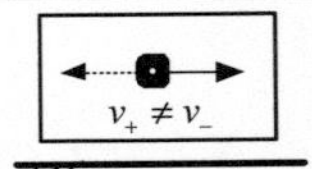 (F)

The creation of a relative displacement between a system and environment can be divided into two principles:

- the contact points of the system to the environment can be moved relative to each other: (A), (C), and (E); or
- the inertias of internal masses, which have no contact to the environment and vibrate relative to the main body of the system, are used: (B), (D), and (F).

The principle (A) comprises systems which realize periodic contact forces in the contact points to the environment (Lobontiu et al. 2001). The changes of the forces are synchronized with the periodic relative displacement of the contact points to each other. It is illustrated in Table 1 by the well-known Inchworm principle. Walking and serpentine motion can be categorized in group (A) too.

Systems of category (B) create horizontal actuation forces by the vibrations of internal masses. The synchronized periodic change of contact forces, e.g. through the vertical vibration of an internal mass, leads to the overall displacement of the system (Bolotnik and Figurina 2008; Zhan et al. 2015).

The principles (C) and (D) are based on the frictional design of the contact areas of the system. The motion in one direction is promoted. It can be modeled with anisotropic friction coefficients. Worm-like locomotion systems (C) are inspired by the peristaltic locomotion of earthworms (Steigenberger and Behn 2012; Gidoni et al. 2014; Bolotnik et al. 2016). They are consisting of two or more segments which can be displaced relative to each other in the direction of motion. If the horizontal force is created by the vibration of an internal mass (D) the system can be constructed without vibrating elements in contact to the environment.

The principles (E) and (F) are based on asymmetric actuation forces. It means that over a period of vibration the values of the acceleration in forward and backward direction differ. Examples are the stick-slip-drive and the inertia drive. Both mechanisms are realized in the field of microrobotics (Eigoli and Vossoughi 2010).

The energetic costs of different locomotion principles are an important point for their application. Especially compared to the rolling motion, the presented principles cannot compete in terms of efficiency. The vibration-driven locomotion is based on differences in the friction properties over a period of the actuation. Periodic sliding motions in forward and backward direction are the usual case. Rolling systems do not have these disadvantages. The potential of the discussed principles lies in the application in environments, which are inappropriate for rolling locomotion, e.g. small maneuvering space or micro scale environments.

3 Locomotion Based on Low and High Frequency Vibration Drives

Robots of micro and mini dimensions for two-dimensional locomotion on flat solid ground are state of the art. Piezoelectric (bending) elements are suitable actuation solutions within the proposed dimensions (1 cm^3–1 dm^3).

Mobile robotics and drive technology are related subjects. Some of the principles discussed in Table 1 are used in piezoelectric motors (Uchino 1998; Janocha 2013; Peng et al. 2015). Examples are inchworm drive, LEGS™ motor, stick-slip principle, and ultrasonic motors. The characteristic conflict of objectives in drive technology between positioning accuracy and actuation speed is present in mobile robotics too.

Inspired by drive technology, mobile micro robots can be divided into quasi-static and resonant systems.

3.1 Quasi-static Mobile Robots

The frequency of the internal actuation vibration of quasi-static robots is considerably smaller than the first natural frequency f_0 of the actuator or a mechanical transmission element. The advantage of such design is the high positioning accuracy: "atomic resolution" (Besocke 1987). Compared to resonant systems, such robots are slow due to the small step size and an excitation with a frequency in the range of $f = 0.01f_0 \dots 0.1f_0$.

Typical quasi-static mobile robots deploy the following principles: walking (Baisch et al. 2011; Hoffman and Wood 2011; Li et al. 2011), inchworm (Yan et al. 2002; Lee et al. 2015), earthworm (Kim et al. 2005), stick-slip and inertia drive (Rembold and Fatikow 1997; Farahani et al. 2012). Within the European project MiCRoN (Estaña and Woern 2007), miniaturized autonomous robots have been developed. Using the stick-slip principle, the MiCRoN-robots can move with a step size down to 10 nm and a speed of 0.4 mm/s.

3.2 Resonant Mobile Robots

Resonant robots are excited by vibration frequencies close to the natural frequencies of elastic mechanical systems, i.e. actuators or transmission elements. Typical realizations are based on piezoelectric bending elements, which excite elastic beams or plates to perform forced vibrations. The internal vibrations are transformed by frictional periodic contacts into motion. The main aim of using the resonance in mobile robots is to achieve an advanced mobility. The systems reported in the literature follow two different vantage points:

1. use of a single natural mode to maximize the robot velocity while minimizing the energy consumption,
2. use of different natural modes to create multidirectional locomotion abilities while minimizing the number of actuators.

Maximizing velocity
The deflection of an actuator or a mechanical transmission element is maximized because of the excitation with a frequency close to a natural frequency. Usually the second natural mode of elastic beams, or the third up to sixth natural mode of a thin elastic plate are used. Anyway, the step size of robots in the considered dimension depends on the concrete design, but it lies typically in the range from nano meters up to micro meters. Using a high excitation frequency from 1 to 100 kHz, fast systems can be designed. Disadvantages can be found in the positioning accuracy and control properties.

Examples of robots based on resonant vibrations are given in Asano et al. (1995), Daugela (1995), DeAmbroggi et al. (1997), Ferreira and Minotti (1997), Woern et al. (2006), Ferreira and Fontaine (2003), Son et al. (2006), Nguyen and Martel (2007), Snis (2008), Avirovik et al. (2014), Hariri et al. (2015), Rios et al. (2015). Within the European I-SWARM project (Woern et al. 2006), autonomous robots for swarm applications have been developed using piezoelectric elastomers (Edqvist 2009). The contributions show that it is rather difficult to achieve a pre-defined frequency dependent locomotion characteristic. In Snis (2008) it is reported that within the I-SWARM project the frequency dependent behavior of every agent needed to be obtained experimentally, as the used second natural frequency could differ around 2 kHz from on agent to another.

Realizing multidirectional locomotion
Early contributions considering the use of different natural modes of a single elastic system to perform multidirectional locomotion are given in Asano et al. (1995), Ferreira and Minotti (1997). The actuator presented in Ferreira and Minotti (1997) consists of a passive metal plate with a fixed bar which creates the frictional contact to the solid ground. A matrix of 4 × 5 piezoelectric ceramic patches is fixed on the system. A synchronized excitation of the piezoelectric elements excites forced vibrations. Using different normal modes, the two-dimensional locomotion can be performed. Degenerated modes allow the rotation of the system. In Ferreira and

Fontaine (2003), a miniaturized system on that basis is presented which consists of 48 such elastic coupled elements. A similar system for one-dimensional locomotion is studied by Son et al. (2006). It uses six piezoelectric patches. In Hariri et al. (2015) it is shown that the idea of Son et al. (2006) can be realized using only one actuator.

4 Multidirectional Resonant Robots

The robots presented in Chap. "A Biologically Inspired Sensor Mechanism for Amplification of Tactile Signals Based on Parametric Resonance" use several actuators to perform two-dimensional locomotion. The deployed principle is based on periodic changes in the frictional contacts between a robot and a solid ground. Similar systems using only one actuator for two-dimensional locomotion are presented in Abaza (2007), Becker et al. (2012). In this paper, the vibration driven locomotion principle is discussed by means of an exemplary prototype.

4.1 Prototype Design

The prototype shown in Fig. 1 consists of a circular piezoelectric plate with three fixed legs made from copper. The legs have contact points to the flat solid ground, i.e. they are acting as vibration transducer. Characteristic values of parameters of the prototype and its actuator are given in Table 2. The robot is driven by a high frequency AC voltage, which is applied to the piezoelectric bending element.

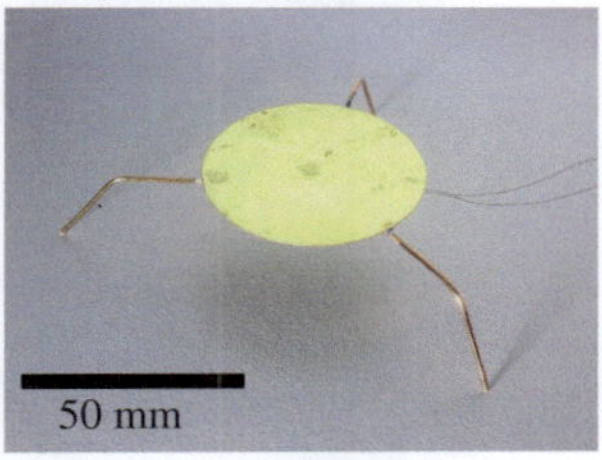

Fig. 1 Robot prototype consisting of a circular piezoelectric bending element and three legs

Table 2 Characteristic data of the prototype and actuator

Robot prototype		Actuator	
Diameter Ø × height	80 mm × 15 mm	Piezo patch	Ø 30 mm × 0.2 mm
Mass	6 g	Brass plate	Ø 50 mm × 0.2 mm
Max. velocity	25 mm/s (on glass)	AC voltage	24 V (0.5–20 kHz)

4.2 Motion Principle and Control

The piezoelectric element vibrates with the frequency of the applied AC voltage and excites the legs to perform bending vibrations. Due to arising complex trajectories of their endpoints the contact and friction properties periodically change. The actuator and the legs with their slightly different geometry form a non-symmetrical vibration system. The endpoint trajectories of each leg differ from each other. Therefore, the resonance characteristic of the elastic overall system can be used to influence on the vibration behavior of every endpoint individually. The locomotion of the system can be controlled by the applied excitation frequency.

To study this phenomenon experimentally, the robots' motion is filmed from above for different frequencies. Two motion markers fixed on the robot are tracked automatically. Results are presented in Fig. 2. The diagram shows dependencies of the translational and rotational velocity on the robots' excitation frequency. The translational velocity is calculated from the trajectory length of the center point of the robot divided by the motion time. The rotational speed gives the change of the robots' orientation in the plane over time. To illustrate the locomotion, three subfigures are added in Fig. 2. For three example frequencies, they illustrate the motion of the prototype by means of an overlay of video frames. The added arrow indicates the locomotion direction.

The motion trajectories of the robots are reproducible. However, our studies (Becker et al. 2012; Becker 2015) and the review of the literature show that it is rather difficult to create robots with predefined locomotion characteristics. Disparities in the used actuators and manual fabrication methods, e.g., soldering and bending, lead to robots with slide structural differences and the varying locomotion behavior of the individual prototypes for the same frequencies. Furthermore,

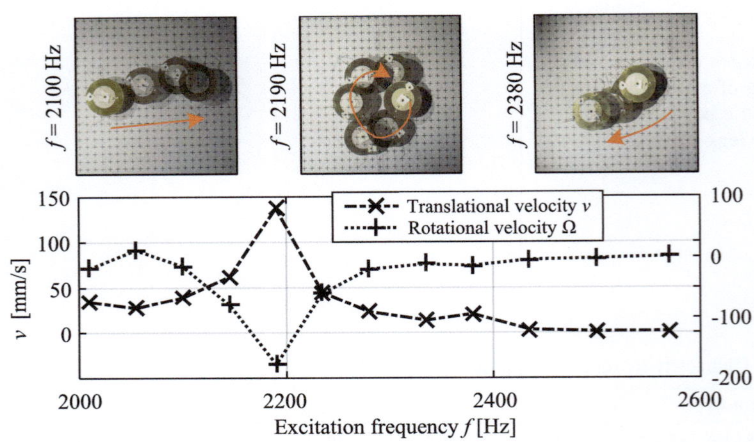

Fig. 2 Locomotion of the prototype excited with a frequencies close to the second natural frequency of a leg $f_2 = 2335$ Hz

the locomotion behavior depends on the substrate of the ground on which the robot is moving on (Becker et al. 2012) and the current mechanical load of the system (Becker 2015). To achieve an automatic control, a teaching process for every single robot is needed.

4.3 Modeling and Simulation

The vibration behavior of the legs has a major influence on the locomotion. The second resonance frequency of an idealized leg, which is considered as a straight Euler-Bernoulli beam with fixed-free support conditions, can obtained with

$$f_2 = \frac{3.507}{L^2} \cdot \sqrt{\frac{EI_z}{\rho A}}. \tag{1}$$

For the parameters of a leg of the presented robot, it follows $f_2 = 2335$ Hz. If the actuator is excited with a frequency close to it, the robot shows a local maximum in translational and rotational speed, as it is presented in Fig. 2. These results confirm the statements in DeAmbroggi et al. (1997), Snis (2008).

To study the described behavior numerically, the finite element method (FEM) is used. The robot is modeled by volume and line elements with and without piezoelectric abilities respectively. Rheonomic conditions are used to fix the center nodes of the actuator. An AC voltage is applied on the boundary of the piezoelectric volume elements. The results presented in Fig. 3 are obtained by the harmonic analysis of the FEM model. The second natural mode of the model is given from two perspectives, and two characteristic points are marked. Additionally, the diagram shows the vibration amplitudes of these points. The local maximum of the amplitude can be found close to the resonance frequency of a single leg. The results

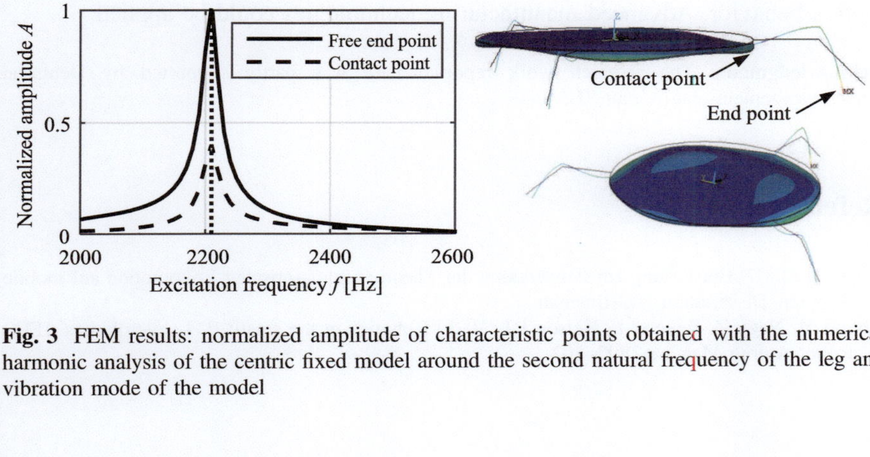

Fig. 3 FEM results: normalized amplitude of characteristic points obtained with the numerical harmonic analysis of the centric fixed model around the second natural frequency of the leg and vibration mode of the model

obtained numerically and experimentally agree with each other. Resonance behavior can be used to maximize the vibration amplitude of the legs' endpoints and, therefore the speed of robots.

5 Conclusions

Vibration-driven locomotion principles are extensively applied in microrobotics. They can be classified by the deployed mechanisms to create the needed asymmetry in the system characteristics or, inspired by drive technology, by the magnitude of the excitation frequency. Quasi-static robots are driven with a frequency considerably smaller than the first natural frequency. A high positioning accuracy, good control properties and a small speed can be achieved. Resonant robots use the frequency-dependent behavior of elastic mechanical elements and actuators to realize fast systems or a multidirectional control. The number of actuators can be minimized.

Robots for multidirectional locomotion can be designed with a single actuator deploying the principle of vibratory locomotion. The dynamics of the legs of a robot has a major influence on the locomotion. The motion trajectories are reproducible, high velocities are achieved. The direction of motion of resonant robots is controlled by the frequency-dependent vibration modes of elastic systems. This control strategy causes special difficulties as the high frequency vibration behavior strongly depends on the boundary conditions and properties of the elastic system. Using prototype manufacturing technologies, it is rather difficult to create robots with a predefined behavior. The frequency-locomotion characteristics need to be obtained individually for every single robot and all combinations of environmental conditions and mechanical loadings. If these characteristics are obtained, it is possible to create locomotion in a precisely desired fashion.

Future extensions of this work will be oriented to increase the robustness of the systems to parameter changes of the environment and robot manufacturing. It should be possible to create robots with a predefined frequency-dependent locomotion behavior. Advanced manufacturing technologies could be applied.

Acknowledgments The research work reported here was partly supported by Deutsche Forschungsgemeinschaft Grant ZI 540/19-1.

References

Abaza K (2007) Ein Beitrag zur Anwendung der Theorie undulatorischer Lokomotion auf mobile Roboter. Dissertation, TU Ilmenau

Abbott JJ, Nagy Z, Beyeler F, Nelson BJ (2007) Robotics in the small: P. 1. microbotics. IEEE Robot Autom Mag 14(2):92–103

Asano M, Matsuoka T, Okamoto H et al (1995) Study for micro mobile machine with piezoelectric driving force actuator. In: Proceedings of the IEEE international conference on robotics and automation (ICRA), vol 3, pp 2955–2960

Avirovik D, Butenhoff B, Priya S (2014) Millipede-inspired locomotion through novel U-shaped piezoelectric motors. IOP Smart Mater Struct 23:1–5

Baisch AT, Heimlich Ch, Karpelson M, Wood RJ (2011) HAMR^3: an autonomous 1.7 g ambulatory robot. In: Proceedings of the IEEE/RSJ international conference on intelligent robots and systems (IROS), pp 5073–5076

Becker F (2015) Zur Mechanik vibrationsgetriebener Roboter für terrestrische und aquatische Lokomotion. Dissertation, TU Ilmenau

Becker F, Minchenya V, Zimmermann K et al (2012) Single piezo actuator driven micro robot for 2-dimensional locomotion. In: Micromechanics and microactuators, vol 2. Springer, Berlin, pp 1–10

Besocke K (1987) An easily operable scanning tunneling microscope. Surf Sci 181:145–153

Blekham II (2000) Vibrational mechanics: nonlinear dynamic effects, general approach, application. World Scientific, Singapore

Bogue R (2015) Miniature and microrobots: a review of recent developments. Ind Robot 42(2): 98–102

Bolotnik N, Figurina T (2008). Vibration-driven systems with movable internal masses: control and optimization. Proceedings of the 53nd international scientific colloquium, Ilmenau, 8–12 Sept 2008, pp 31–32

Bolotnik N, Pivovarov M, Zeidis I et al (2016) The motion of a two-body limbless locomotor along a straight line in a resistive medium. Z Angew Math Mech 4:429–452

Caprari G (2003) Autonomous micro-robots: applications and limitations. Dissertation, École polytechnique fédérale de Lausanne

Daugela A, Fujii H, Jeronymo CE et al (1995) Piezo ceramic based locomotive drive. In: Proceedings of the 6th IEEE international symposium of micro machine and human science, pp 187–192

DeAmbroggi F, Fortuna L, Muscato G (1997) PLIF: piezo light intelligent flea-new micro-robots controlled by self-learning techniques. In: Proceedings of the IEEE international conference on robotics and automation (ICRA), vol 2, pp 1767–1772

Diller E, Sitti M (2013) Micro-scale mobile robotics. Found Trends Robot 2(3):143–259

Driesen W (2008) Concept, modeling and experimental characterization of the modulated friction inertial drive (MFID) locomotion principle: application to mobile microrobots. Dissertation, École polytechnique fédérale de Lausanne

Edqvist E, Snis N, Mohr R et al (2009) Evaluation of building technology for mass producible millimetre-sized robots using flexible printed circuit boards. J Micromech Microeng 19(7)

Eigoli AK, Vossoughi GR (2010) Dynamic modeling of stick-slip motion in a legged piezoelectric driven microrobot. Int J Adv Robotic Syst 7(3):201–208

Estaña R, Woern H (2007) The MiCRoN robot project. Informatik akutell, Autonome mobile systeme. Springer, Berlin, pp 334–340

Farahani AA, Suratgar AA, Talebi HA (2012) Optimal controller design of legless piezo capsubot movement. Int J Adv Robot Syst 10(126):1–7

Ferreira A, Fontaine J (2003) Dynamic modeling and control of a conveyance microrobotic system using active friction drive. IEEE-ASME T Mech 8(2):188–202

Ferreira A, Minotti P (1997) Control of a multidegree of freedom standing wave ultrasonic motor driven precise positioning system. Rev Sci Instrum 68(4):1779–1786

Gidoni P, Noselli G, DeSimone A (2014) Crawling on directional surfaces. Int J Non-Linear Mech 61:65–73

Hariri HH, Soh GS, Foong SH et al (2015) Miniature piezoelectric mobile robot driven by standing wave. Proceedings of the 14th world congress in mechanism and machine science, pp 1–6

Hoffman KL, Wood RJ (2011) Passive undulatory gaits enhance walking in a myriapod millirobot. In: Proceedings of the IEEE/RSJ international conference on intelligent robots and systems (IROS), pp 1479–1486

Janocha H (2013) Unkonventionelle Aktoren: Eine Einführung. Oldenbourg Verlag, München

Kim B, Park S, Jee CY et al (2005) An earthworm-like locomotive mechanism for capsule endoscopes. In: Proceedings of the 2005 IEEE/RSJ international conference on intelligent robots and systems, 2–6 Aug 2005, pp 2997–3002

Lee K, Kim Y, Park JK et al (2015) Clawed miniature inchworm robot driven by electromagnetic oscillatory actuator. J Bionic Eng 12(4):519–526

Lepora NF, Verschure P, Prescott TJ (2013) The state of the art in biomimetics. Bioinspir Biomm 8(1):013001

Li W, Li J, Hu, H et al (2011) Analysis and experiment of stick-slip motion principle in a legged microrobot. In: Proceeding of the 6th forum on strategic technology, vol 1, pp 328–332

Lobontiu N, Goldfarb M, Garcia EA (2001) A piezoelectric-driven inchworm locomotion device. Mech Mach Theory 36(4):425–443

Lysenko V, Minchenya W, Zimmermann K (2007) Minimization of the number of actuators in legged robots using biological objects. In: Proceedings of the 52nd international scientific colloquium, Ilmenau, 10–13 Sept 2007, pp 483–488

Naguyen AT, Martel S (2007) Locomotion of a miniature robot based on synchronized vibrating actuation mechanisms. In: Proceedings of the IEEE/ASME international conference on advanced intelligent mechatronics (AIM), pp 1–6

Ostrowski J, Burdick J, Lewis AD et al (1995) The mechanics of undulatory locomotion: the mixed kinematic and dynamic case. In: Proceedings of the IEEE international conference on robotics and automation (ICRA), vol 2, pp 1945–1951

Rembold U, Fatikow S (1997) Autonomous microrobots. J Intell Robot Syst 19(4):375–391

Rios SA, Fleming AJ, Yong YK (2015) Design of a two degree of freedom resonant miniature robotic leg. In: Proceedings of the IEEE international conference on advanced intelligent mechatronics (AIM), pp 318–323

Sahu B, Taylor CR, Leang KK (2010) Emerging challenges of microactuators for nanoscale positioning, assembly, and manipulation. J Manuf Sci Eng 132(3):030917-1–16

Sitti M, Ceylan H, Hu W et al (2015) Biomedical applications of untethered mobile milli/microrobots. Proc IEEE 103(2):205–224

Snis N (2008) Actuators for autonomous microrobots. Digital Comprehensive Summaries of Uppsala Dissertations from the Faculty of Science and Technology 431, Uppsala University

Son KJ, Kartik V, Wickert JA et al (2006) An ultrasonic standing-wave-actuated nano-positioning walking robot: piezoelectric-metal composite beam modeling. J Vib Control 12(12):1293–1309

Steigenberger J (2011) Some theory towards a stringent definition of 'locomotion'. Multibody Syst Dyn 26(1):81–90

Steigenberger J, Behn C (2012) Worm-like locomotion systems: an intermediate theoretical approach. Oldenbourg Verlag, München

Uchino K (1998) Piezoelectric ultrasonic motors: overview. Smart Mater Struct 7:273–285

Woern H, Szymanski M, Seyfried J (2006) The I-SWARM project. In: Proceedings of the 15th IEEE international symposium on robot and human interactive communication, pp 492–496

Yan G, Lu, Q, Ding G et al (2002) The prototype of a piezoelectric medical microrobot. In: 13th IEEE international symposium of micromechtronics and human science, pp 73–77

Yuxin P, Yulong P, Gu X et al (2015) A review of long range piezoelectric motors using frequency leveraged method. Sensor Actuat A-Phys 235:240–255

Zhan X, Xu J, Fang H (2015) Planar locomotion of a vibration-driven system with two internal masses. Appl Math Model. http://dx.doi.org/10.1016/j.apm.2015.06.016

Zimmermann K, Zeidis I, Behn C (2009) Mechanics of terrestrial locomotion. Springer, Berlin

Exploration of Carbon-Filled Carbon Nanotube Vascular Stents

Darrell J. Skousen, Kristopher N. Jones, Takami Kowalski, Anton E. Bowden and Brian D. Jensen

Abstract The purpose of this research was to design, fabricate, and test coronary stent designs composed of carbon-infiltrated carbon nanotubes (CI-CNTs). Coronary stents currently have two major complications: restenosis and thrombosis. CI-CNT stents may provide improved clinical outcomes for both of these issues. Multiple stent design concepts were generated, evaluated, and two stent designs were selected: one with a semi-auxetic nature, and one designed for maximum force. These designs were further developed and optimized using analytical tools along with finite element analysis. Planar versions of the stent designs were manufactured and mechanically tested to verify performance. The performance of the cylindrical stent configurations was analyzed using finite element modeling. A sample cylindrical stent was also fabricated. This research demonstrates that feasible coronary stent designs can be manufactured from CI-CNTs. However, a major challenge for CI-CNT stent designs is meeting the design requirement of sufficient radial force.

Keywords Carbon nanotubes · Vascular stent · Microfabrication

D.J. Skousen
Merit Medical Systems, Inc., South Jordan, UT, USA
e-mail: darrellsk@gmail.com

K.N. Jones
Ford Motor Company, Dearborn, MI, USA
e-mail: kris.jones13@gmail.com

T. Kowalski · A.E. Bowden · B.D. Jensen (✉)
Department of Mechanical Engineering, Brigham Young University, Provo, UT, USA
e-mail: bdjensen@byu.edu

T. Kowalski
e-mail: kowalski.takami@gmail.com

A.E. Bowden
e-mail: abowden@byu.edu

L. Zentner et al. (eds.), *Microactuators and Micromechanisms*,
Mechanisms and Machine Science 45, DOI 10.1007/978-3-319-45387-3_10

1 Introduction

The leading cause of death in the United States is heart disease, claiming approximately 24 % of all deaths (Hoyert and Xu 2012; Gutman and Golomb 2012). Coronary artery disease is a condition where coronary arteries narrow due to fatty plaque build up, reducing the blood flow to the heart, which can lead to heart failure and death (Gaziano et al. 2011). A minimally invasive procedure called percutaneous coronary intervention (PCI), including balloon and stent angioplasty has been developed to treat coronary artery disease. There are two major complications associated with PCI: restenosis (renarrowing) and thrombosis (blood clots). Restenosis is caused by a combination of early elastic recoil, negative remodeling, and neointimal formation. Early elastic recoil occurs immediately, and is due to the elastic properties of the arteries. Late lumen loss in balloon angioplasty is caused by neointimal formation (tissue in-growth) (20–30 %) and negative remodeling (arterial shrinking) (70–80 %) (Mintz et al. 1996; Kimura et al. 1997; Weintraub 2007).

In stent angioplasty, a cylindrical scaffold wire mesh (stent) often made of stainless steel is implanted in the artery to prevent restenosis. These stents prevent elastic recoil and negative remodeling; however, neointimal formation can still lead to restenosis (Fig. 1). Stents have proven to reduce the rates of restenosis more than angioplasty alone. Drug-eluting stents have further reduced restenosis rates, but there is a concern for their ability to prevent late-term thrombosis (Moses et al. 2003; Stone et al. 2004; Luscher et al. 2007). New stent materials that can improve these two complications will be advantageous in stent development.

Pyrolytic carbon (PyC) is a biomaterial that has excellent biocompatibility, and has been used in over 4 million heart valves. The blood compatibility of PyC has been tested extensively (Goodman et al. 2007; Yang et al. 1996; Cenni et al. 1995; Slaughter et al. 2011; Stary et al. 2003; Ratner et al. 2004; Antoniucci et al. 2000). Medical implants (including vascular stents) constructed from stainless steel or titanium have been coated with PyC to improve biocompatibility (Visconti 2011). Carbon-infiltrated carbon nanotube (CI-CNT) materials produced in this research are a form of PyC which can be fabricated into coronary stent designs.

CNTs can be used as a framework in producing precise geometries including MEMS by performing controlled growth of VACNTs (vertically aligned carbon nanotubes). In this process, CNTs are grown vertically and can therefore produce complicated structural geometry with excellent manufacturing tolerances (1–3 μm) (Hutchison et al. 2010). CI-CNTs are patterned in a 2-dimensional geometry, which can then be infiltrated with PyC using chemical vapor deposition to create a structure. Typical ceramics have undesirable properties for use in compliant (flexible) designs, but the CI-CNT material differs from typical ceramics by tolerating high elastic strains of up to 2.3 % (Fazio et al. 2011).

Many of the issues affecting existing vascular stents including restenosis, thrombosis, and the need for anti-coagulant drugs therapy (blood thinners) could be improved or eliminated by using a more biocompatible material. Significant challenges exist for the design of stents from ceramic materials such as CI-CNTs. Most

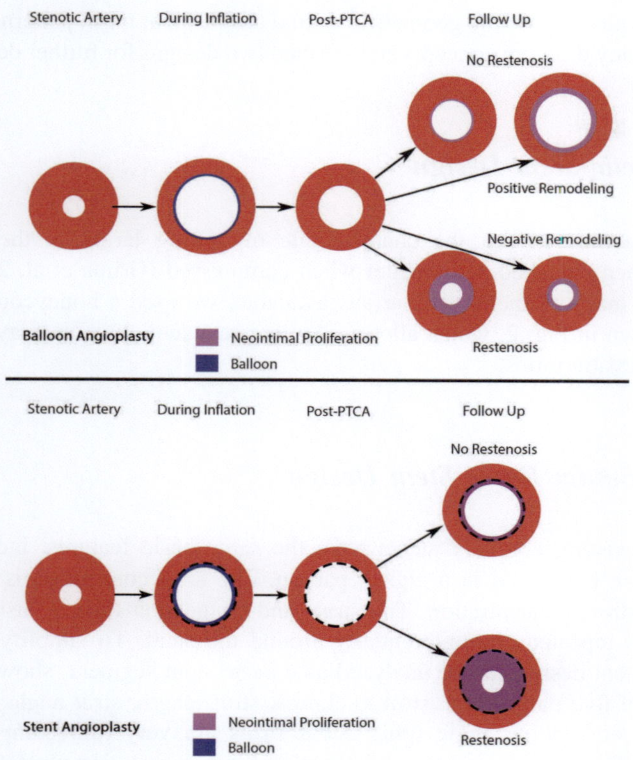

Fig. 1 *Upper* Pathway for balloon angioplasty without a stent. *Lower* Pathway following stent implantation (Falk et al. 2009)

notably, compliant, expanding and contracting mechanisms can be difficult to design because of their brittle nature. If these challenges are overcome, then a biocompatible CI-CNT coronary stent could be developed to improve treatment. The purpose of this research was to design, fabricate, and test coronary stent designs composed of CI-CNTs. The designs were fabricated and tested as planar patterns, and a sample pattern was rolled and assembled into a cylindrical stent.

2 Design Exploration

Vascular stents undergo a large amount of deflection during insertion. Due to the relatively high strain of CI-CNTs, coronary stents made of this material can be fabricated in their expanded state and elastically compressed for insertion into the body. We explored possible geometries to find usable stent mesh patterns. Our team generated many different concepts and selected two designs for further development.

2.1 Auxetic Stent Design

Auxetic materials exhibit the characteristic of getting larger in the transverse direction when stretched and smaller when compressed (Grima et al. 2008). After considering many of the auxetic arrays available, we used a honeycomb-like pattern, as shown in Fig. 2, which allows us to create a semi-auxetic array capable of high compression ratios.

2.2 Maximum Force Stent Design

Nearly all existing coronary stents have the same basic features, including thin struts that are connected in a zigzag pattern. The strut connection is rounded to reduce the stress concentration. The struts and connection form a basic stent segment that is repeated circumferentially around the stent. To simplify the design process, a stent design can be analyzed as a basic stent segment, shown in Fig. 3, consisting of five parameters: strut thickness, strut length, strut angle (theta), end radius, and end radius angle (phi). Stent struts are very thin compared to the diameter of the stent, making an analysis of a planar stent design almost identical to a cylindrical design.

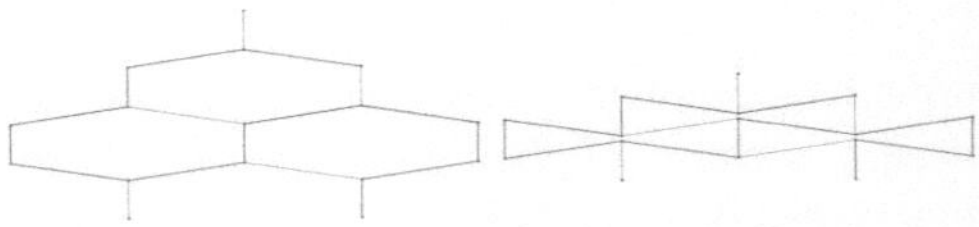

Fig. 2 The honeycomb-like pattern expanded (*left*) and compressed (*right*)

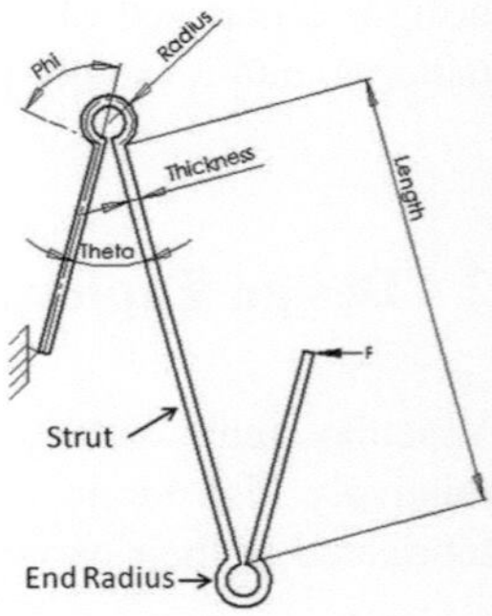

Fig. 3 A basic stent segment described by 5 parameters: strut thickness, angle, and length, and end radius and end radius angle

3 Optimization

Both stent designs were further developed and optimized into a usable stent geometry. The use of engineering fundamentals as well as finite-element analysis (FEA) were critical in optimizing the designs.

3.1 Auxetic Stent Optimization

The semi-auxetic honeycomb provided the rigid link pattern to design a CI-CNT stent. Because the bowtie design is a repeated cell array, we optimized a single cell. Further, a single cell consists of four identical beams which undergo the same deflection. We know from compliant mechanism theory that thinner beams can undergo larger deflections with lower stresses (Howell 2001). The diagonal beams from Fig. 2 needed to be thin and flexible while the vertical beams remained thick and rigid to provide structure, shape, and support. The stent was designed to have an expanded diameter of 3 mm and a compressed diameter of 1 mm.

The connection between the rigid and flexible beams results in a stress concentration. Instead of using a fillet to reduce stress, a more effective spline geometry of varying thickness was used to optimize the stress distribution throughout the material. Using the optimizer in ANSYS, a routine was created which built the geometry to optimize the stress. For spline creation, four horizontally spaced control points were chosen that were allowed to move vertically, thus dictating the shape of the spline. The code built the geometry with a specified set of spline control points, performed the deflection and stress analysis, and, using gradient based optimization, changed the vertical component of each control point, performing the analysis again until an optimum was reached. Figure 4 demonstrates how ANSYS was able to optimize a spline geometry within the given parameters and constraints to minimize and distribute stress along the beam.

3.2 Maximum Force Stent Optimization

CI-CNT stent designs inherently provide lower radial force than metallic stents due to the lower elastic modulus and brittle nature of CI-CNT's. The second stent design was optimized to provide the maximum possible radial force without

Fig. 4 Example of optimized beam taper showing deflection and stress

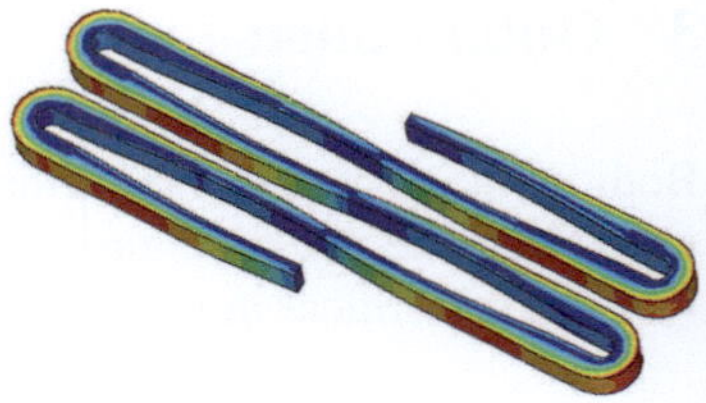

Fig. 5 FEM of the TMF design with first principal stresses distributed fairly uniformly. The design does not clash even at 65 % compression

exceeding the allowable stress. By reducing the stent strut thickness, the stresses could be lowered as needed; however, a trade-off occurs as radial stiffness/force decreases with decreased thickness. The stresses and reaction forces for the basic stent segment were calculated using mechanics of materials equations as well as the pseudo-rigid-body model for compliant mechanisms (Howell 2001). Due to the symmetric nature of the basic stent segment, only a quarter of that segment was needed for optimization.

The basic stent segment was optimized using an exhaustive search with discrete values for continuous variables. The constraints included a tensile stress less than 80 MPa, a compressive stress less than 120 MPa, and no physical contact (clash) between the stent segments. The optimal design had the largest possible thickness without exceeding the allowable stress. Also, the optimal strut angle was the smallest possible angle that could be achieved without the segments clashing before reaching a 2/3 compression.

The basic stent segment was further analyzed and optimized by finite element modeling (FEM) using ANSYS. 21,700 8-node brick (Solid185) elements were used. A tapered beam was used instead of a constant thickness beam. The struts were changed from being straight to slightly curved to avoid clashing. The stent was designed to have a depth radial depth of 100 μm, with 12 circumferentially repeating segments. A manual approach was used to optimize the tapering and curvature of the stent design. After approximately 30 analyses were performed, a final design was selected (Fig. 5). The improved design resulted in stresses that were much more uniformly distributed, and adjacent stent segments that did not clash together. The improvements tripled the reaction force compared to the original design while only slightly increasing the maximum tensile stress. The final design was named the tapered maximum force (TMF) design.

4 Fabrication and Testing

Planar versions of both stent designs were fabricated for mechanical testing, and a cylindrical stent was also demonstrated. Samples were also analyzed using an FE atherosclerotic arterial model.

4.1 Fabrication

The same general method for CI-CNT fabrication described in detail by Jones et al. (2013) was used in the fabrication of the stent designs. This process included standard photolithography using a layer of alumina, a layer of iron for CNT growth, growth of CNTs and infiltration by chemical vapor deposition, and finally removal of the sample from the substrate. Planar samples of each design were fabricated.

4.2 Fixture Design and Testing Procedure

Each sample was tested in an Instron tabletop tensile testing machine. A specially designed fixture and a force transducer were used to gradually compress the stent and record the resulting force and displacement. Figure 6 shows a fabricated semi-auxetic stent in the test set-up. Each sample was tested by securing the pattern in the fixture and compressing at a constant rate of 3 mm/min. Figure 7 shows a typical test sample at the start and end of the test. Figure 8 shows the force-deflection curves for the semi-auxetic design. The plot reveals that the data

Fig. 6 Photograph of testing fixture in the tabletop Instron. A semi-auxetic stent pattern is loaded in the fixture

Fig. 7 Example CI-CNT semi-auxetic stent mesh before and after compression

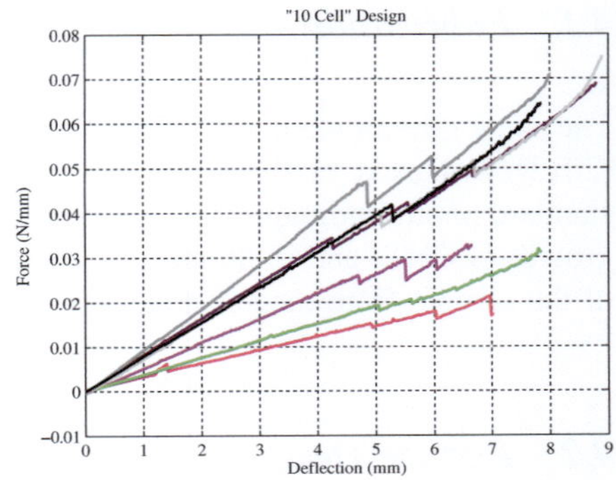

Fig. 8 Force deflection plots for several instances of the semi-auxetic design

are nearly linear and do not have plastic deformation before failure. Jagged points on the data represent buckling or fracturing of a single beam.

The Young's modulus was extracted from the test data using a finite element model. The average Young's modulus of the semi-auxetic samples was 18.6 GPa, while that of the TMF samples was 10.6 GPa. This difference may be caused because the semi-auxetic designs have smaller strut thicknesses (25 μm) compared to the TMF design (75 μm). The smaller thicknesses may result in more complete infiltration during fabrication, which results in a higher material stiffness. However, as shown in Fig. 8, there was significant difference in the measured stiffness even from sample to sample for the same design. In addition, the average percent compression for both designs was about 36 %.

4.3 Cylindrical Stent Fabrication

To create a cylindrical stent, we added keys and locks to either end of the pattern. The resulting parts were taken and prepared for manipulation by using cyanoacrylate to fix small stainless steel rods to the stent to allow a micromanipulator to attach to the stent. A micromanipulator was used to bring and rotate the end attached to the micromanipulator to the opposite side to connect the locks. Another micromanipulator was used to drip cyanoacrylate onto the lock and key to seal the connection. A sample of such a cylindrical stent is shown in Fig. 9.

Fig. 9 A cylindrical stent fabricated by curling a flat pattern into a cylinder

4.4 Arterial Response

In order to predict the performance of a stent before it is used in clinical trials, FE models of an atherosclerotic coronary artery have been used in stent analyses, and were also created for this research (Migliavacca et al. 2004; Liang et al. 2005; Lally et al. 2005). This arterial model was validated by finite-element analysis of an existing stainless steel stent, and then used to analyze the CI-CNT stent designs.

A normal sized coronary artery with a diameter of 3 mm was modeled (Hermiller et al. 2009). A simple model consisting of an artery and plaque, each 0.5 mm thick hollow concentric cylinders, was developed. The typical arterial size in clinical trials is a diameter of around 3 mm with a pre-dilation minimal lumen diameter of about 1 mm. During dilation the balloon permanently deforms the plaque and artery lining, leaving a minimal diameter of roughly 2 mm (Suryapranata et al. 1998).

The finite element arterial model had the following dimensions at rest: Artery OD—3.75 mm, Artery ID—2.75 mm, Plaque OD—2.75 mm, Plaque ID—1.75 mm. When applying an internal pressure of 100 mm Hg, the plaque ID expands to 2.02 mm, and the artery ID expands to 2.95 mm, similar to clinical trial arteries. The material parameters used are given by Migliavacca (2004) in their analysis, which used parameters calculated from the testing of biological tissues by Salunke (2001).

Due to the symmetric nature of the artery and plaque, only a quarter cylinder was modeled. To validate our model, we simulated a common stainless steel stent (Cordis BX Velocity) and predicted the artery's response (Migliavacca et al. 2005). The material properties of the stent used are those of Liang and include: elastic modulus = 201 GPa, poisson ratio = 0.3, yield stress = 330 MPa, limit stress = 750 MPa (Liang et al. 2005).

The stent was expanded outward to a diameter of 3.30 mm. The stent then experienced elastic recoil reducing its diameter to 3.19 mm, and leaving the final minimum lumen diameter of the artery at 2.71 mm, a result similar to those seen in clinical trials (Fig. 10) (Moses et al. 2003).

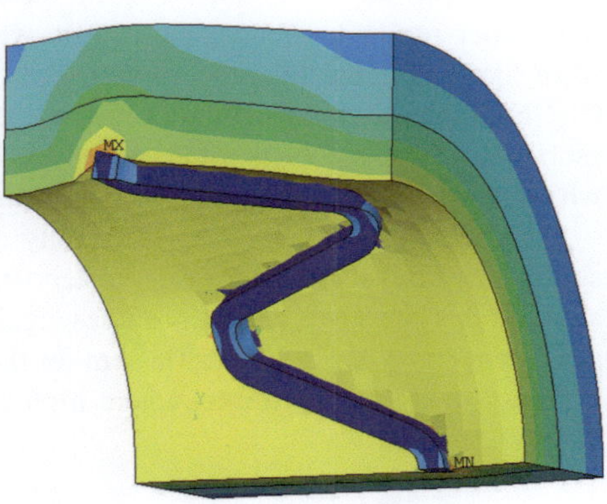

Fig. 10 Arterial response to stent angioplasty using a Cordis BX Velocity stent

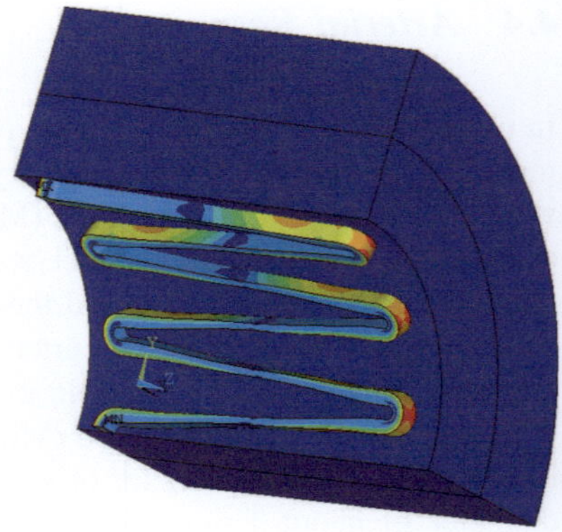

Fig. 11 Arterial response to stent angioplasty using a TMF stent design (1st principal stress)

We also analyzed the performance of a quarter section of the TMF stent design using the FE arterial model. Material properties for the CI-CNT material were taken from the test data described in Sect. 4.2. The stent was compressed, and then moved into the artery and allowed to spring back, pushing against the artery wall. The artery was pressurized at 100 mmHg, and had an initial minimum lumen diameter of 2.00 mm. After the stent was released the artery expanded to have a minimum lumen diameter of 2.05 mm (Fig. 11). The semi-auxetic stent design was also analyzed using this method and expanded the artery from an initial diameter of 2.006–2.011 mm.

5 Discussion

Mechanical testing of the manufactured planar stent samples revealed that the designs can fracture during compression earlier than desired (36 % average compression). Even though the semi-auxetic designs performed largely as anticipated, there was significant variability in the compression testing results. Several factors contributed to the variation. First, the fabrication equipment used was also shared with other research groups; these tests were conducted over several months and were likely affected by process variation. Also, due to the ceramic nature of the CI-CNT material, failure often occurred due to defects in the material caused during fabrication. These defects often caused the stent designs to fail early in one location while the rest of the design could withstand greater compression.

Another area of concern is the ability of the CI-CNT stent designs to provide sufficient radial force to prevent restenosis. It is clear from the arterial response analysis that the TMF design provides significantly less radial force than a stainless steel stent (final diameter of 2.05 mm for the TMF design compared to 2.71 mm for stainless steel). In conditions where high radial force is needed to prevent arterial recoil and negative remodeling, the CI-CNT stents may have difficulty preventing restenosis. However, in situations where restenosis is driven by negative remodeling and intimal hyperplasia, the CI-CNT designs may outperform existing stents.

6 Conclusions and Recommendations

This research has produced two promising coronary stent designs. Research was conducted to validate the use of CI-CNTs to create planar stent geometries. Two stent designs were developed and optimized. Samples were fabricated and tested in compression to demonstrate the material's flexibility and to confirm its mechanical properties. 3-dimensional designs and analyses were performed to predict stent performance. We confirmed that stents composed of CI-CNTs can deflect sufficiently without fracturing and could be used in stent applications, though the deflection was not as great as desired. However, due to a relatively small Young's modulus, the CI-CNT stents produce significantly less radial force than stainless steel stents.

Acknowledgments The research work reported here was partially supported by Clarke Capital Partners. We thank Jordan Tanner, Melanie Ferguson, and Warren Robison for their help with testing.

References

Antoniucci D, Bartorelli A et al (2000) Clinical and angiographic outcome after coronary arterial stenting with the carbostent. Am J Cardiol 85(7):821–825

Cenni E, Granchi D, Arciola CR et al (1995) Adhesive protein expression on endothelial cells after contact in vitro with polyethylene terephthalate coated with pyrolytic carbon. Biomaterials 16 (16):1223–1227

Falk E, De Feyter PJ, Shah PK (2009) Ischemic heart disease. Manson

Fazio WC, Lund JM, Wood TS et al (2011) Material properties of carbon-infiltrated carbon nanotube-templated structures for microfabrication of compliant mechanisms. In: Proceedings of the ASME 2011 international mechanical engineering congress & exposition, no. 64168

Gaziano J, Ridker O, Libby P (2011) Braunwald's heart disease: a textbook of cardiovascular medicine. Saunders Elsevier

Goodman SL, Scranton VL, Brendzel AM (2007) Platelet responses to silicon-alloyed pyrolytic carbons. J Biomed Mater Res A 83(1):64–69

Grima JN, Farrugia PS, Caruana C et al (2008) Auxetic behavior from stretching connected squares. J Mater Sci 43:5962–5971

Gutman D, Golomb G (2012) Liposomal alendronate for the treatment of restenosis. J Controlled Release 161:619–627

Hermiller JB, Fergus T, Pierson W et al (2009) Clinical and angiographic comparison of everolimus-eluting and paclitaxel-eluting stents in small coronary arteries: a post hoc analysis of the spirit iii randomized trial. Am Heart J 158(6):1005–1010

Howell LL (2001) Compliant mechanisms. Wiley, New York

Hoyert DL, Xu J (2012) Deaths: preliminary data for 2011. National Vital Statistics Reports, 61

Hutchison DN, Morrill NB, Aten Q et al (2010) Carbon nanotubes as a framework for high-aspect-ratio MEMS fabrication. J Microelectromech Syst 19:75–82
Jones K, Jensen BD, Bowden A (2013) Fabrication and testing of planar stent mesh designs using carbon-infiltrated carbon nanotubes. J Nanotechnol Eng Med 4(2)
Kimura T, Kaburagi S, Tamura T et al (1997) Remodeling of human coronary arteries undergoing coronary angioplasty or atherectomy. Circulation 96(2):475–483
Lally C, Dolan F, Prendergast PJ (2005) Cardiovascular stent design and vessel stresses: a finite element analysis. J Biomech 38(8):1574–1581
Liang DK, Yang DZ, Qi M, Wang WQ (2005) Finite element analysis of the implantation of a balloon-expandable stent in a stenosed artery. Int J Cardiol 104(3):314–318
Luscher TF, Steffel J, Eberli FR et al (2007) Drug-eluting stent and coronary thrombosis: biological mechanisms and clinical implications. Circulation 115(8):1051–1058
Migliavacca F, Petrini L, Massarotti P et al (2004) Stainless and shape memory alloy coronary stents: a computational study on the interaction with the vascular wall. Biomech Model Mechanobiol 2(4):205–217
Migliavacca F, Petrini L, Montanari V et al (2005) A predictive study of the mechanical behaviour of coronary stents by computer modelling. Med Eng Phys 27(1):13–18
Mintz GS, Popma JJ et al (1996) Intravascular ultrasound assessment of the mechanisms and predictors of restenosis following coronary angioplasty. J Invasive Cardiol 8(1):1–14
Moses JW, Leon MB et al (2003) Sirolimus-eluting stents versus standard stents in patients with stenosis in a native coronary artery. N Engl J Med 349(14):1315–1323
Ratner BD, Hoffman AS, Shoen FJ, Lemons JE (2004) Biomaterials science: an introduction to materials in medicine. Elsevier Academic Press
Salunke NV, Topoleski LD, Humphrey JD, Mergner WJ (2001) Compressive stress-relaxation of human atherosclerotic plaque. J Biomed Mater Res 55(2):236–241
Slaughter MS, Pederson B, Graham JD et al (2011) Evaluation of new forcefield technology: reducing platelet adhesion and cell coverage of pyrolytic carbon surfaces. J Thorac Cardiovasc Surg 142(4):921–925
Stary V, Bacakova L, Hornik J, Chmelik V (2003) Bio-compatibility of the surface layer of pyrolytic graphite. Thin Solid Films 433:191–198
Stone GW, Ellis SG, Cox DA et al (2004) A polymer-based, paclitaxel-eluting stent in patients with coronary artery disease. N Engl J Med 350(3):221–231
Suryapranata H, van't Hof AW, Hoorntje JC et al (1998) Randomized comparison of coronary stenting with balloon angioplasty in selected patients with acute myocardial infarction. Circulation 97(25):2502–2505
Visconti G, Focaccio A, Tavano D et al (2011) The CID Chrono cobalt-chromium alloy carbofilm-coated coronary stent system. Int J Cardiol 149(2):199–204
Weintraub WS (2007) The pathophysiology and burden of restenosis. Am J Cardiol 100(5):S3–S9
Yang Y, Franzen SF, Olin CL (1996) In vivo comparison of hemocompatibility of materials used in mechanical heart valves. J Heart Valve Dis 5(5):532–537

A Novel Gripper Based on a Compliant Multistable Tensegrity Mechanism

S. Sumi, V. Böhm, F. Schale, R. Roeder, A. Karguth and K. Zimmermann

Abstract Within this paper a novel gripper is introduced. The gripper is based on a tensegrity mechanism with multiple states of self-equilibrium. These tensegrity mechanism is built upon a mechanical compliant tensegrity structure, consisting of tensile and compressive members. The existence, shape and stability of the different states of self-equilibrium depend on the parameters of the members. After investigating this dependency with a form-finding algorithm the structure is extended with additional members to obtain a two-finger-gripper. The equilibrium configurations represent the opened and the closed state of the gripper. Due to the several equilibrium configurations, the control of the gripper can be realised easily and is discussed with theoretical analyses. The working principle of the gripper and the mechanical compliance is verified by a prototype. In addition, as an outlook for further works, a miniaturised prototype of the tensegrity mechanism with two stable states of self-equilibrium has been built.

Keywords Compliant tensegrity structure · Compliant gripper · Multiple states of self-equilibrium · Gripper application

S. Sumi (✉) · A. Karguth
TETRA GmbH Ilmenau, Ilmenau, Germany
e-mail: Susanne.Sumi@tetra-ilmenau.de

A. Karguth
e-mail: Andreas.Karguth@tetra-ilmenau.de

V. Böhm · F. Schale · R. Roeder · K. Zimmermann
Department of Mechanical Engineering, Technical Mechanics Group, Ilmenau University of Technology, Ilmenau, Germany
e-mail: Valter.Boehm@tu-ilmenau.de

F. Schale
e-mail: Florian.Schale@tu-ilmenau.de

R. Roeder
e-mail: Richard.Roeder@tu-ilmenau.de

K. Zimmermann
e-mail: Klaus.Zimmermann@tu-ilmenau.de

L. Zentner et al. (eds.), *Microactuators and Micromechanisms*,
Mechanisms and Machine Science 45, DOI 10.1007/978-3-319-45387-3_11

1 Introduction

In the field of robotics the use of prestressed mechanically compliant mechanisms is a recently discussed topic. Tensegrity mechanisms, based on mechanical compliant tensegrity structures, are an example of such mechanisms. Tensegrity structures consist of groups of not directly connected compressive members, connected by a continuous net of tensile members (Motro 2003; Pugh 1976).

Robots based on tensegrity mechanisms are lightweight, deployable, have a simple system design, shock absorbing capabilities and a very high strength to weight ratio (Ashwear 2015; Böhm and Zimmermann 2013; Caluwaerts et al. 2014; Rieffel et al. 2010). These are great advantages for the use of tensegrity mechanisms in robot applications. But these properties result in a large control effort. An overview about recent developments can be found in Boehler et al. (2015), Böhm et al. (2015), Mirats Tur and Hernandez (2009), Skelton and De Oliveira (2009).

Known robots based on tensegrity structures are using conventional tensegrity structures with only one state of self-equilibrium. A special property of tensegrity mechanisms is that they may have multiple states of self-equilibrium (equilibrium configurations). These tensegrity structures are called multistable tensegrity structures. In literature only few structures of this kind were previously reported (Defossez 2003; Micheletti 2013; Xu and Luo 2011). If the members have a pronounced elasticity it may be possible to change between the equilibrium configurations. This means to change the shape of the tensegrity structure with small control effort, which is an advantage compared to the conventional tensegrity structures. The consideration of multistable compliant tensegrity structures is an attractive new research topic.

In the present article a planar multistable tensegrity structure is discussed and the possible use of these structures in gripper applications is introduced. Some of the results are already discussed in Sumi et al. (2016). In this article the tensegrity structure and the gripper-application are investigated more detailed.

The presented multistable tensegrity structure has two stable equilibrium configurations. These two equilibrium configurations correspond to the opened and closed states of the gripper. Due to the design, the gripping force results only from the prestress of the structure.

A first, in Sect. 2, the topology of the tensegrity mechanism is considered and the equilibrium configurations in dependency of the parameters of the members are discussed. Then gripper arms are added to the tensegrity mechanism and different kinds of actuation are introduced. In Sect. 3, experimental results are presented and finally Sect. 4 is devoted to conclusions and further research directions.

2 Theory

The tensegrity structure consists of three members of high axial stiffness (members 1–3) and six members of low axial stiffness (members 4–9). The topology is shown in Fig. 1.

The following theoretical analyses (including the form-finding procedure) are performed with the static Finite Element Method including geometric nonlinearity.

All members are massless 2D linear spring elements with constant axial stiffness (parameters of the members: k_i—axial stiffness, L_{0i}—free length). For the element stiffness matrix see Guest (2006).

The characteristic equation is

$$\{F(\{u\})\} = [K(\{u\})]\,\{u\}, \tag{1}$$

with $\{F(\{u\})\}$ vector of nodal forces, $\{u\}$ vector of nodal displacements, $[K(\{u\})]$ tangent stiffness matrix. This Eq. (1) is solved with an incremental-iterative procedure with the geometric boundary conditions: $u_{x1} = u_{y1} = u_{y2} = 0$ mm. The static stability of the equilibrium configurations is determined with the help of the eigenvalues of $[K(\{u\})]$, see Zhang and Ohsaki (2015).

If the members' parameters are properly selected, the tensegrity structure has three stable equilibrium configurations, see Fig. 1 A, B and C. For more detailed information how these parameters are selected, see Böhm et al. (2016). All

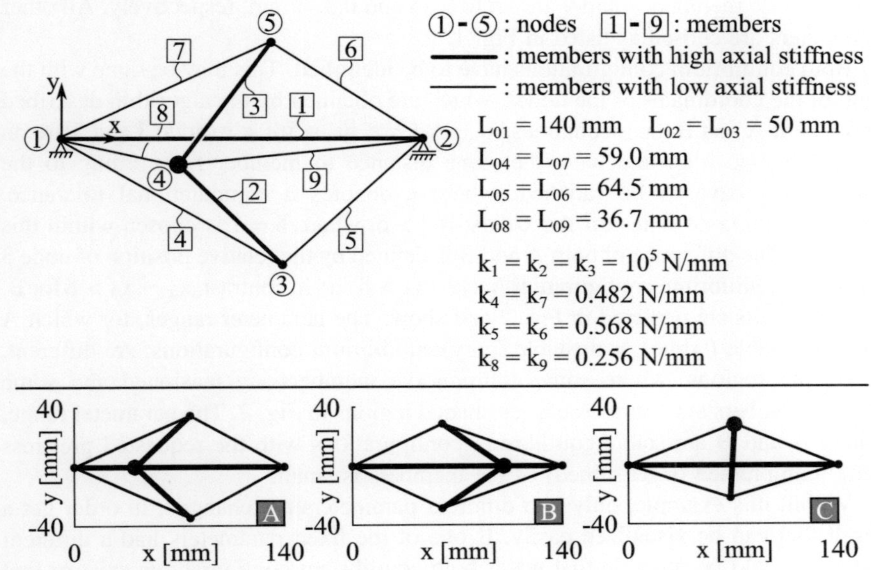

Fig. 1 Topology, members' parameters and three possible stable equilibrium configurations of the considered planar tensegrity structure

members of high axial stiffness are compressed and all members of low axial stiffness are tensioned, in all three configurations. There are two groups of compressed members (member 1 as the first, members 2 and 3 as the second), which are only indirectly connected with tensile members, so it is a tensegrity mechanism, indeed. Since there are member 2 and 3 directly connected, it's a class 2 tensegrity structure (Skelton et al. 2001).

In the following only equilibrium configurations A and B are considered. These two configurations are needed for the specific application.

2.1 Equilibrium Configurations

A study of the equilibrium configurations is presented in this section. The equilibrium configurations, their existence, shape and stability depend on the parameters of the members (stiffness and free length). This is shown by an example, where some of the parameters of the members are changed and the equilibrium configurations are investigated.

The considered tensegrity structure, with the parameters from Fig. 1, has three stable equilibrium configurations A, B and C. As mentioned above, only A and B are considered in the following.

Within this example the free lengths of the tensile members 4 and 7 and of the tensile members 5 and 6 are varied: $L_{04} = L_{07} = 40, \ldots, 80$ mm, $L_{05} = L_{06} = 40, \ldots,$ 80 mm with a step size of 0.3 mm. In order to disturb the symmetry, the stiffnesses of the tensile members 8 and 9 are set to 0.25 and 0.3 N/mm, respectively. All other parameters are chosen as listed in Fig. 1.

Both equilibrium configurations have to be identified. This may be done with the help of the coordinates of the nodes, which are obtained by the algorithm described above. For A and B the inequalities $|x_5 - x_3| < \varepsilon$, $|y_3 + y_5| < \varepsilon$ (node 3 and 5 are on the same x-position and have the same distance to member 1 according to the y-direction) have to be satisfied, where ε denotes a computational tolerance. Furthermore, node 3 lies off the x-axis: $|y_3| > \delta$, where $\delta = 1$ is chosen within this example. The difference of both A and B is defined by the relative position of node 3 and 4: for equilibrium configuration A $x_3 - x_4 > \delta$ and in contrast, $x_4 - x_3 > \delta$ for B.

The results are depicted in Fig. 2 and show: The parameter ranges, for which A or B are stable (white) or unstable (grey) equilibrium configurations, are different. There are regions, where some compressive members are tensioned and some tensile members are compressed, see shaded regions in Fig. 2. The parameter range, where A and B are stable equilibrium configurations with the requested prestress state (compressed or tensioned) of the members is small.

Within this example, only two different parameters are changed, in order get a result that can be visualized easily. If one of the fixed parameters had a different value, it could be possible that never both equilibrium configurations exist or that never both equilibrium configurations are stable. A corresponding example is presented in Fig. 3. There is only one modification of the example above, the

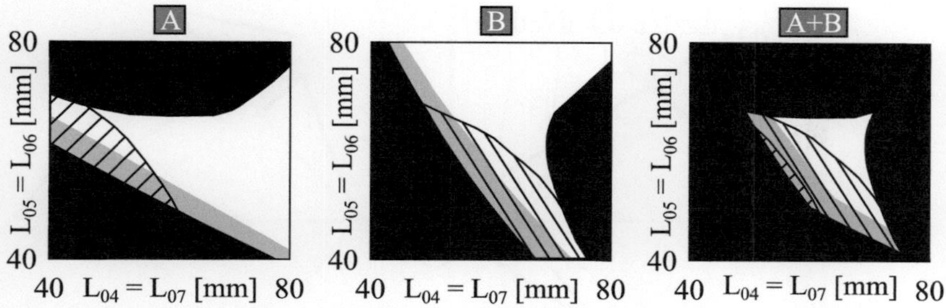

Fig. 2 Existence and stability of equilibrium configurations A and B, depending on parameters of the members. *Black* Equilibrium configuration does not exist or no convergent result; *White/grey* Stable/unstable equilibrium configuration; *Shaded* At least one member hasn't the requested prestress state

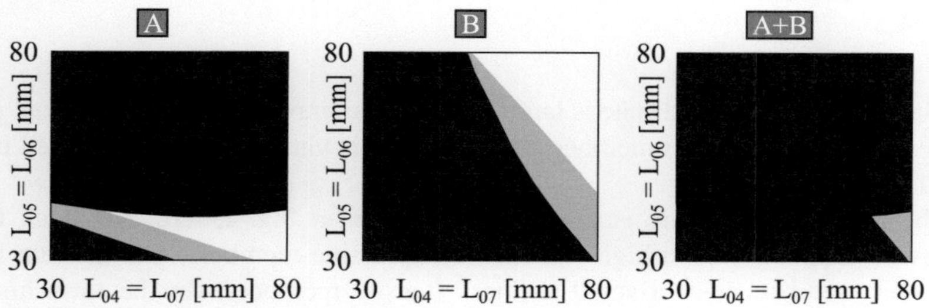

Fig. 3 Existence and stability of equilibrium configurations A and B, depending on parameters of the members. *Black* Equilibrium configuration does not exist or no convergent result; *White/grey* Stable/unstable equilibrium configuration

stiffnesses of the tensile members 8 and 9 are set to 0.1 and 0.7 N/mm, respectively. As a result, there are values for $L_{04} = L_{07}$ and $L_{05} = L_{06}$, where equilibrium configuration A exists and is stable, and there are values, where B exists and is stable. But it's not possible to have values for $L_{04} = L_{07}$ and $L_{05} = L_{06}$ within the interval [30 mm, 80 mm] so that both A and B are existing and stable.

In conclusion: The parameters of the members have to be chosen carefully to get the two stable equilibrium configurations A and B.

2.2 Gripper Arm Design

According to the mentioned application, there may be gripper arms added to the tensegrity structure. Then the equilibrium configurations A and B represent the opened and the closed state of the gripper, respectively.

In a simple case stiff gripper arms are fixed connected to the compressive members 2 and 3. In Fig. 5 (first row) the opened and the closed state and the change between the states are depicted. It can be observed, that the space

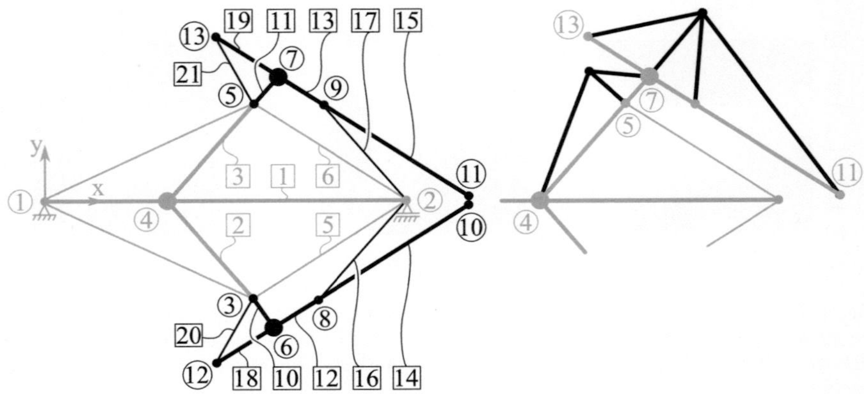

Fig. 4 Construction of the gripper arms and the truss frameworks that are needed for the calculation

requirement in the opened state is large. This can be prevented, if an additional joint and two additional tensile members are added. This variant will be discussed more detailed, in the following.

The tensegrity structure is extended, as shown in Fig. 4, left. The thick lines represent members with a high stiffness (compressive members) and the thin lines represent members with a small stiffness (tensile members). In the form-finding procedure, all members are connected with pin-joints. In the considered design variant the connections of members 2 and 10, and of members 3 and 11 have to be stiff. Additional the connection of members 14, 12 and 18 and of the three members 15, 13 and 19 has to be stiff. For that reason, there have to be added truss frameworks to the structure for the used algorithm. The truss frameworks are depicted in Fig. 4, right. The added members are members with high stiffness.

The parameters of the members are determined, so that member 14 (15) is parallel to member 5 (6), in the closed state of the gripper and so that member 14 and 15 are parallel to member 1, in the opened state of the gripper.

To achieve a larger gripping-force, there may be pin-joints with a limited rotational range used in node 6 and 7. Using these modified joints the angle between member 10 and 18 (11 and 19) cannot get smaller than needed in the closed state of the gripper. With this modification the gripping-force depends only on the underlying tensegrity structure. Another positive aspect of these modified joints is, that members 16, 17, 20 and 21 may have lower stiffnesses than the other members with a low stiffness in the underlying tensegrity structure. They have to be only as strong as needed to compensate the weight force of the gripper arms.

The opening of the gripper is depicted in Fig. 5, lower row. It can be observed, that the space requirement is lower than in the first variant (first row).

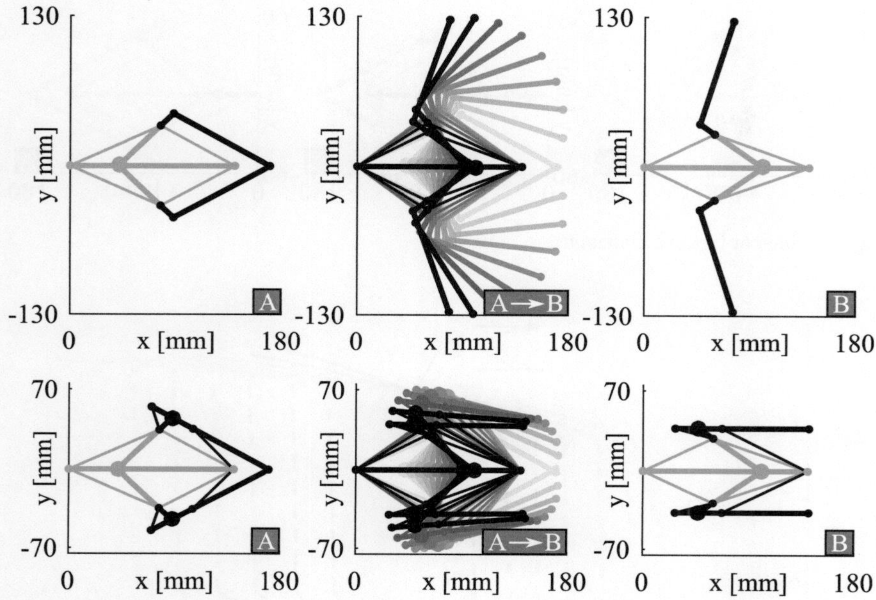

Fig. 5 Two variants to realise the gripper arms and the transition between the closed and the opened state of the gripper

2.3 Actuation

Actuation of the gripper means to open and to close the gripper, which is the transition between equilibrium configuration A and B. There are many different possible principles to actuate the gripper:

- Change the relative position between selected members
- Change the prestress of selected members (by changing stiffnesses and/or lengths)
- Change the position of selected members.

For all three principles are examples depicted in Fig. 6. In the left variant, the angle between the members 2 and 3 is changed. In the second variant, the stiffnesses of member 8 (to close the gripper) and 9 (to open the gripper) are changed. And in the right variant, node 4 is moved along the x-axis.

The third variant of actuation is considered in the following, as an example. While altering the position of node 4 along the x-axis with an actuation force, there is a snap-through between the equilibrium configurations (both from A to B and from B to A). The relation between the actuation force at node 4 (F_{4x}) and the position of node 4 (x_4) is depicted in Fig. 7. It is shown for seven different states of prestress. For the coloured line, the parameters of the members are chosen as listed in Fig. 1 and described in Sect. 2.2. For the other plots is satisfied: Brighter lines

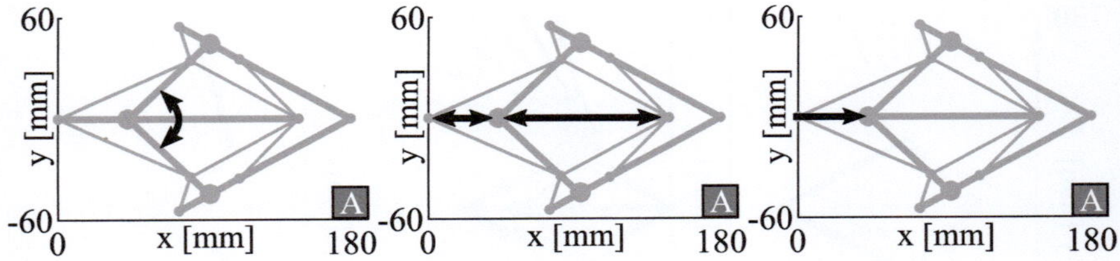

Fig. 6 Different types of actuation

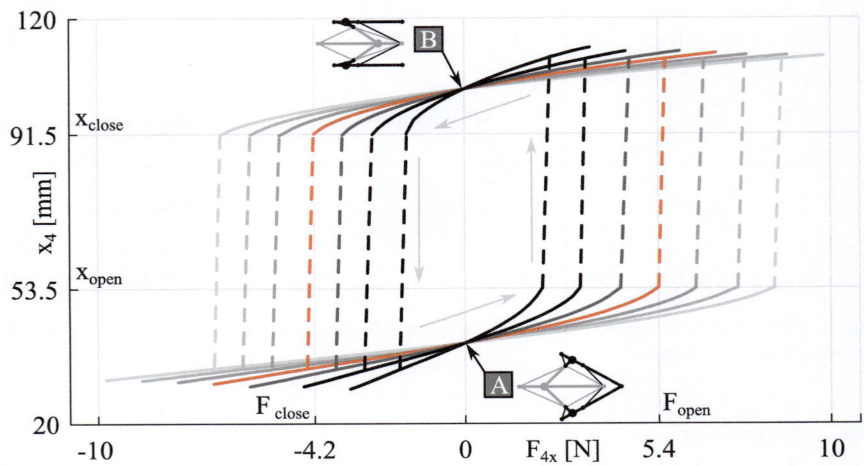

Fig. 7 Relation between the actuation force at node 4 (F_{4x}) and the position of node 4 (x_4). For the coloured line the members' parameters from Fig. 1 and from Sect. 2.2 are used. Brighter lines stand for a higher prestress and darker lines for a lower prestress

mean a higher prestress of the system and darker lines stand for a lower prestress. To change the prestress, the stiffnesses of all members are scaled by the same factor.

It can be seen (Fig. 7), that the change of the prestress does not change the form of the equilibrium configurations A and B (see $F_{4x} = 0$ N). The position of node 4, where the snap-through takes place, x_{open}, x_{close}, is the same for the different states of prestress. But the amount of force (F_{open}, F_{close}) that is needed to reach these positions (x_{open}, x_{close}) depends on the state of prestress. If the prestress is higher (brighter lines in Fig. 7), the amount of force is larger.

This shows, that the mechanical compliance of the gripper can be controlled by the level of prestress of the system.

To maintain the mechanical compliance of the gripper during actuation, the actuator (e.g. a linear servomotor) is coupled with node 4 by a connecting piece, containing an elongated hole. During actuation at the snap-though-position (x_{open} or x_{close}) the actuator may stop and the gripper snaps into the other state and node 4 snaps to the other side of the elongated hole. This process is schematically depicted

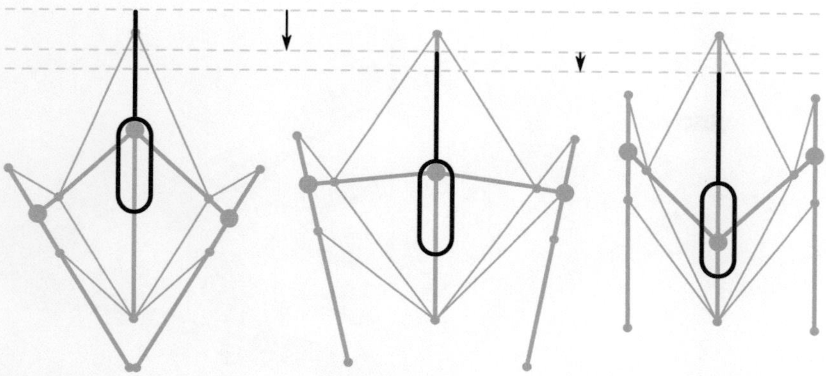

Fig. 8 Actuation of the gripper realised by a connecting piece with an elongated hole to preserve the snap-through-effect

in Fig. 8. The elongated hole shall be longer than needed, to ensure that node 4 does not clash with the connecting piece.

Due to this realisation, the maximal force needed for the actuation is 5.4 N and the minimal force is −4.2 N. To hold the gripper in its opened or closed state no energy is needed. The mechanical compliance is preserved, although there is an actuator used, which is not compliant.

3 Experiments

The introduced gripper has been built to verify the theoretical results. The parameters from Fig. 1 are used (free lengths and axial stiffnesses of the springs, lengths of the compressive members). Since it is a planar tensegrity structure, it has to be slightly modified to be able to build it (e.g. member 1 and 8 lie in the same plane, which is not possible within a prototype) and to ensure spatial stability. For this reason, member 2–9 have been doubled, see Fig. 9.

The actuation is realised, like presented in Fig. 8, with a Linear DC-Servomotor (LM 2070-080-01, Faulhaber Inc.), with max. peak/continuous forces of 27.6 N/9.2 N, with a stroke length of 80 mm and with a max. speed of 2.3 m/s. The connecting piece has an elongated hole with a length of 30 mm (see Fig. 9).

The functionality of the system has been proven by preliminary tests. The gripper may grip objects with a width up to 60 mm and has a maximal gripping force of approximately 2 N.

The process to open and to close the gripper is shown in Fig. 10.

The gripper has a high mechanical compliance, which is depicted in Fig. 11. Due to this property it is possible to grip objects that are eccentrically positioned. While moving the gripper, small obstacles don't harm the movement.

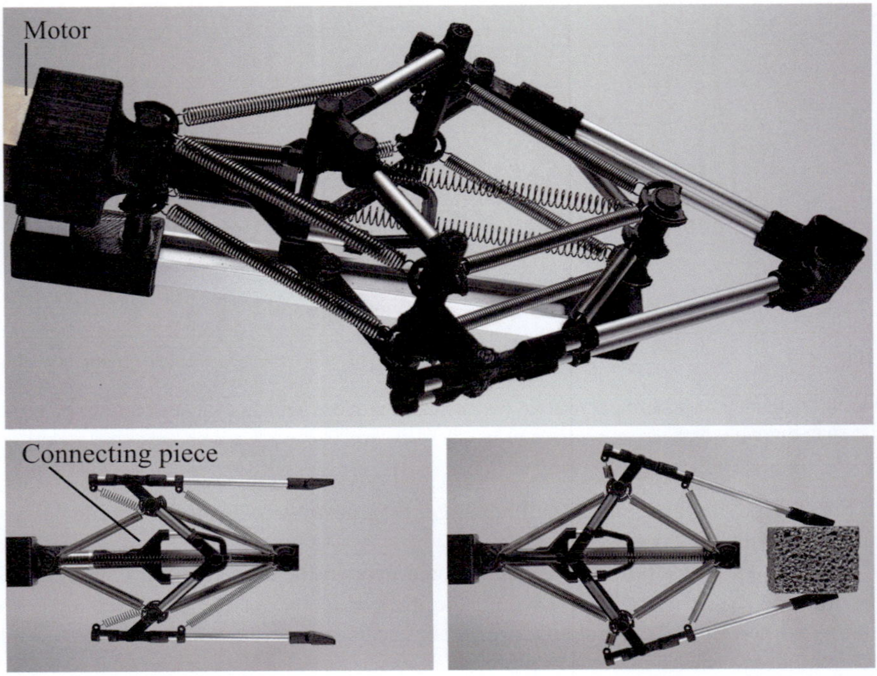

Fig. 9 *Top* Isometric view of the prototype; *Bottom* The gripper in opened and closed state with a gripped object (40 mm × 54 mm × 49 mm)

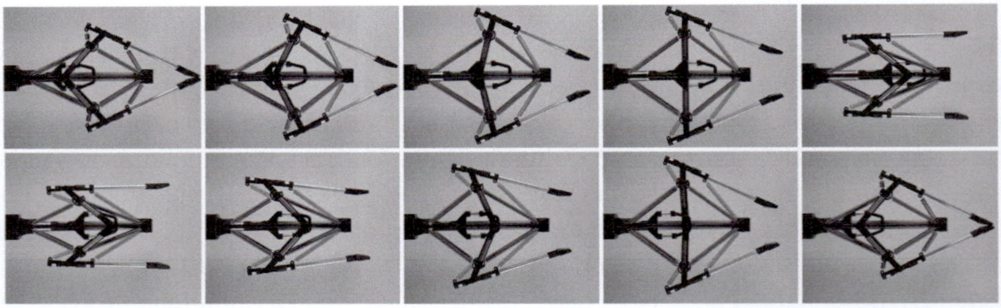

Fig. 10 Process to open (*top* row, *left* to *right*) and to close (*bottom* row, *left* to *right*) the gripper

As a first step for further development in the direction of miniaturisation, another prototype of the tensegrity structure has been build. It is shown in Fig. 12. Member 1 has a length of approximately 70 mm and members 2 and 3 of 30 mm. In this prototype the tensile elements are merged to a continuous monolithic net made of silicone elastomer. This design method enables the further realisation of miniaturised grippers based on the introduced principle.

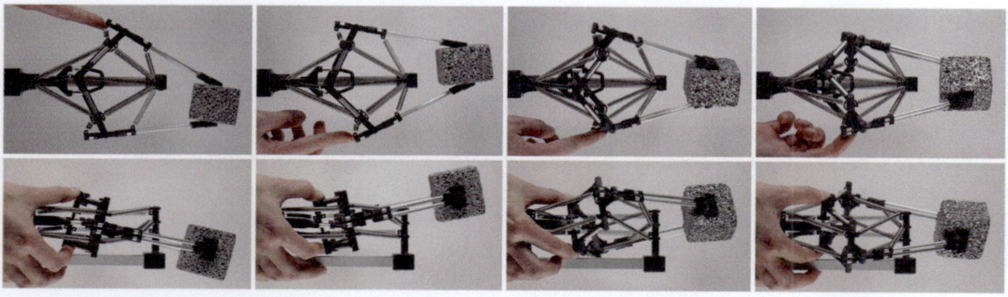

Fig. 11 Demonstration of the mechanical compliance of the gripper

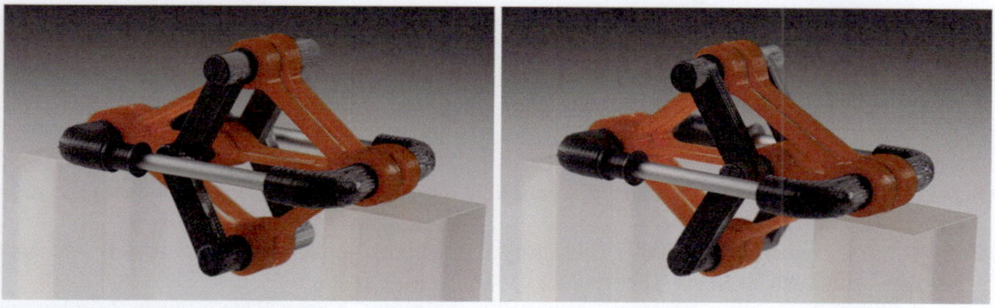

Fig. 12 Two stable equilibrium configurations of a multistable tensegrity structure with a monolithic net of tensile elements made of silicone elastomer

4 Conclusions

In this article a new design principle of compliant grippers based on tensegrity mechanisms with multiple states of self-equilibrium is presented. It is possible to build compliant grippers based on structures with pronounced elasticity. Grippers based on tensegrity mechanisms with multiple states of self-equilibrium have several advantages, e.g. simple system design and small control effort. To meet the needs of a selected application only few system-parameters have to be customized.

Additional a smaller version of the tensegrity mechanism has been built to show the possibilities of further miniaturisation.

Future work will focus on the miniaturisation of tensegrity mechanisms (for gripper applications and robotic applications). Further topologies for tensegrity mechanisms with multiple states of self-stress will be investigated. The realisation of systems with tuneable performance (compliance, gripping force) by adjusting their prestress will be considered. Alternative actuation techniques will be tested.

Acknowledgments This work is supported by the Deutsche Forschungsgemeinschaft (DFG project BO4114/2-1).

References

Ashwear N, Tamadapu G, Eriksson A (2015) Optimization of modular tensegrity structures for high stiffness and frequency separation requirements. Int J Solids Struct 80:297–309

Boehler Q, Charpentier I, Vendrines MS et al (2015) Definition and computation of tensegrity mechanism workspace. J Mech Robot 7(4):4, paper No: JMR-14–1168,

Böhm V, Zimmermann K (2013) Vibration-driven mobile robots based on single actuated tensegrity structures. In: Proceedings of the IEEE international conference on robotics and automation, pp 5455–5460

Böhm V, Zeidis I, Zimmermann K (2015) An approach to the dynamics and control of a planar tensegrity structure with application in locomotion systems. Int J Dyn Control 3(1):41–49

Böhm V, Sumi S, Kaufhold T et al (2016) Compliant multistable tensegrity structures with simple topologies. In: Proceedings of the 6th European conference on mechanism science 2016

Caluwaerts K, Despraz J, Iscen A et al (2014) Design and control of compliant tensegrity robots through simulation and hardware validation. J Roy Soc Interf, 11

Defossez M (2003) Shape memory effect in tensegrity structures. Mech Res Commun 30(4):311–316

Guest S (2006) The stiffness of prestressed frameworks: a unifying approach. Int J Solids Struct 43(3–4):842–854

Micheletti A (2013) Bistable regimes in an elastic tensegrity system. In: Proceedings of the royal society A, vol 469, no 2154

Mirats Tur JM, Hernàndez S (2009) Tensegrity frameworks: dynamic analysis review and open problems. Int J Mech Mach Theory 44(1):1–18

Motro R (2003) Tensegrity: structural systems for the future. Kogan Page Science, London and Sterling

Pugh A (1976) An introduction to tensegrity. University of California Press

Rieffel JA, Valero-Cuevas FJ, Lipson H (2010) Morphological communication: exploiting coupled dynamics in a complex mechanical structure to achieve locomotion. J Roy Soc Interf 7:613–621

Skelton RE, De Oliveira MC (2009) Tensegrity systems. Springer, Berlin

Skelton RE, Adhikari R, Pinaud JP et al (2001) An introduction to the mechanics of tensegrity structures. In: The mechanical systems design handbook: modeling, measurement, and control. CRC Press, Boca Raton

Sumi S, Böhm V, Schale F et al (2016) Compliant gripper based on a multistable tensegrity structure. In: Proceedings of the 6th European conference on mechanism science 2016

Xu X, Luo Y (2011) Multistable tensegrity structures. J Struct Eng 137(1):117–123

Zhang JY, Ohsaki M (2015) Tensegrity structures—form, stability, and symmetry. Springer, Berlin

Selection of the Optimal Rigid-Body Counterpart Mechanism in the Compliant Mechanism Synthesis Procedure

N.T. Pavlović, N.D. Pavlović and M. Milošević

Abstract The main problems appearing in the compliant mechanism synthesis procedure are the realization of the accuracy of the pre-defined function as best as possible, as well as the realization of the motion range of the mechanism as large as possible. One of the most frequent compliant mechanism synthesis procedure is to develop the compliant mechanism on the basis of the rigid-body linkage being able to realize pre-defined function of the compliant mechanism. There are many advantages of using the compliant joints in the mechanism structure. On the other hand, the mechanisms with compliant joints can realize relatively small motion range, that is, their mobility is limited. Another limitation to their use is fatigue failure at the compliant joints. In each particular compliant joint the angular deflection of a flexure hinge is limited in depending of the contour and hinge dimensions as well as the type of the material being used for the production of the compliant mechanism. Therefore, if there are several rigid-body counterparts to be used in compliant mechanism synthesis procedure, it should be necessary to choose the particular rigid-body counterpart with the smallest values of the angular deflection of the compliant joints. This approach will be presented in the synthesis procedure of the compliant mechanism for rectilinear guiding.

Keywords Compliant mechanisms · Compliant joints · Mobility · Synthesis · Rectilinear guiding

---

N.T. Pavlović (✉) · N.D. Pavlović · M. Milošević
Faculty of Mechanical Engineering, University of Niš, Niš, Serbia
e-mail: pnenad@masfak.ni.ac.rs

N.D. Pavlović
e-mail: pavlovic@masfak.ni.ac.rs

M. Milošević
e-mail: mmilos@masfak.ni.ac.rs

L. Zentner et al. (eds.), *Microactuators and Micromechanisms*,
Mechanisms and Machine Science 45, DOI 10.1007/978-3-319-45387-3_12

1 Introduction

Compliant mechanisms gain some or all of their mobility from the relative flexibility of their joints rather than from rigid-body joints only. There are many advantages of using the compliant joints in the mechanism structure: a mechanism can be built in one piece, the weight can be reduced and wear, clearance, friction, noise and need for lubrication can be eliminated. Therefore they are suitable to be applied in micromachining. On the other hand, the mechanisms with compliant joints can realize relatively small displacements, that is, their mobility is limited.

There are many papers considering the structure and function of the compliant joints and compliant mechanisms. The paper (Midha et al. 1994) has established basic nomenclature and classification for the components of compliant mechanisms. The paper (Howell and Midha 1994) has introduced a method to aid in the design of a class of compliant mechanisms wherein the flexible sections (flexural pivots) are small in length compared to the relatively rigid sections. The paper (Ananthasuresh and Kota 1995) has presented a formal structural optimization technique called the homogenization method in order to design flexible structures (compliant mechanisms). The paper (Böttcher et al. 2001) has introduced new ideas of technically realizable joints from nature and their integration into elastically movable structures for motion tasks in positioning and manipulating engineering.

Some papers have analyzed the influence of the geometry, as well as the material type of the compliant joints on guiding accuracy of the point of the compliant mechanisms (Pavlović and Pavlović 2002, 2003, 2004, 2013).

The paper (Pavlović and Pavlović 2009) has presented the original designs of the compliant parallel-guiding mechanisms.

The frequent method to synthesize a compliant mechanism is to design it as the counterpart of the rigid-body linkage being able to realize pre-defined function of the compliant mechanism (rigid-body replacement method). This paper deals with a selection of the optimal rigid-body counterpart mechanism in the compliant mechanism synthesis procedure. In each particular compliant joint there is a limitation of the angular deflection of a flexure hinge, in depending of the contour and hinge dimensions as well as type of the material being used for the production of the compliant mechanism. Therefore, if there are several rigid-body counterparts to be used in compliant mechanism synthesis procedure, it should be necessary to choose the particular rigid-body counterpart with the smallest values of the angular deflection of the compliant joints. The aim of this paper is to present the above mentioned approach in the synthesis procedure of the compliant mechanism for rectilinear guiding.

2 Rigid-Body Linkages for Rectilinear Guiding

There are many designs of the rigid-body linkages for realization of rectilinear guiding. Five well-known designs will be presented in the next sections.

2.1 Hoecken Four-Bar Linkage

The coupler point C of the Hoecken four-bar linkage (Fig. 1), located at the end of the coupler, can be guided on an approximate rectilinear path segment (Dizioglu 1965).

The link lengths of this mechanism, as well as the position of the input crank, that enable realizing of horizontal displacement of the coupler point C of Δx_C = 5 mm, with the small angular displacement of the input crank ($\Delta\varphi = 5^{\circ}$) and minimal difference between realized and exact rectilinear path (Δy_C), have been shown in Fig. 1 (Pavlović and Pavlović 2003, 2013):

The angles defining the relative positions of the neighboring links, that is, the links being connected with particular revolute joint, can be defined as follows:

$$\begin{aligned} \angle xA_0A &= \phi \\ \angle A_0AB &= \pi - \phi + \kappa = \alpha \\ \angle ABB_0 &= \psi - \kappa = \beta \\ \angle xB_0B &= \psi \end{aligned} \tag{1}$$

The values of the rotation angle in the revolute joints caused by the rotation of the input crank of $\Delta\varphi = 5^{\circ}$ ($\varphi = 177^{\circ} \div 182^{\circ}$) have been shown in Table 1.

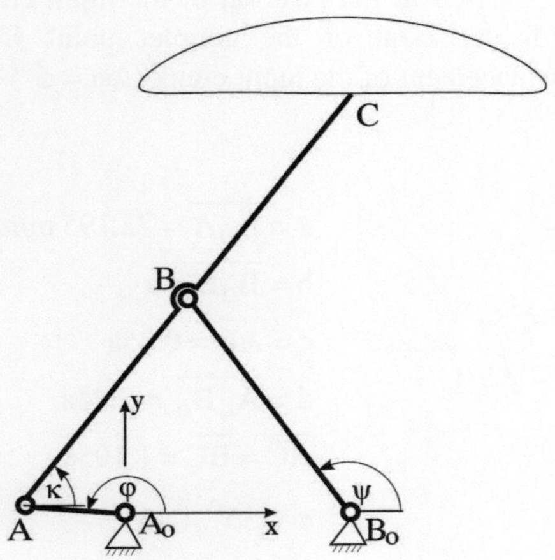

$a = \overline{A_0A} = 42.971\,\text{mm}$

$b = \overline{B_0B} = 2.5a$

$c = \overline{AB} = 2.5a$

$d = \overline{A_0B_0} = 2a$

$r = \overline{AC} = 5a$

$\varphi = 177^{o} \div 182^{o}$

$\Delta y_C = 3.3\,\mu m$

Fig. 1 Hoecken four-bar linkage (Dizioglu 1965)

Table 1 The rotation angles in all joints of the Hoecken four-bar linkage

$\Delta\varphi$ (°)	$\Delta\alpha$ (°)	$\Delta\beta$ (°)	$\Delta\psi$ (°)
5	3.34	0.01	1.67

2.2 Roberts-Tsebysev Four-Bar Linkage

The coupler point C of the Roberts-Tsebysev four-bar linkage (Fig. 2), located in the corner of the coupler as a ternary link, can be guided on an approximate rectilinear path segment (Bloch 1951).

The link lengths of this mechanism, as well as the position of the input crank, that enable realizing of horizontal displacement of the coupler point C of $\Delta x_C = 5$ mm, with the small angular displacement of the input crank ($\Delta\varphi = 5°$) and minimal difference between realized and exact rectilinear path (Δy_C), have been shown in Fig. 2 (Pavlović and Pavlović 2003, 2013):

The angles defining the relative positions of the neighboring links, that is, the links being connected with particular revolute joint, can be defined by Eq. 1.

The values of the rotation angle in the revolute joints caused by the rotation of the input crank of $\Delta\varphi = 5°$ ($\varphi = 35° \div 40°$) have been shown in Table 2.

2.3 Watt Four-Bar Linkage

The coupler point C of the Watt four-bar linkage (Fig. 3), located in the middle of the coupler, can be guided on an approximate rectilinear path segment (Bloch 1951).

The link lengths of this mechanism, as well as the position of the input crank, that enable realizing of horizontal displacement of the coupler point C of $\Delta x_C = 5$ mm, with the small angular displacement of the input crank ($\Delta\varphi = 5°$) and

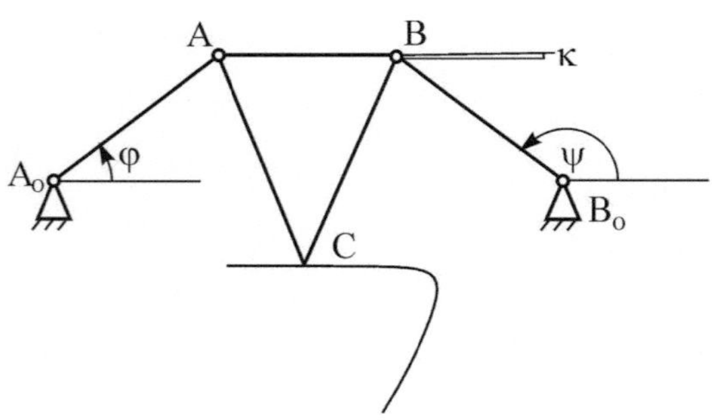

$$a = \overline{A_0A} = 22.195\,\text{mm}$$
$$b = \overline{B_0B} = a$$
$$c = \overline{AB} = 0.85a$$
$$d = \overline{A_0B_0} = 2.44a$$
$$\overline{AC} = \overline{BC} = 1.105a$$
$$\varphi = 35^0 \div 44^0$$
$$\Delta y_C = 0.8\,\mu\text{m}$$

Fig. 2 Roberts-Tsebysev four-bar linkage (Bloch 1951)

Table 2 The rotation angles in all joints of the Roberts-Tsebysev four-bar linkage

$\Delta\varphi$ (°)	$\Delta\alpha$ (°)	$\Delta\beta$ (°)	$\Delta\psi$ (°)
5	14.63	14.89	5.26

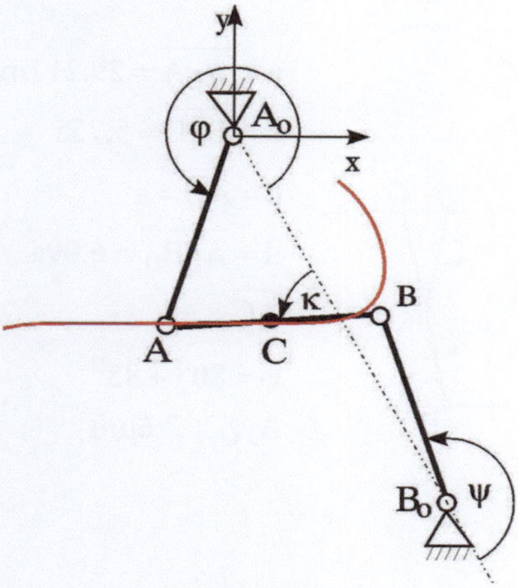

$a = \overline{A_0A} = 61.406\text{mm}$

$b = \overline{B_0B} = a$

$c = \overline{AB} = 1.04a$

$d = \overline{A_0B_0} = 2.1a$

$\overline{AC} = \overline{BC} = \frac{c}{2}$

$\varphi = 311^{\circ} \div 316^{\circ}$

$\Delta y_C = 1.8\mu\text{m}$

Fig. 3 Watt four-bar linkage (Bloch 1951)

Table 3 The rotation angles in all joints of the Watt four-bar linkage

$\Delta\varphi$ (°)	$\Delta\alpha$ (°)	$\Delta\beta$ (°)	$\Delta\psi$ (°)
5	2.26	7.5	4.76

minimal difference between realized and exact rectilinear path (Δy_C), have been shown in Fig. 3 (Pavlović and Pavlović 2003, 2013).

The angles defining the relative positions of the neighboring links, that is, the links being connected with particular revolute joint, can be defined by Eq. 1.

The values of the rotation angle in the revolute joints caused by the rotation of the input crank of $\Delta\varphi = 5°$ ($\varphi = 311° \div 316°$) have been shown in Table 3.

2.4 Evans Four-Bar Linkage

The coupler point C of the Evans four-bar linkage (Fig. 4), located at the end of the coupler, can be guided on an approximate rectilinear path segment (Dizioglu 1965).

The link lengths of this mechanism, as well as the position of the input crank, that enable realizing of horizontal displacement of the coupler point C of $\Delta x_C = 5$ mm, with the small angular displacement of the input crank ($\Delta\varphi = 5°$) and minimal difference between realized and exact rectilinear path (Δy_C), have been shown in Fig. 4 (Pavlović and Pavlović 2013).

The angles defining the relative positions of the neighboring links, that is, the links being connected with particular revolute joint, can be defined by Eq. 1.

The values of the rotation angle in the revolute joints caused by the rotation of the input crank of $\Delta\varphi = 5°$ ($\varphi = 80° \div 85°$) have been shown in Table 4.

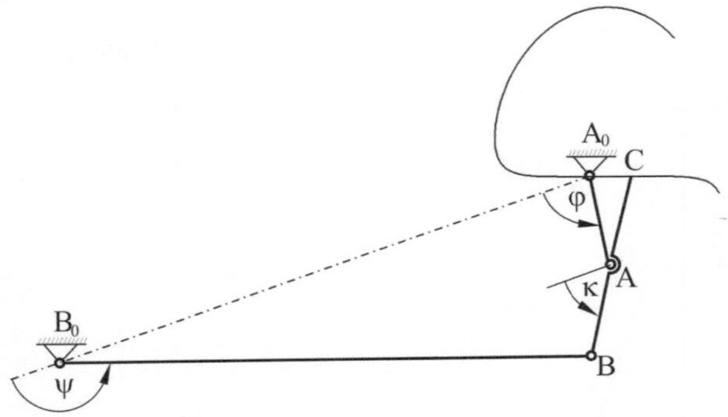

$$a = \overline{A_0A} = 29.217\text{mm}$$
$$b = \overline{B_0B} = 5.75a$$
$$c = \overline{AB} = a$$
$$d = \overline{A_0B_0} = 6.09a$$
$$\overline{AC} = a$$
$$\varphi = 80^o \div 85^o$$
$$\Delta y_C = 2.6\mu\text{m}$$

Fig. 4 Evans four-bar linkage (Dizioglu 1965)

Table 4 The rotation angles in all joints of the Evans four-bar linkage

$\Delta\varphi$ (°)	$\Delta\alpha$ (°)	$\Delta\beta$ (°)	$\Delta\psi$ (°)
5	10.42	5.58	0.37

2.5 *Scott-Russel Mechanism*

The coupler point C of the Scott-Russel mechanism (the isosceles slider–crank mechanism), located at the end of the coupler (Fig. 5), can be guided on an exact rectilinear path segment.

The link lengths of this mechanism, as well as the position of the input crank, that enable realizing of vertical displacement of the coupler point C of $\Delta x_C = 5$ mm, with the small angular displacement of the input crank ($\Delta\varphi = 5°$) and minimal difference between realized and exact rectilinear path (Δy_C), have been shown in Fig. 5.

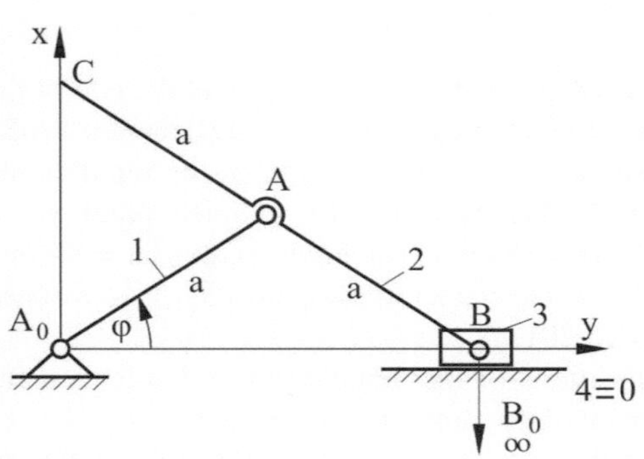

$$a = \overline{A_0A} = 38.869\text{mm}$$
$$c = \overline{AB} = a$$
$$\overline{AC} = a$$
$$\varphi = 40^o \div 45^o$$
$$\Delta y_C = 0\mu\text{m}$$

Fig. 5 Scott-Russel mechanism

Table 5 The rotation angles in all joints of the Scott-Russel mechanism

$\Delta\varphi$ (°)	$\Delta\alpha$ (°)	$\Delta\beta$ (°)
5	10	5

The angles defining the relative positions of the neighboring links, that is, the links being connected with particular revolute joint, can be defined as follows:

$$\begin{aligned} \angle xA_0A &= \phi \\ \angle A_0AB &= \pi - 2\phi = \alpha \\ \angle ABB_0 &= \phi + \frac{\pi}{2} = \beta \end{aligned} \tag{2}$$

The values of the rotation angle in the revolute joints caused by the rotation of the input crank of $\Delta\varphi = 5°$ ($\varphi = 40° \div 45°$) have been shown in Table 5.

2.6 Discussion

The rectilinear displacement of the coupler point C of $\Delta x_C = 5$ mm as well as the rotation angle of the input crank of $\Delta\varphi = 5°$ have been adopted for all above-mentioned rigid-body linkages. Using the method of position analysis of mechanism we have calculated respective link lengths of the mechanisms in order to satisfy the above-mentioned demands.

The values of the rotation angle in the revolute joints caused by the rotation of the input crank of $\Delta\varphi = 5°$, as well as the difference between realized and exact rectilinear path (Δy_C) have been shown in Table 6 in order to be compared.

If we take into consideration only guiding accuracy, the Scott-Russel mechanism or Roberts-Tsebysev four-bar linkage are to be suggested as a counterpart mechanism to be applied in compliant mechanism synthesis procedure. However, the mechanisms with compliant joints can realize relatively small motion range, that is, their mobility is limited. Another limitation to their use is fatigue failure at the compliant joints. Therefore the influence of the values of the angles defining the relative positions of the neighboring links (the rotation angles in the joints) on the mobility and fatigue failure has to be taken into consideration. Due to the smallest

Table 6 The rotation angles in all joints and the differences between realized and exact rectilinear path of the linkages

	$\Delta\varphi$ (°)	$\Delta\alpha$ (°)	$\Delta\beta$ (°)	$\Delta\psi$ (°)	Δy_C (mm)
Hoecken	5	3.34	0.01	1.67	3.3
Roberts	5	14.63	14.89	5.26	0.8
Watt	5	2.26	7.5	4.76	1.8
Evans	5	10.42	5.58	0.37	2.6
Scott-Russel	5	10	5	–	0

values of the rotation angles in the joints (Table 6), the Hoecken four-bar linkage has been chosen as a suitable counterpart mechanism to be applied in the synthesis procedure of compliant mechanism for rectilinear guiding (the rotation angles in all joints are not greater than 5°).

3 Compliant Hoecken Mechanism

We have designed the compliant mechanism for rectilinear guiding by using the Hoecken four-bar linkage as a counterpart mechanism.

Figure 6 shows two of the many different shapes of the compliant joints (Pavlović and Pavlović 2013): film joint (Fig. 6a) and notch joint (Fig. 6b).

The compliant mechanisms with film and notch joints can be made in one piece. A film joint (Fig. 6a) has been fully determined by three parameters (the width of relatively rigid segments w_R, the width of relatively elastic segments w_E and the length of the elastic segments l), while a semi-circular notch joint (Fig. 6b) has been fully determined by two parameters (w_R and w_E).

Figure 7 shows a Hoecken compliant mechanism with semi-circular notch joints. The results of the position analysis, performed by ANSYS Software, show that the greatest accuracy (the smallest Δy_C on rectilinear path segment of $\Delta x_C = 5$ mm) have been obtained for the mechanisms with the great rigidity ratio w_R/w_E ($w_R = 10$ mm, $w_E = 0.75$ mm $\Rightarrow \Delta y_C = 18.5$ μm). The limit positions of the "input crank" A_0A of the mechanism, determined by the permissible maximal bending stress, enable the maximal realizable angular deflection of $\Delta\varphi_{max} = 16.2°$. In other words, the mobility of this compliant mechanism is defined by angular deflection of $\Delta\varphi_{max} = 16.2°$, with respective maximal acting force causing the appearance of the maximal permissible bending stress $F_{max} = \pm 2.54$ N.

We have also designed a Hoecken compliant mechanism with film joints (Fig. 8), with similar geometrical properties of the compliant joints ($w_R = 10$ mm,

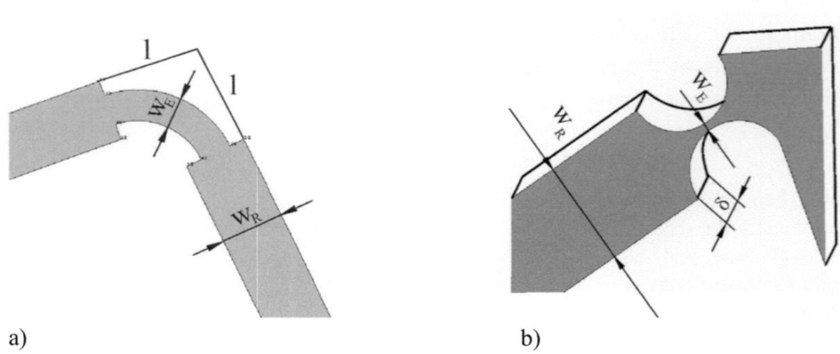

Fig. 6 The compliant: **a** film joint; **b** notch joint (Pavlović and Pavlović 2013)

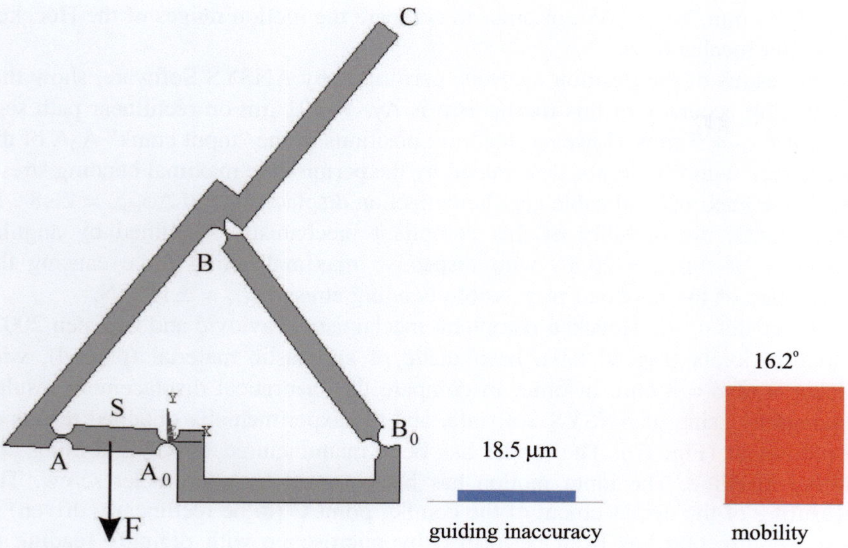

Fig. 7 Hoecken compliant mechanism with semi-circular notch joints (Pavlović and Pavlović 2013)

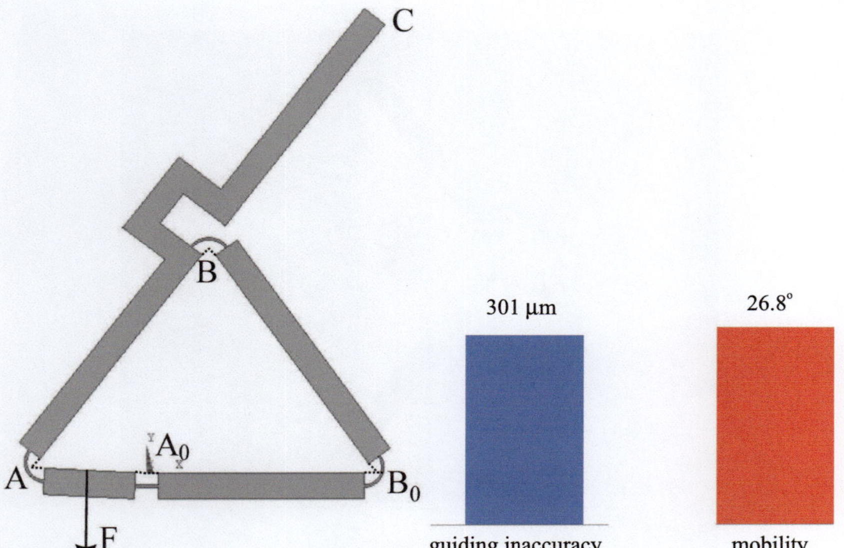

Fig. 8 Hoecken compliant mechanism with film joints (Pavlović and Pavlović 2013)

$w_E = 0.75$ mm, $l/a = 0.05$), in order to compare the motion ranges of the Hoecken compliant mechanisms.

The results of the position analysis, performed by ANSYS Software, show that the guiding accuracy of this mechanism is $\Delta y_C = 301$ μm on rectilinear path segment of $\Delta x_C = 5$ mm. However, the limit positions of the "input crank" A_0A of the mechanism with film joints, determined by the permissible maximal bending stress, enable the maximal realizable angular deflection displacement of $\Delta\varphi_{max} = 26.8°$. In other words, the mobility of this compliant mechanism is defined by angular deflection of $\Delta\varphi_{max} = 26.8°$, with respective maximal acting force causing the appearance of the maximal permissible bending stress $F_{max} = \pm 1.77$ N.

The model of the Hoecken compliant mechanism (Pavlović and Christen 2002) with film joints (Fig. 9) have been made of an plastic material (piacryl), with thickness of $\delta = 4$ mm, in order to compare the theoretical displacement results, obtained by using of ANSYS Software, and the experimentally obtained results of displacement (Fig. 10). The model has been manufactured by CNC drilling and milling machine. The input motion has been caused by micrometer screw. The measuring of the displacement of the coupler point C (to be rectlinearly driven) in x- and y-direction has been performed by microscope with digitally reading of results. Experimental results correspond to the results obtained by ANSYS Software (Fig. 10).

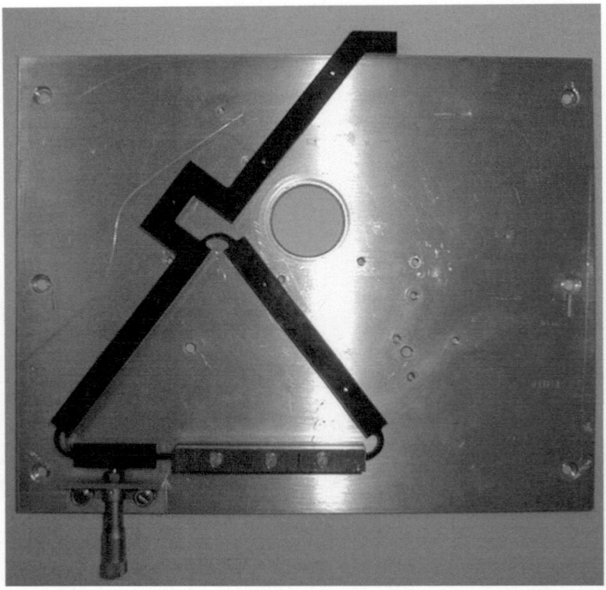

Fig. 9 The model of the Hoecken compliant mechanism with film joints (Pavlović and Christen 2002)

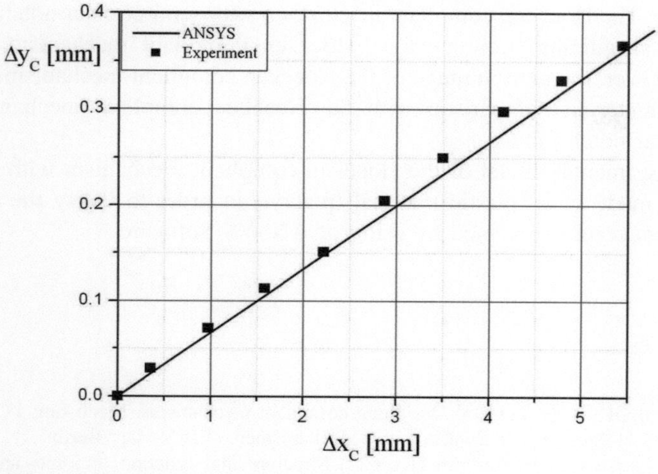

Fig. 10 Theoretical and experimental displacement results

4 Conclusions

The frequent method to synthesize a compliant mechanism is to design it as the counterpart of the rigid-body linkage being able to realize pre-defined function of the compliant mechanism (rigid-body replacement method). This paper has dealt with a selection of the optimal rigid-body counterpart mechanism in the compliant mechanism synthesis procedure regarding a limitation of the rotational deflection in each particular compliant joint.

The joint angular rotation as well as the difference between realized and exact rectilinear path have been analyzed and compared for five rigid-body linkages being suitable to realize the rectilinear guiding of the coupler point (Hoecken, Roberts-Tsebisev, Watt, Evans and Scott-Russel mechanism).

If we take into consideration only guiding accuracy, the Scott-Russel mechanism or Roberts-Tsebysev four-bar linkage are to be suggested as a counterpart mechanisms to be applied in compliant mechanism synthesis procedure. However, the mechanisms with compliant joints can realize relatively small motion range, that is, their mobility is limited. Another limitation to their use is fatigue failure at the compliant joints. Therefore the influence of the values of the angles defining the relative positions of the neighboring links (the rotation angles in the joints) on the mobility and fatigue failure has to be taken into consideration. Due to the smallest values of the rotation angles in the joints, the Hoecken four-bar linkage has been chosen as a suitable counterpart mechanism to be applied in the synthesis procedure of compliant mechanism for rectilinear guiding.

The compliant Hoecken mechanisms with film and semi-circular notch joints have been designed by using of the Hoecken four-bar linkage as a counterpart mechanism. The Hoecken compliant mechanism with semi-circular notch joints can realize better guiding accuracy than Hoecken compliant mechanism with film joints. However, the motion range of the Hoecken compliant mechanisms with film joints is greater in comparison with the Hoecken compliant mechanisms with semi-circular notch joints.

The experimental model of the Hoecken compliant mechanism with film joints have been made of an plastic material (piacryl) in order to verify the theoretical displacement results, obtained by using of ANSYS Software.

References

Ananthasuresh GK, Kota S (1995) Designing compliant mechanisms. Mech Eng 117(11):93–96

Bloch SS (1951) Angenäherte Synthese von Mechanismen.VEB Verlag, Berlin

Böttcher F, Christen G, Pfefferkorn H (2001) Structure and function of joints and compliant mechanism. In: Motion Systems (Collected Short Papers of the Innovationskolleg "Bewegungssysteme" Friedrich-Schiller Universität Jena), Technische Universität Jena, Technische Universität Ilmenau, ShakerVerlag, Aachen, 2001, pp. 30–35

Dizioglu B (1965) Getriebelehre: Ein Lehrbuch für Studierende und Ingenieure, Teil 1: Grundlagen. Braunschweig

Howell LL, Midha A (1994) A method for the design of compliant mechanisms with small-length flexural pivots. ASME J Mech Des 116(1):280–290

Midha A, Norton TW, Howell LL (1994) On the nomenclature, classification and abstractions of compliant mechanisms. ASME J Mech Des 116(1):270–279

Pavlović TN, Christen G (2002) Experimental research of the compliant four-bar linkage for rectilinear guiding. In: Proceedings of the 47th Internationales Wissentschaftliches Kolloquium, Ilmenau, September 2002, pp 320–321

Pavlović TN, Pavlović DN (2002) Stress analysis and guiding accuracy of the compliant four-bar linkages for rectilinear guiding. In: Proceedings of the 47th Internationales Wissentschaftliches Kolloquium, Ilmenau, September 2002, pp 345–346

Pavlović TN, Pavlović DN (2003) Motion characteristics of the compliant four-bar linkages for rectilinear guiding. J Mech Eng Des (Yugoslav Society for Machine Elements and Design) 6 (1):20–27

Pavlović TN, Pavlović DN (2004) Rectilinear guiding accuracy of Roberts-Чебышев compliant four-bar linkage with siliconejoints. Механика на машините 12(4):53–56

Pavlović TN, Pavlović DN (2009) Compliant mechanism design for realizing of axial link translation. Mech Mach Theory 44(5):1082–1091

Pavlović TN, Pavlović DN (2013) Compliant mechanisms (in Serbian). Faculty of Mechanical Engineering, University of Niš, Serbia

Design and Experimental Characterization of a Flexure Hinge-Based Parallel Four-Bar Mechanism for Precision Guides

P. Gräser, S. Linß, L. Zentner and R. Theska

Abstract This paper presents the investigation of the influence of the flexure hinge contour in compliant linkage mechanisms for precision engineering applications. Especially the influence on the precision of the path of motion and the stroke of the compliant mechanism is reflected. Based on previous results on optimized single polynomial flexure hinges, the validity of proposed guidelines is analyzed for a combination of several flexure hinges in one compliant mechanism. A parallel crank mechanism is used as an example for a compliant rectilinear guiding mechanism with constant link orientation. The parameters of the approximated linear motion are investigated for the rigid-body model, a compliant analytic model and a FEM model. Finally these results are compared with measurement results taken at manufactured prototypes.

Keywords Compliant mechanism · Parallel crank mechanism · Linear motion · Flexure hinge · Contour optimization · Precision guide

P. Gräser (✉) · R. Theska
Department of Mechanical Engineering, Institute of Design and Precision Engineering, Precision Engineering Group, Technische Universität Ilmenau, Ilmenau, Germany
e-mail: philipp.graeser@tu-ilmenau.de

R. Theska
e-mail: rene.theska@tu-ilmenau.de

S. Linß · L. Zentner
Department of Mechanical Engineering, Mechanism Technology Group, Technische Universität Ilmenau, Ilmenau, Germany
e-mail: sebastian.linss@tu-ilmenau.de

L. Zentner
e-mail: lena.zentner@tu-ilmenau.de

L. Zentner et al. (eds.), *Microactuators and Micromechanisms*, Mechanisms and Machine Science 45, DOI 10.1007/978-3-319-45387-3_13

1 Introduction

In precision engineering linkage mechanisms with optimized flexure hinges gain in importance. They are distinguished from common mechanisms with conventional hinges by numerous positive properties. Since they are based on elastic bending there is no external friction, no wear and no need of maintenance or lubrication. Therefore compliant linkage mechanisms are suitable for the use under vacuum and clean room conditions. They are preferred in many applications in the semiconductor industry, astronautics or precision measurement.

Compliant mechanisms can be subdivided based on different attributes (Zentner 2014). One option is the distribution of the compliance in the hinge. A distinction is made between distributed and lumped compliance based on the ratio of the length and the smallest thickness of the hinge (Zentner 2014). This article will only address fully compliant mechanisms with lumped compliances.

In precision engineering, decisive for the use of compliant mechanisms are the reproducibility, the precision of the path of motion, and the maximum stroke. These parameters are affected by the characteristics of flexure hinges. Based on the nature of flexure hinges there is no stationary rotation axis, causing a deviation of the path of motion compared to the rigid-body mechanism. The maximum stroke of the compliant mechanism is limited by the admissible strain in the flexure hinges. First investigations of the authors in generalized guidelines for the synthesis of compliant mechanisms for the optimization via geometric parameters of the flexure hinges have shown positive results. Therefore optimized flexure hinges with polynomial contours have been investigated. The validity of the derived guidelines will be demonstrated on a planar compliant linkage mechanism. The example used in the presented investigations is a compliant parallel crank with the task of motion of approximate linear planer guidance.

2 Motion Task

The starting point is a motion task which should be realized by a compliant mechanism via synthesis. Based on this approach, firstly guidelines for the combination of several flexure hinges need to be formulated. Secondly existing design rules for polynomial contours need to be validated. Therefore a procedure is recommended which leads from a given motion task to a suitable compliant mechanism. The procedure includes the selection of an appropriate rigid-body mechanism, the constructional realization of the compliant mechanism and the goal-oriented design of the geometric parameters of the flexure hinges. These synthesis steps are discussed in detail in Chap. 3.

A frequent task in precision engineering is to guide each point of a plane along a straight line. The deviation of this line should be minimal and highly reproducible. Furthermore, the smallest possible rotation of the moved link should occur during the motion. Such a task is necessary for adjusting and positioning of elements in precision manufacturing or in metrology. Therefore, a parallel crank mechanism is chosen due to the practical relevance, demonstrated in a large number of publications e.g. in (Liaw and Shirinzadeh 2008; Luo et al. 2015; Yang et al. 2010). In these cases, the trajectory of each coupler point of the parallel crank mechanism is equivalent to a circular translation. However, the motion can be approximated as a straight line with deviations within the set boundaries.

After the determination of the motion task and selection of a rigid-body model the requirements for the subsequent implementation of the mechanism are specified. The following three requirements are defined which are related directly to the output motion: The range of motion along the guiding line v_x, the maximum error in the orthogonal direction v_y and the maximum rotation angle of the guided coupler link δ (cf. Sect. 3.1). Another requirement arises due to the flexure hinges. According to the state of the art and preliminary investigations the rotation angle φ^* for each hinge is limited to 10 degrees because otherwise the maximum admissible strain ε_{zul} is exceeded.

3 Synthesis of the Compliant Linkage Mechanism

The used synthesis approach is the rigid-body replacement method (Howell et al. 2013). Therefore, the rigid-body hinges are replaced by prismatic flexure hinges in the compliant mechanism. The approach includes the following synthesis steps (Linß 2015):

- Determination of a suitable rigid-body mechanisms for the required path of motion,
- Design and implementation of the compliant mechanism,
- Geometric design and optimization of the flexure hinges,
- Verification of results, and of the need for further adjustment and optimization.

The selection of the rigid-body mechanism includes the choice of an appropriate model and the dimensioning of associated geometrical parameters. This is done with respect to required path of motion. Based on the specified concrete model the path of motion can be determined.

A further possibility of the analytical calculation of the motion parameters is the use of mathematical models for the compliant mechanism. In this way, the deformations can be determined by equations. An example for this is a model of the parallel spring guide derived by Nönnig (1980) with and without stiffened crank arms. Higher accuracy in the calculation of the motion path can be achieved with FEM simulations based on the CAD models.

After the choice of the kinematic parameters and the constructional realization of the compliant mechanism the geometric design and optimization of the four flexure hinges is a crucial step in the mechanism synthesis. For a simplified determination of suitable flexure hinge contours in comparison to computerized optimization

FEM-based guidelines and design graphs can be used (Linß 2015). Therefore at least the relative rotation angle in the rigid-body model φ^* for each hinge must be analyzed. It has been shown that hinge contours based on polynomial functions are particularly suitable in this case (Gräser et al. 2015).

3.1 Rigid-Body Mechanism

The rigid-body mechanism describes an ideal path of motion which is the reference in the following qualitative studies in addition to the exact straight line. The path of motion corresponds to an analytically predictable function, which results from the geometric parameters of the mechanism. The function can be characterized by the displacement in x and y direction of one particular point on the coupler of the mechanism. In the chosen example of the parallel crank, these parameters are the coupler length c and the crank length a. The point which is analyzed during motion is the point K on the coupler. This point is called coupler point, see Fig. 1.

For dimensioning of a compliant parallel crank mechanism the following requirements for the guiding of a plane along a straight line are given:

- Stroke of motion $|v_x^*| \geq 10\,\text{mm}$,
- Guiding deviation in y direction $|v_y^*| \leq 1\,\text{mm}$,
- Rotation angle of the coupler $\delta \leq 90''$.

The requirements are defined based on the addressed application in precision engineering. Additionally the angle of rotation should not exceed the permissible angle of the used plane mirror interferometer for the length measurement, see Sect. 5. With the given requirements and in consideration of the maximum deflection angle of the hinges φ^* suitable values for the parameter a of the parallel crank can be chosen with the help of the created design graph in Fig. 2.

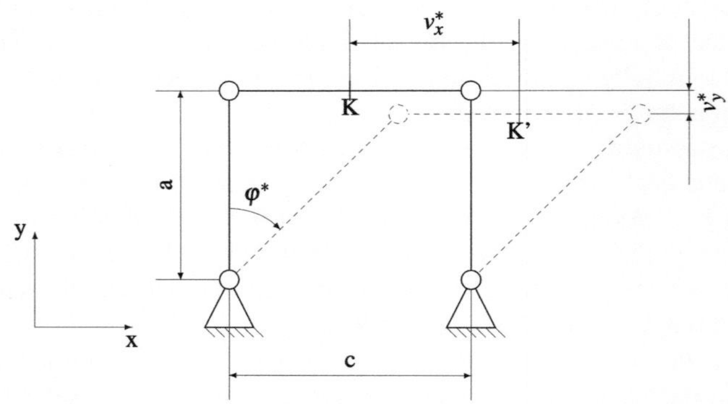

Fig. 1 Rigid-body model of a parallel crank mechanism with the coupler point K respectively K', the geometrical parameters (a, c) and the parameters of motion (v_x^*, v_y^*)

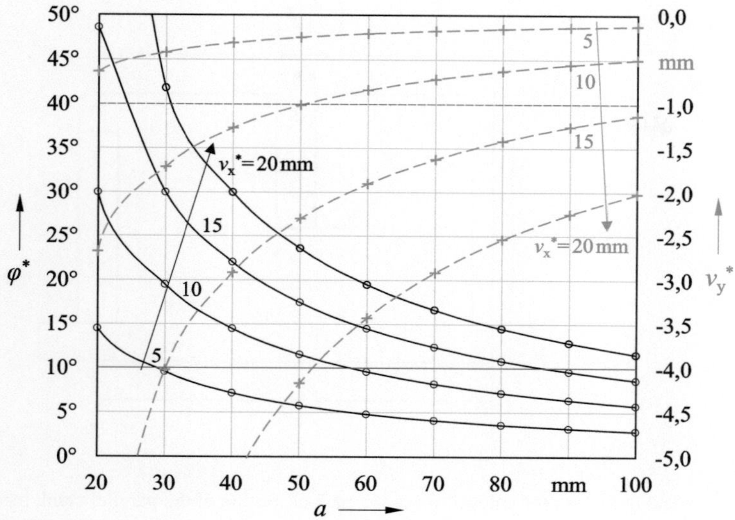

Fig. 2 Design graph for the determination of parameters for the rigid-body mechanism

In the design graph, the functions of the angle of the hinges φ^* and the straight line deviation v_y^* are given in dependence of the crank length a for different maximum strokes v_x^*. With the condition $v_x = 10$ mm and the selected crank length of $a = 80$ mm a maximum hinge angle of $\varphi^* = 7.1°$ and a maximum straight line deviation of $v_y^* = -0.6$ mm result. The basis for this design graph is the following equation for the rigid-body mechanism:

$$v_y^* = \frac{v_x^*}{\tan(\varphi^*)} - a, \text{ with } \varphi^* = \arcsin\left(\frac{v_x^*}{a}\right). \tag{1}$$

3.2 Compliant Mechanism

After the dimensioning of the rigid-body mechanism the implementation of the compliant mechanism is discussed in this section. First the idealized rigid-body joints are replaced by flexure hinges. These are connected by links which have a significant higher stiffness than the hinges. Thus, the deformation of these connecting link segments can be neglected. The path of motion of the coupler point can be calculated in various ways. In this work, the compliant parallel crank mechanism is calculated by the model of a concentrated parallel spring guide according to Nönnig (1980), see Fig. 3 and Eq. 2 and 3.

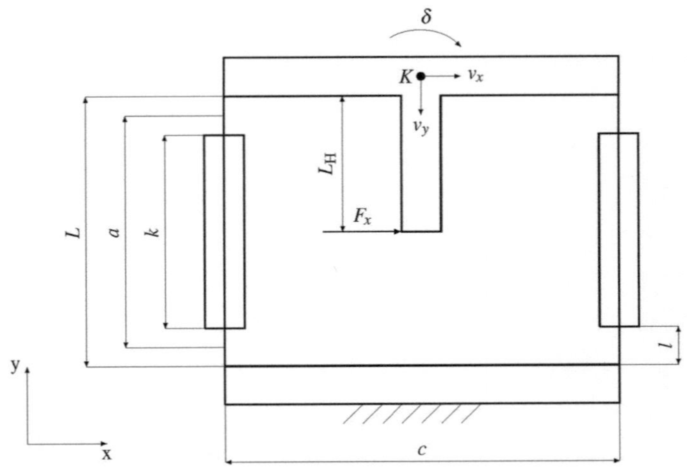

Fig. 3 Analytical model for the calculation of the path of motion of the parallel crank mechanism with concentrated compliance

$$v_y = \frac{3}{5}\frac{v_x^2}{L}\eta, \quad \eta = \frac{1 - \frac{5}{2}m^3 + \frac{3}{2}m^5}{(1 - m^3)^2}, \quad m = \frac{k}{L} \tag{2}$$

$$\delta = \left(16 \cdot 10^{-5}\frac{L^3\frac{L-2L_H}{2c}}{H_0 c}F_x + \frac{L^7\frac{L-2L_H}{2c}}{201600 H_0^3 c}F_x^3\right)\psi \tag{3}$$

The parameter η is a path factor in the case of stiffened model and m is the stiffening ratio. The parameter ψ is a correction factor which depends on the stiffening ratio m. With v_x and v_y the motion of the coupler point K of the compliant parallel crank mechanism can be described. In addition to the rigid-body model without coupler rotation, in the case of the compliant counterpart the rotation angle of the coupler δ must be regarded, because a minimal rotation occurs.

With a given maximum rotation δ and the guiding deviation v_y suitable values for the coupler length c can be determined as shown in Fig. 4. For the example of the investigated parallel crank mechanism a value of c = 100 mm results.

3.3 Deployment of Design Guidelines for Flexure Hinge Contour Optimization

The polynomial function derived by (Linß et al. 2015) allows the description of contours from elementary to very complex shape. Since all flexure hinges in the compliant parallel crank mechanism are equally shaped they all can be described by

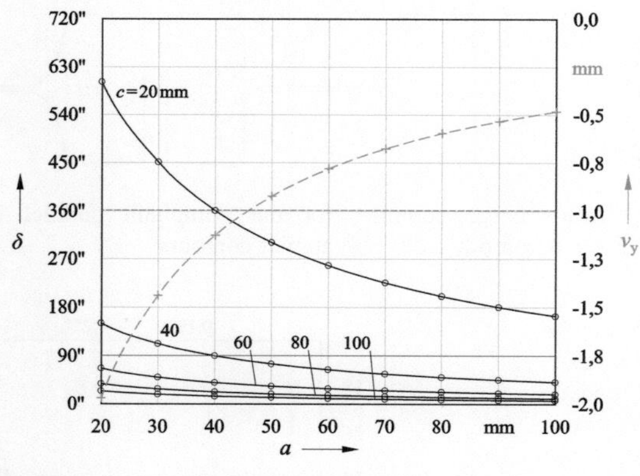

Fig. 4 Design graph for the determination of the coupler length c of the compliant parallel crank mechanism for the listed requirements ($v_x = 10\,\text{mm}, v_y < 1\,\text{mm}, \delta < 90''$)

Fig. 5 Geometric parameters of a flexure hinge for deflection with the angle φ

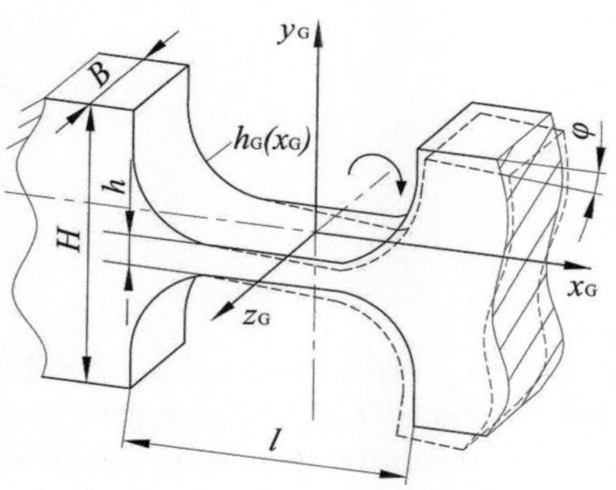

the variable hinge height $h_G(x_G)$ based on each hinge coordinate system, see Fig. 5. To determine the geometric design it needs the definition of the basic shape of the hinge contour and the geometric parameters.

Compared to common circular or corner-filleted contours, special polynomial contours offer a great potential for optimization deploying a comparably simple contour modeling, see Fig. 6. For the investigations symmetrical flexure hinges with the minimum hinge height h in the joint center height are considered.

With regard to the precision and the possible stroke of the compliant mechanism the following geometric parameters have to be considered:

- the hinge length ratio l/H,
- the hinge height ratio h/H and
- the used order n of the polynomial function.

To determine appropriate parameter values for the flexure hinges with polynomial contours derived guidelines can be used. For the simultaneous decrease of the two objectives path deviation and maximum strain (which limits the motion range

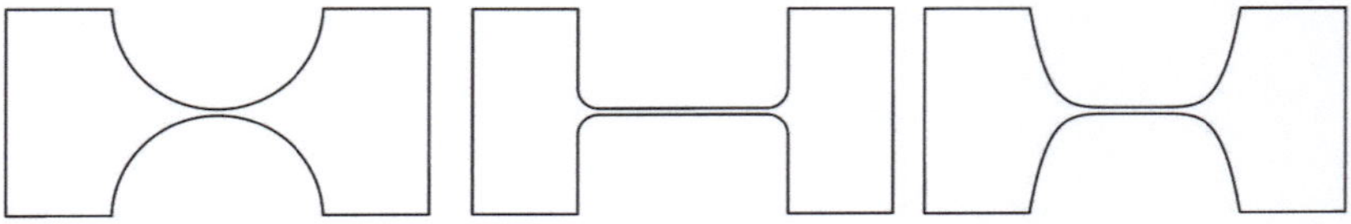

Fig. 6 Investigated flexure hinge contours for the compliant parallel crank mechanism: semi-circular, corner-filleted and 6th-order polynomial contours

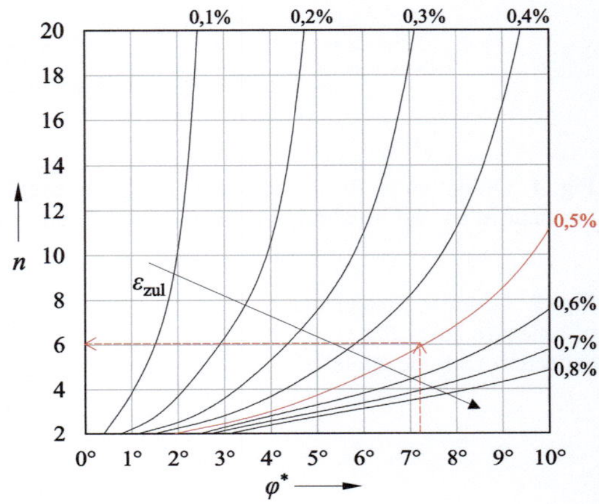

Fig. 7 Design graph for the determination of the order n of the polynomial function for the contour of the flexure hinge in the mechanism with given admissible strain ε_{adm} and rotation angle $\varphi^* \left(\frac{l}{H} = 1, \frac{h}{H} = 0.03\right)$

respectively possible stroke) the use of the hinge length ratio $l/H = 1$ and the hinge height ratio $h/H = 0.03$ is suggested (Linß 2015). The mechanism width B can be chosen arbitrarily, because it affects only the necessary force to deflect the mechanism. In the presented example the width is chosen as $B = 6\,\text{mm}$.

After defining the basic hinge dimensions $l = H = 10\,\text{mm}$ and $h = 0.03\,\text{mm}$, a suitable polynomial order n can be determined by means of the created design graph in Fig. 7.

For instance the admissible maximum strain of the utilized aluminum alloy EN AW 7075 is $\varepsilon_{adm} = 0.5\,\%$. Together with the maximum relative rotation angle of the hinges φ^* of a polynomial order of at least $n = 6$ is needed.

4 Simulative Investigations of the Parallel Mechanism

4.1 Rigid-Body Simulation

To investigate the path of motion of the rigid-body model the regarded mechanism is simulated with the software SAM. The parameters for the link lengths and the position of the hinges are the input for the program. The result is a model of the rigid-body mechanism, see Fig. 8. The motion in the simulation of the parallel

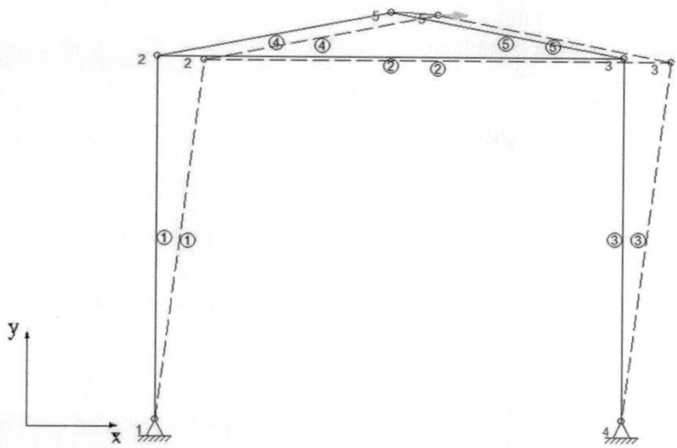

Fig. 8 Simulation of the motion path of the rigid-body mechanisms with the software SAM (coupler point K and K′ in point 5)

crank is specified by a deflection of a point of the coupler in positive x direction $v_x^* = 10\,\mathrm{mm}$. An additional parameter for the evaluation of the path of motion is the displacement of the point K in y direction v_y^*. A rotation of the coupler δ^* does not occur in this case.

4.2 FEM Simulation

To proof the potential of the synthesis method based on suitable polynomial contours, the influence of the variation of the hinge contour on the mechanism properties is investigated by means of a static-mechanical FEM analysis. For this purpose, any flexure hinge is modeled with the same polynomial order $n = 6$, while the centers of all flexure hinges correspond to the coordinates of each revolute joint in the rigid-body mechanism.

An additional parameter that results due to the constructional implementation is the orientation of the hinges. This is initially chosen for the flexure hinges so that the longitudinal axes of the flexure hinges are aligned parallel to the crank. As a reference two compliant mechanisms, one with usual semicircular and one with corner-filleted hinge contours, are considered.

The simulation model was created using the software ANSYS Workbench 16.2 in consideration of large deformations. The deflection of the mechanism with a linear actor is given by a force with constant x direction at the coupler. The coupler displacement in y direction and the rotation around the z axis are free. The compliant parallel crank mechanism with the used polynomial contours of 6th-order is shown in Fig. 9 for the non-deformed and deformed state at full input $|v_\mathrm{x}| = 10\,\mathrm{mm}$. To obtain accurate simulation results comparable to the metrological investigation for the selected non upright position the following three aspects are considered in the FEM model too: The influence of gravity on the deformation in the z direction, a boundary condition to simulate the weight compensation

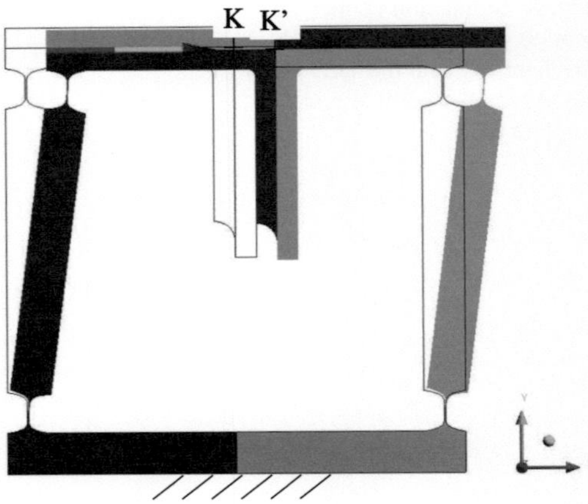

Fig. 9 Simulation of path of motion of the compliant mechanism with ANSYS Workbench

implemented in the measurement setup by means of a guided ball and the weight of the measuring mirrors used for interferometric length measurement.

After specifying the input displacement $v_x = v_x^*$ the guiding properties of the coupler link can be determined. Thus, the position of the regarded coupler point K in the compliant mechanism results in comparison to the ideal straight line (straight line deviation v_y) and in comparison to the rigid-body model (lateral path deviation Δv_y). In addition, the angle of rotation δ of the coupler is calculated based on the coordinates of two points. The analysis of the maximum strain according to von Mises ε_V in the mechanism allows the calculation of the realizable range of motion of the manufactured mechanisms according to the admissible strain for the chosen material. The mechanisms with corner-filleted and polynomial contours can be examined for full deflection of $|v_x| = 10\,\text{mm}$, while the mechanism with semi-circular contours can only be deflected with an input displacement in x direction of $|v_x| = 4\,\text{mm}$.

5 Measurements of Parallel Mechanism Prototypes

To verify the presented model prototypes which were fabricated by means of wire-cut EDM are measured. Three parallel crank mechanisms with different flexure hinge contour have been investigated: semi-circular, corner-filleted and polynomial function of 6th-order. The shapes of the contours are the semi-circular, the corner-filleted and the polynomial function of 6th-order. The hard aluminum alloy EN AW 7075 is used, as well as in the simulations.

The metrological investigation of the path of motion of the prototypes is made with a specially constructed test bench. The investigated parameters are corresponding to those of the simulation, the displacement v_x, the displacement v_y and

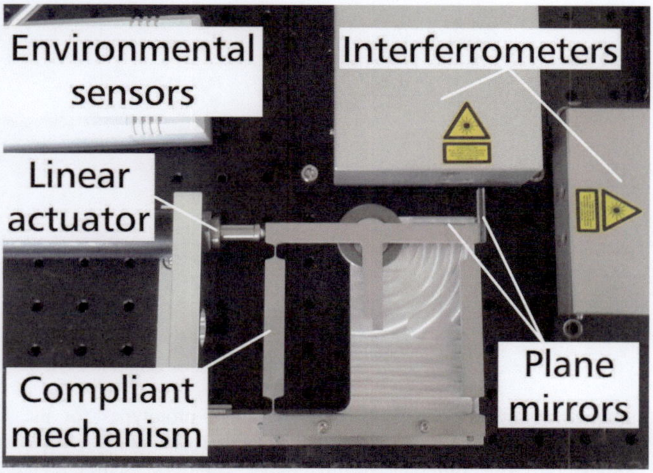

Fig. 10 Setup for the high precision measurement of the path of motion

the coupler rotation δ. The two displacements can be determined directly be means of length measurements. To measure the rotation, it is necessary to use two parallel linear measurements carried out at a known distance and from their difference to determine the angle. The length measurements are made by plane mirror interferometers, which have a resolution of 0.1 nm. For realizing the described three length measurements, a single-beam interferometer and orthogonal a two-beam interferometer are used, see Fig. 10.

The initial driving motion is directly applied by a precision linear actuator at the coupler of the parallel crank, for all mechanisms see Fig. 10. The resolution of the linear actuator is 0.1 μm. To compensate the dead load of the mechanism and the mirrors a rolling ball is used additionally, which is placed below the coupler and is moved along during deflection.

To ensure a statistically firm analysis, multiple measurements for each prototype have been made. The mean values are presented as functions in the following diagrams, while the indication of the confidence intervals is not shown due to a better clarity.

5.1 Linear Deviation and Path Deviation of the Coupler

The evaluation of the measurement results and the verification of the different models are made under consideration of two different motion parameters. They differ by the desired trajectory of the compliant mechanism. A first option is to refer directly to an ideal straight line. The resulting error is called here straight line deviation v_y (cf. Fig. 11 left).

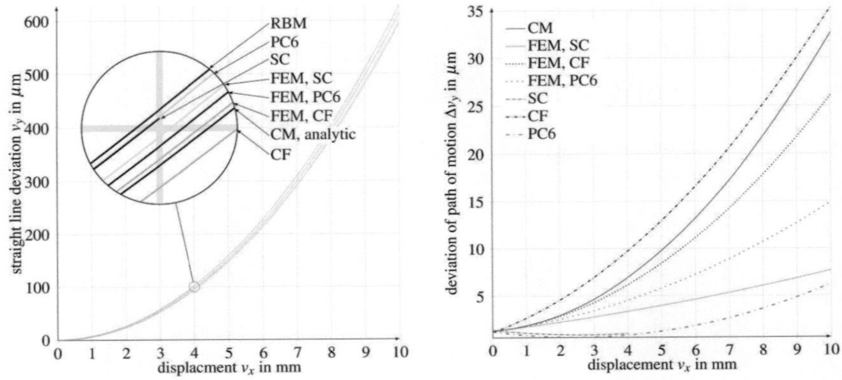

Fig. 11 Measurement results for the different models and prototypes for the parallel crank mechanism: rigid-body mechanism (RBM); analytically calculation of compliant mechanism (CM); FEM simulation with different flexure hinge contours: semi-circular (FEM, SC), corner-filleted (FEM, CF), 6th-order polynomial (FEM, PC6); manufactured prototypes with different flexure hinge contours: semi-circular (SC), corner-filleted (CF), 6th-order polynomial (PC6)

Secondly, the motion path is compared to the result derived with the rigid-body model. This comparison shows how accurately the constructed compliant mechanism realizes the motion path of the rigid body mechanism. For the parallel crank mechanism this path deviation is called Δv_y (see Fig. 11 right).

In the diagram of the straight line deviation (Fig. 11 left) the trajectories of the various models and prototypes are shown. The individual motion patterns have the same characteristic, but differ in their distances from the straight line. Differences between the parallel crank mechanisms with different notch contours occur. The arrangement in the diagram is the same for the simulation and the measurements. Differences between simulation and measurement can be attributed to several reasons. Either, the simulation model inaccuracies as well as tolerances of the manufactured mechanisms compared with the CAD models, lead to deviations. The trajectory of the prototype with semicircular contour ends already at $v_x = 4$ mm, because from this value the maximum admissible strain is exceeded in the flexure hinges.

Compared with the trajectories of the rigid-body model (Fig. 11 right) the semi-circular and also the polynomial contours lead to better values than the corner-filleted contours. But the advantage of the polynomial contour in comparison with the semi-circular contours is the larger range of motion. The mechanism with 6th-order polynomial flexure hinges can be deflected without exceeding the strain limit at $v_x = 10$ mm. Thus, the polynomial flexure hinges represent a good compromise between semi-circular and corner-filleted contours regarding the precision and the stroke.

5.2 *Rotation of the Coupler*

A rotation of the coupler δ occurs due to the replacement of the rigid-body mechanism with the compliant mechanism. The rotation δ results from transversal forces and moments generated by the flexure hinges. For the analytical model of the compliant mechanism and the FEM simulation the maximum rotation of the coupler for the predetermined range of motion results.

For the manufactured prototypes the rotation angle δ of the coupler is examined according to the resolution of the proposed measurement system. The resolution is mainly limited by the surface quality of the mirror and it is $\pm 4''$. Thus, it can be concluded that the rotation of the manufactured prototypes is below $\delta = 10''$. Compared with the state of the art this minimal angle of rotation is very low for a straight line guiding of a link and sufficient for a variety of applications in precision engineering.

6 Conclusion and Outlook

The investigations in this paper show that the design of the notch geometry of flexure hinges in compliant linkage mechanisms has an influence on both, the accuracy of the path of motion as well as the maximum range of motion. A flexure hinge contour with polynomial functions offer an advantage compared to conventional contours like semi-circular contours, as these allow an increased stroke for the same precision. For the selection of suitable polynomial orders the presented design graphs can be used. These provide recommendations regarding the geometric parameters of the compliant mechanism and its flexure hinge design.

The metrological investigation confirms that analytical calculations allow only an approximation of the motion. For the detailed design of compliant mechanisms a numerical solution for example by means of FEM simulation is necessary. The increase in accuracy results from the consideration of the exact nonlinear deflection of the flexure hinges, which is neglected in the analytic calculations. Thus, a simplified design with respect to precision and range of motion for compliant linkage mechanisms with optimized flexure hinge contours is possible by means of special design graphs and guidelines.

Another parameter for validating the FEM model is the characteristic stiffness of the mechanism. This can be evaluated by monitoring the driving force of the actuator as a function of the displacement. Simulations show the dependence on the hinge contour. This will be demonstrated by existing prototypes in further investigations. Furthermore, there is an initial investigation of the hinge orientation in the compliant mechanism, i.e. the spatial orientation of each hinge. The consideration of the different deflection angles of each hinge leads to the conclusion to use different notch geometries for flexure hinges deflected with different rotation angle. Through a combination of hinges with different orders of the polynomial function it is possible to increase the range of motion of a mechanism without a loss in precision of the path of motion. It is even possible to improve it. More aspects to be examined are manufactured-based effects and the use of different materials.

Further investigations should validate existing synthesis methods and provide new guidelines for the mechanisms synthesis with a focus on high precision and reproducibility of the path of motion.

Acknowledgments The development of this project is supported by the Deutsche Forschungsgemeinschaft (DFG) under Grant No. TH 845/5-1.

References

Gräser P, Linß S, Zentner L, Theska R (2015) Ultraprecise linear motion generated by means of compliant mechanisms. In: Proceedings of euspen's 15th International Conference & Exhibition, Leuven, Belgium, 01–05 June 2015, pp 241

Howell LL, Magleby SP, Olsen BM (2013) Handbook of compliant mechanisms. Wiley, New York

Liaw HC, Shirinzadeh B (2008) Robust generalised impedance control of piezo-actuated flexure-based fourbar mechanisms for micro-nano manipulation. Sens Actuators, A 148(2):443

Linß S (2015) Ein Beitrag zur geometrischen Gestaltung und Optimierung prismatischer Festkörpergelenke in nachgiebigen Koppelmechanismen. Dissertation, TU Ilmenau

Linß S, Milojevic A, Pavlovic ND et al (2015) Synthesis of compliant mechanisms based on goal-oriented design guidelines for prismatic flexure hinges with polynomial contours. In: Proceedings of the 14th World Congress in Mechanism and Machine Science, Taipei, 25–30 Oct 2015, pp 6

Luo Y, Liu W, Wu L (2015) Analysis of the displacement of lumped compliant parallel-guiding mechanism considering parasitic rotation and deflection on the guiding plate and rigid beams. Mech Mach Theory 91:50–68

Nönnig R (1980) Untersuchungen an Federgelenkführungen unter besonderer Berücksichtigung des räumlichen Verhaltens. Dissertation, Ilmenau: Technische Hochschule Ilmenau

Yang X, Li W, Wang Y, Ye G (2010) Output displacement analysis for compliant single parallel four-bar mechanism. In: Proceedings of the International Conference on Mechatronics and Automation (ICMA) 2010. International Conference on Mechatronics and Automation (ICMA). Xian, China. 04–07 Aug 2010. pp 1354

Zentner L (2014) Nachgiebige Mechanismen. De Gruyter Oldenbourg, München

Dynamic Model of a Compliant 3PRS Parallel Mechanism for Micromilling

A. Ruiz, F.J. Campa, C. Roldán-Paraponiaris and O. Altuzarra

Abstract The objective of this work is to develop a manipulator of 5 degrees of freedom for micromilling. It consists of a XY stage under a 3PRS compliant parallel mechanism, obtaining the advantages of the compliant joints as are higher repetitiveness, smoother motion and a higher bandwidth, due to the high precision demanded from the process, under 0.1 μm. In this work, the dynamics of the compliant stage will be developed. The modelling approach is based on the use of the Principle of Energy Equivalence combined with the Boltzmann-Hamel equations to analyze the rotational dynamics of the platform. A pseudo-rigid model has been assumed for the compliant joints, calculating the flexural and torsional stiffness by FEA. Finally, a prototype has been built and some preliminary results are shown comparing the simulation and the measurements.

Keywords Compliant mechanism · Parallel mechanism · Dynamics

1 Introduction

Nowadays, micromilling applications have a great demand in several fields: scientific, medical, metrology and communication. The requirements are very restrictive in terms of precision and surface finish. For several parts, the machines

A. Ruiz · F.J. Campa (✉) · C. Roldán-Paraponiaris · O. Altuzarra
Department of Mechanical Engineering, Compmech Research Group,
University of the Basque Country UPV/EHU, Leioa, Spain
e-mail: fran.campa@ehu.es

A. Ruiz
e-mail: antonio.ruiz@ehu.es

C. Roldán-Paraponiaris
e-mail: constantino.roldan@ehu.es

O. Altuzarra
e-mail: oscar.altuzarra@ehu.es

L. Zentner et al. (eds.), *Microactuators and Micromechanisms*,
Mechanisms and Machine Science 45, DOI 10.1007/978-3-319-45387-3_14

used in this sector are too large compared to the demanded workspace. For that reason, the design of smaller and more compact machines is desirable, which is the objective of the present work.

Here, the development of a compact micromilling machine based on a compliant parallel manipulator is proposed. The workpiece will be supported on a 3PRS compliant parallel mechanism (Gao et al. 2010) in series with an XY stage whereas the high speed spindle will remain on a fixed gantry. The 3PRS solution is a fusion of a compliant mechanism (Howell 2001) and a parallel mechanism (Merlet 2000), so the advantages of both devices can be obtained: no friction, not need lubrication and zero backlash from the flexure devices and the higher mechanical stiffness, higher loading capacity, and higher positioning accuracy from the parallel mechanism.

The use of compliant mechanisms in high precision applications is becoming very common. For example, in (Choi and Kim 2012), a mechanism for a single-axis flexure-based nano-positioning stage (1dof) with a range of motion up to a millimeter and a compact stage size is presented. For mechanisms with 2 dof, some different configurations have been developed. A planar motion stage design based on flexure elements is shown in (Choi et al. 2011). Another example of this type of device is resented in (Aphale et al. 2009), where a device with a relatively large range and high scanning speed is shown. Also, 3 dof mechanisms have been proposed. For example, an ultra-precision XYθz flexure stage with nanometer accuracy is presented by Ahn et al. (2010). A high-performance three-axis serial-kinematic nano-positioning stage for high-bandwidth applications is developed by Kenton and Leang (2012). Other design of a XYZ compliant parallel manipulator is shown in Hao and Kong (2012), where the structure is composed by identical spatial double four-beam modules.

On the other hand, the manufacturing of microlenses consists of milling a matrix of NxN concave aspherical cavities with the negative of the lens on one of a cylindrical workpiece, which diameter ranges from 10 to 20 mm, see Fig. 1. Different shapes can be developed, ranging diameters from 0.5 to 2 mm and sagittal depths less than 1 mm. The kinematic requirements in terms of needed displacements have been obtained by means of the design of a standard process, where a matrix of 4×4 cavities in a cylindrical mould of 18 mm is achieved. The cavities are spherical with a diameter of 2 mm, and are machined in down-milling with a spiral down strategy in counter-clock direction. The offset between the part and the tool tip for the motions between cavities is of 1 mm. The resulting displacements needed in the workpiece are ± 10 mm in X and Y.

To perform this operation, a XY3PRS hybrid mechanism has been selected. It is based on a 3PRS compliant parallel kinematics manipulator mounted in series on a XY stage, see Fig. 1. The moving platform is connected to the XY stage by each leg, where a P joint, a R joint and a S joint are placed in sequence and the P joint is actuated by a linear actuator. Thus, three identical PRS linkages attach the moving platform to the base (Li and Xu 2007). This configuration allows meeting the requirements of the process and offers the option of performing rotations around the

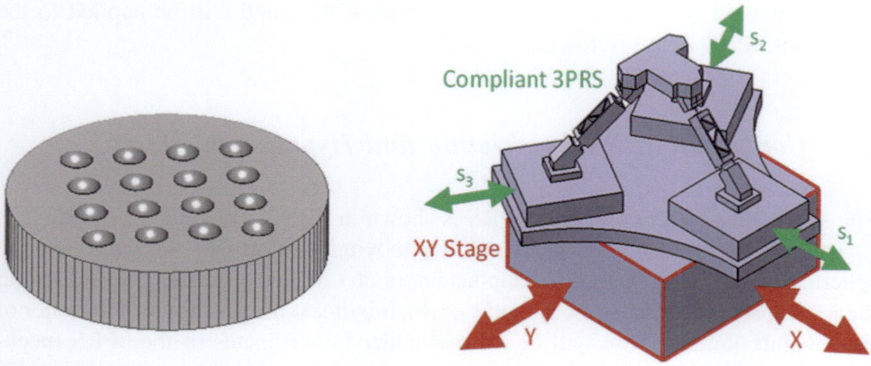

Fig. 1 *Left* CAD model of a microlenses mould. *Right* Schematic of a XY3PRS manipulator

X and Y axis. this offers the possibility of performing rotations around the X and Y axis, which would allow a better orientation of the ball end mill with respect to the workpiece.

The present study is focused on the dynamic analysis of the 3PRS compliant manipulator, whose rigid body kinematics has been previously solved following the work from (Li and Xu 2007). That dynamic model will be used for integrated simulation of the mechatronics of the manipulator during the milling process.

2 Dynamic Analysis

The dynamic problem has been solved using the Principle of Energy Equivalence, that is, the mechanism will be divided in b open-chain subsystems where the Lagrange equations can be applied with their local generalized coordinates q_b. The condition that the b individual subsystems move like belonging to an assembled mechanism implies that q_B, the set of all the generalized coordinates q_b is a function of the generalized coordinates q of the assembled mechanism. Thus, the virtual displacements can be related by means of the Jacobian. Furthermore, as the movement is the same, the virtual work performed by the assembly system and by the set of subsystems must be the same:

$$\delta W = \delta W_B \Rightarrow \boldsymbol{\delta q}^{\mathbf{T}} \boldsymbol{\tau} = \boldsymbol{\delta q}_{\mathbf{B}}^{\mathbf{T}} \boldsymbol{\tau}_{\mathbf{B}} = \boldsymbol{\delta q}^{\mathbf{T}} \mathbf{J}^{\mathbf{T}} \boldsymbol{\tau}_{\mathbf{B}} \tag{1}$$

Hence, the actuating forces on the assembled mechanism can be calculated as:

$$\boldsymbol{\tau} = \mathbf{J}^{\mathbf{T}} \boldsymbol{\tau}_{\mathbf{B}} = \sum_{b=1}^{B} \mathbf{J}_{\mathbf{b}}^{\mathbf{T}} \boldsymbol{\tau}_{\mathbf{B}} \tag{2}$$

This method was covered in Altuzarra et al. (2015) and will be applied to the compliant 3PRS in the following sections.

2.1 Compliant 3PRS Description and Hypotheses

The schematics of the compliant 3PRS is shown in Fig. 2. The length of the legs is L, and the radius from the center of the moving platform to the center of the spherical joints is b. Three prismatic actuators at 120° have been employed, being the joint space coordinates s_1, s_2, and s_3, which indicate the position of the center of the revolute joints. Those will be the generalized coordinates of the 3PRS mechanism. The angles between the legs and the XY stage are α_i, whose values are 45° in the default position. As a result, the platform is able to perform two rotations around X and Y axis, ψ and θ, respectively, as well as movements in Z direction. Parasitic motions in X, Y and a rotation around Z, ϕ, also appear.

For the study of the dynamics, all the elements will be considered as rigid bodies, whereas the joints will be treated as conventional using torsional springs to model their stiffness. That is, a pseudo-rigid model will be considered.

2.2 Actuated Plates

The base of each leg performs a translation measured by s_i under the force imposed by the actuators F_i where i = 1, 2, 3, see Fig. 2. Applying the Lagrange equations, the following equation of motion can be obtained, where $\mathbf{M_{act}}$ is the inertial matrix containing the mass of the plates.

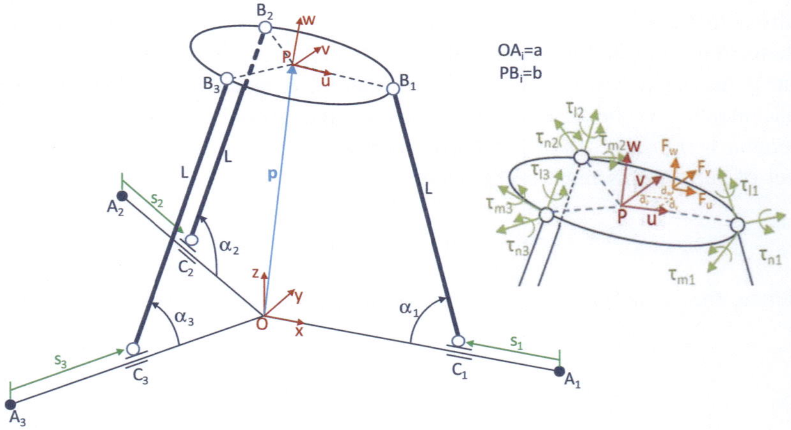

Fig. 2 Schematic of the compliant 3PRS manipulator and loads acting on the platform

$$\mathbf{M_{act}}\ddot{\mathbf{q}} = \mathbf{f_{act}} \Rightarrow \mathbf{M_{act}}\{\, \ddot{s}_1 \quad \ddot{s}_2 \quad \ddot{s}_3 \,\}^T = \{\, F_1 \quad F_2 \quad F_3 \,\}^T \tag{3}$$

2.3 Platform Translational Dynamics

To solve the dynamic problem of the platform, the translational and rotational dynamics of the platform have been decoupled. The generalized coordinates are defined by the position of the platform center $\mathbf{p_t} = \{\, x_P \quad y_P \quad z_P \,\}^T$. Considering the gravity $\mathbf{g}$ and the cutting forces due to the milling process as a force $\mathbf{f}$ applied in a position defined by vector $\mathbf{d}$ in the moving frame UVW, see Fig. 3, the equations of motion of the platform translational dynamics are:

$$\mathbf{Rf} = \mathbf{M}\ddot{\mathbf{p}}_\mathbf{t} + \mathbf{g} \tag{4}$$

Where R is the rotation matrix that relates the moving frame of the platform UVW with the fixed XYZ.

$$\mathbf{R} = \mathbf{R_y}(\theta)\mathbf{R_x}(\psi)\mathbf{R_z}(\phi) = \begin{bmatrix} c\theta c\phi + s\psi s\theta s\phi & -c\theta s\phi + s\psi s\theta c\phi & c\psi s\theta \\ c\psi s\phi & c\psi c\phi & -s\psi \\ -s\theta c\phi + s\psi c\theta s\phi & s\theta s\phi + s\psi c\theta c\phi & c\psi c\theta \end{bmatrix} \tag{5}$$

Projecting to the manipulator coordinates, the contribution to the global dynamics is:

$$\mathbf{J}^T_{\mathbf{plat_t}}\mathbf{f} = \mathbf{J}^T_{\mathbf{plat_t}}\mathbf{M}\mathbf{J}_{\mathbf{plat_t}}\ddot{\mathbf{q}} + \mathbf{J}^T_{\mathbf{plat_t}}\mathbf{M}\dot{\mathbf{J}}_{\mathbf{plat_t}}\dot{\mathbf{q}} + \mathbf{J}^T_{\mathbf{plat_t}}\mathbf{g} \tag{6}$$

Where $\mathbf{J_{platt}}$ is the Jacobian that relates the platform center coordinates $\mathbf{p_t}$ with the generalized coordinates $\mathbf{q}$. The Jacobian is obtained analyzing the kinematics of the 3PRS. Taking into account the three loop closure equations and the three restrictions due to the planar motion of the legs in the planes at 120°, the following relations can be written:

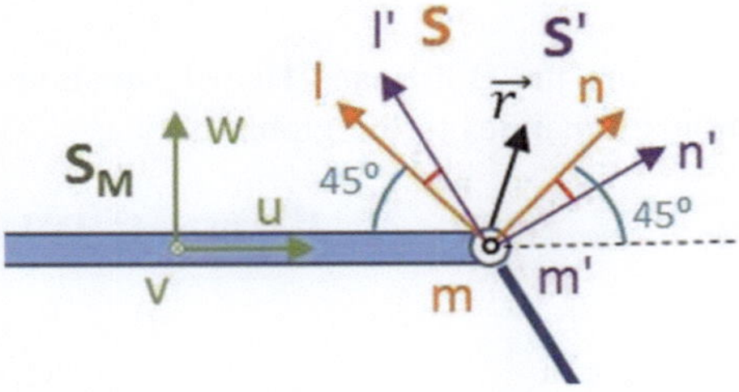

Fig. 3 Coordinate frames used to obtain the spherical joint rotations

$$\mathbf{OP} + \mathbf{PB_i} = \mathbf{OA_i} + s_i \cdot \mathbf{s_{io}} + L \cdot \mathbf{l_{io}} \quad \mathbf{OB}_i \cdot \mathbf{u}_i = (\mathbf{OP} + \mathbf{PB}_i) \cdot \mathbf{u}_i = 0 \tag{7}$$

After derivation, the 3PRS generalized coordinates $\mathbf{q}$ are related to the platform center velocity and the platform angular speed $\boldsymbol{\omega}$ components, regarded here as quasi-velocities, by means of $\dot{\mathbf{p}}^{\mathbf{T}} = \left\{ \dot{x}_p \quad \dot{y}_p \quad \dot{z}_p \quad \omega_x \quad \omega_y \quad \omega_z \right\} = \left\{ \dot{\mathbf{p}}_{\mathbf{t}}^{\mathbf{T}} \quad \boldsymbol{\omega}^{\mathbf{T}} \right\}$.

$$\left.\begin{array}{r} \mathbf{l_{io}} \cdot \dot{\mathbf{p}}_{\mathbf{t}} + (\mathbf{PB_i} \wedge \mathbf{l_{io}}) \cdot \boldsymbol{\omega} = \mathbf{s_{io}} \cdot \mathbf{l_{io}} \cdot \dot{s}_i \\ \mathbf{u_i} \cdot \dot{\mathbf{p}}_{\mathbf{t}} + (\mathbf{PB_i} \wedge \mathbf{u_i}) \cdot \omega = \mathbf{0} \end{array}\right\} \Rightarrow \left\{ \begin{array}{l} \mathbf{J}_{\mathbf{p1}i}\dot{\mathbf{p}} = \mathbf{J}_{\mathbf{q}i}\dot{\mathbf{q}} \\ \mathbf{J_{p2}}\dot{\mathbf{p}} = 0\dot{q} \end{array}\right. \tag{8}$$

Finally, the Jacobian of the platform is obtained as in Eq. (9), and divided in its translational and rotational part.

$$\begin{aligned} \begin{bmatrix} \mathbf{J_{p1}} \\ \mathbf{J_{p2}} \end{bmatrix} \dot{\mathbf{p}} = \mathbf{J_p}\dot{\mathbf{p}} = \begin{bmatrix} \mathbf{J_{q1}} \\ \mathbf{0} \end{bmatrix} \dot{\mathbf{q}} = \mathbf{J_q}\dot{\mathbf{q}} \Rightarrow \dot{\mathbf{p}} \\ = \mathbf{J_p}^{-1}\mathbf{J_q}\dot{\mathbf{q}} = \mathbf{J_{plat}}\dot{\mathbf{q}} = \begin{bmatrix} \mathbf{J_{plat_t}} \\ \mathbf{J_{plat_r}} \end{bmatrix} \dot{\mathbf{q}} \end{aligned} \tag{9}$$

2.4 Platform Rotational Dynamics

To study the rotational dynamics of the platform, the Boltzmann-Hamel equations will be used, as the quasi-velocities $\boldsymbol{\omega}$ will be the generalized local coordinates that define the orientation change of the platform. The quasi-velocities can be related to the Euler rotations $\dot{\mathbf{e}}$ as:

$$\boldsymbol{\omega} = \mathbf{D^T}\dot{\mathbf{e}} = \begin{bmatrix} c\varphi & s\theta s\varphi & 0 \\ -s\varphi & s\theta c\varphi & 0 \\ 0 & c\varphi & 1 \end{bmatrix} \begin{Bmatrix} \dot{\theta} \\ \dot{\psi} \\ \dot{\varphi} \end{Bmatrix} \tag{10}$$

Quasi-velocities are also related to the 3PRS generalized coordinates by means of the platform rotational Jacobian $\mathbf{J_{platr}}$ as in Eq. (9), so it is possible to relate Euler rotations and the 3PRS generalized coordinates by means of $\mathbf{J_e}$ Jacobian.

$$\boldsymbol{\omega} = \mathbf{D^T}\dot{\mathbf{e}} = \mathbf{J_{plat_r}}\dot{\mathbf{q}} \Rightarrow \dot{\mathbf{e}} = \left[\mathbf{D^T}\right]^{-1}\mathbf{J_{plat_r}}\dot{\mathbf{q}} = \mathbf{J_e}\dot{\mathbf{q}} \tag{11}$$

Using the Boltzmann-Hamel equations, the contribution of the platform rotational dynamics to the global is:

$$\mathbf{J}_{\mathbf{plat_r}}^{T}\mathbf{m} = \mathbf{J_e^T}\left[\mathbf{D^T}\mathbf{I_{plat}}\mathbf{D^T}\right]\mathbf{J_e}\ddot{\mathbf{q}} + \mathbf{c_{plat}} \tag{12}$$

Where $\mathbf{I_{plat}}$ is the inertia tensor in the moving frame, $\mathbf{c_{plat}}$ is a term that depends on the centrifugal and Coriolis forces and $\mathbf{m}$ is the vector of the moments acting on the platform.

$$\begin{aligned}\mathbf{c}_{\mathbf{plat}_r} = \mathbf{J_e^T}\Big[\left[\mathbf{D^T I_{plat} D^T}\right]\dot{\mathbf{J}}_\mathbf{e} + \left[\mathbf{D^T I_{plat} \dot{D}^T}\right]\mathbf{J_e} \\ + \left[2\dot{\mathbf{D}}^\mathbf{T}\mathbf{I_{plat}} - \left(\frac{\partial \dot{\boldsymbol{\omega}}}{\partial \dot{\mathbf{e}}}\right)^T \mathbf{I_{plat}}\right]\mathbf{J}_{\mathbf{plat}_r}\Big]\dot{\mathbf{q}}\end{aligned} \tag{13}$$

Regarding the loads on the platform, see Fig. 2, two sources have been considered, the moments due to the cutting forces and the moments due to the spherical joints deflection. Although their elastic deflection can be considered in the potential energy used of the Lagrangian formulation, instead, here they have been regarded as external moments acting on the platform. Hence, the moments vector is Eq. (14), where $\mathbf{d}$ is the position of the tool center point with respect to the platform center.

$$\mathbf{m} = \mathbf{m_{cut}} + \mathbf{m_{joints}} = \mathbf{R}(\mathbf{d} \wedge \mathbf{f}) + \mathbf{m_{joints}} \tag{14}$$

To compute the torques due to the spherical joints deflections, two coordinate systems have been located in each spherical joint, the first, $\mathbf{S_i}$, fixed to the platform and the second, S_{io}, fixed to the leg, see Fig. 3 for the leg 1. In the zero position, at $\alpha_i = 45°$, both systems are identical. Rotations around the m- and n-axes make reference to the joint deflection and rotation around the l-axis represents torsional deformation. To obtain the rotations between the systems $\mathbf{S_i}$ and $\mathbf{S_{i0}}$, the rotation matrix that relates them must be developed. To do that, the unit vectors of each system $\mathbf{S_{io}}$ regarding to the global system XYZ must be calculated. These systems belong to the different limbs of length L, with $\mathbf{l_{i0}}$ aligned with the leg and $\mathbf{m_{i0}}$ parallel to the revolute joint axis. As a result, the rotation matrici $\mathbf{R_{io}}$ to change from $\mathbf{S_{io}}$ that is placed in each leg to the fixed frame XYZ can be expressed as:

$$\mathbf{R_{i0}} = [\,\mathbf{m_{i0}} \quad \mathbf{n_{i0}} \quad \mathbf{l_{i0}}\,] \quad i = 1, 2, 3 \tag{15}$$

On the other hand, the rotation matrices $\mathbf{R_i}$ that $\mathbf{S_i}$ systems with XYZ are calculated. The unit vectors for each system $\mathbf{S_i}$ can be defined by their fixed position in the platform, being $\alpha_{i0} = 45°$ the default value of α_i. To refer these vectors to the fixed frame, it is necessary to multiply them by the rotation matrix $\mathbf{R}$ that relates the moving frame attached to the platform UVW, with the fixed one XYZ:

$$\mathbf{R_i} = [\,\mathbf{m_i} \quad \mathbf{n_i} \quad \mathbf{l_i}\,] \tag{16}$$

Finally, the rotation matrix $\mathbf{R_{i\text{-}io}}$ that relates frames $\mathbf{S_{io}}$ and $\mathbf{S_i}$ is:

$$\mathbf{R_{i-i0}} = \mathbf{R}_{\mathbf{i0}}^T\mathbf{R_i} \quad i = 1, 2, 3 \tag{17}$$

Furthermore, this matrix can be developed as a function of three rotations around all the axes system $\mathbf{S_{io}}$, (β_{mi}, β_{ni}, β_{li}), where β_{mi} and β_{ni} are the two rotations due to the deflection of the joints, and β_{li} is due to the torsion of the joints. Therefore, the rotation matrix that relates the systems $\mathbf{S_i}$ with $\mathbf{S_{i0}}$ may also be expressed as the product of the above rotation matrices.

$$\begin{aligned}\mathbf{R_{i-i0}} &= \mathbf{R_m}(\beta_m)\mathbf{R_n}(\beta_n)\mathbf{R_l}(\beta_l)\\ &= \begin{bmatrix} c\beta_n c\beta_l & -c\beta_n s\beta_l & s\beta_n \\ s\beta_m s\beta_n c\beta_l + c\beta_m s\beta_l & -s\beta_m s\beta_n s\beta_l + c\beta_m c\beta_l & -s\beta_m c\beta_n \\ -c\beta_m s\beta_n c\beta_l + s\beta_m s\beta_l & c\beta_m s\beta_n s\beta_l + s\beta_m c\beta_l & c\beta_m c\beta_n \end{bmatrix}\end{aligned} \tag{18}$$

Equaling the terms of the matrices in Eq. (17) and (18), the rotations in the spherical joints can be obtained. The resulting expressions are:

$$\beta_{mi} = \mathrm{a}\tan\left(\frac{-\mathbf{R_{i-i0}}(2,3)}{\mathbf{R_{i-i0}}(3,3)}\right) \quad \beta_{ni} = \mathrm{a}\sin(\mathbf{R_{i-i0}}(1,3)) \quad \beta_{li} = \mathrm{a}\tan\left(\frac{-\mathbf{R_{i-i0}}(1,2)}{\mathbf{R_{i-i0}}(1,1)}\right) \tag{19}$$

Knowing the rotations produced in the spherical joints, the torques in each hinge can be obtained multiplying by the corresponding stiffness.

$$\mathbf{m_{joints}} = \sum_{i=1}^{3}\{\boldsymbol{\tau_{mi}} + \boldsymbol{\tau_{ni}} + \boldsymbol{\tau_{li}}\}\begin{cases} \tau_{mi} = k_{\mathrm{fsph}} \cdot \beta_{mi} \\ \tau_{ni} = k_{\mathrm{fsph}} \cdot \beta_{ni} \\ \tau_{li} = k_{\mathrm{tsph}} \cdot \beta_{li} \end{cases} \tag{20}$$

2.5 Legs

The position of the mass centers and the angular position of each limb have been considered as local generalized coordinates, $\mathbf{p_{legi}} = \{x_{Gi} \quad y_{Gi} \quad z_{Gi} \quad \alpha_i\}$, see Fig. 4. After obtaining the equations of motion of each leg, their contribution to the global dynamics of the manipulator is:

$$\mathbf{J_{legi}^T}\,\mathbf{M_{leg}}\,\mathbf{J_{legi}}\ddot{\mathbf{q}} + \mathbf{J_{legi}^T}\,\mathbf{M_{leg}}\,\dot{\mathbf{J}}_{\mathbf{legi}}\dot{\mathbf{q}} + \mathbf{J_{legi}^T}\,\mathbf{gk_{legi}} = \mathbf{0} \quad i = 1,2,3 \tag{21}$$

Where $\mathbf{M_{leg}}$ is the mass matrix of each leg, $\mathbf{J_{legi}}$ are the Jacobians that relate $\mathbf{p_{legi}}$ with the 3PRS generalized coordinates $\mathbf{q}$, and $\mathbf{gk_{legi}}$ is the vector that contains gravitational and elastic terms from the revolute joints.

Developing the closure loop equation for each leg and deriving, it is possible to relate the legs mass center position with the 3PRS generalized coordinates $\mathbf{q}$.

$$\frac{d}{dt}\left(\mathbf{OG_i} = \mathbf{OA}_i + s_i \cdot \mathbf{s_{i0}} + \frac{L}{2} \cdot \mathbf{l_{i0}}\right) \Rightarrow \mathbf{J}_{Gi}\,\dot{\mathbf{p}}_{\mathbf{legi}} = \mathbf{J}_{qGi}\dot{\mathbf{q}} \quad i = 1,2,3 \tag{22}$$

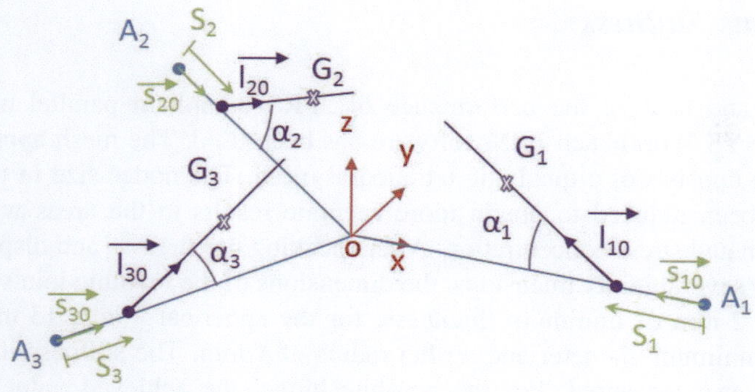

Fig. 4 Schematic representation of the legs variables

Also, to relate the legs angular position α_i with **q**, it is necessary to average the three loop closure equations of the manipulator and derive:

$$\frac{d}{dt}\left(\mathbf{OP} = \frac{1}{3}\sum_{i=1}^{3}\mathbf{OA_i} + s_i\mathbf{s_{i0}} + L\mathbf{l_{i0}} - \mathbf{PB_i}\right) \Rightarrow \mathbf{J}_{\alpha i}\dot{\alpha}_i = \mathbf{J}_{q\alpha i}\dot{\mathbf{q}} \tag{23}$$

Finally, combining Eqs. (22) and (23), it is possible to obtain the three legs Jacobians.

$$\begin{bmatrix}\mathbf{J}_{Gi}\\ \mathbf{J}_{\alpha i}\end{bmatrix}\dot{\mathbf{p}}_{\mathbf{legi}} = \begin{bmatrix}\mathbf{J}_{qGi}\\ \mathbf{J}_{q\alpha i}\end{bmatrix}\dot{\mathbf{q}} \Rightarrow \dot{\mathbf{p}}_{\mathbf{legi}} = \mathbf{J}_{\mathbf{legi}}\dot{\mathbf{q}} \tag{24}$$

2.6 *Dynamic Model of the Manipulator*

Once the dynamic analyses for each of the components of the mechanism have been developed, it is necessary to group all of them in a single expression to model the whole manipulator. To do that, a summation of Eq. (3), (6), (12) and (21) is performed. The resultant expression is:

$$\begin{aligned}&[\mathbf{M_{act}}\ddot{\mathbf{q}}] + \left[-\mathbf{J}^T_{\mathbf{plat_t}}\mathbf{f} + \mathbf{J}^T_{\mathbf{plat_t}}\mathbf{M}\mathbf{J}_{\mathbf{plat_t}}\ddot{\mathbf{q}} + \mathbf{J}^T_{\mathbf{plat_t}}\mathbf{M}\dot{\mathbf{J}}_{\mathbf{plat_t}}\dot{\mathbf{q}} + \mathbf{J}^T_{\mathbf{plat_t}}\mathbf{g}\right]\\ &+\left[-\mathbf{J}^T_{\mathbf{plat_r}}\mathbf{m} + \mathbf{J}^\mathbf{T}_\mathbf{e}\left[\mathbf{D^T I_{plat} D^T}\right]\mathbf{J_e}\ddot{\mathbf{q}} + \mathbf{c_{plat_r}}\right] +\\ &+\sum\nolimits_{i=1}^{3}\left[\mathbf{J}^\mathbf{T}_{\mathbf{leg}i}\mathbf{M_{leg}}\mathbf{J}_{\mathbf{leg}i}\ddot{\mathbf{q}} + \mathbf{J}^\mathbf{T}_{\mathbf{leg}i}\mathbf{M_{leg}}\dot{\mathbf{J}}_{\mathbf{leg}i}\dot{\mathbf{q}} + \mathbf{J}^\mathbf{T}_{\mathbf{leg}i}\mathbf{g}\mathbf{k}_{\mathbf{leg}i}\right] = \mathbf{f_{act}}\end{aligned} \tag{25}$$

2.7 *Joints Stiffness*

To design and analyze the performance of 3PRS compliant parallel mechanism joints, ANSYS Workbench FEM software has been used. The mesh applied to the mechanism consists of a quadratic tetrahedral mesh. The nodes size in the flexure joints has been reduced to obtain more accurate results in the areas with a high deformation and stress concentration. After checking the stresses and displacements reached for several joint dimensions, the dimensions of the revolute joints are 8 mm radius and 2 mm of minimum thickness, for the spherical joints, 13 mm length, 3 mm of minimum diameter and a fillet radius of 4 mm. The stiffness if the joints have been also measured. For the revolute hinge, the achieved value has been $k_{\text{rev}} = 98.37$ Nm/rad. For the spherical joint, the bending stiffness and the torsional stiffness have been $k_{\text{fsph}} = 32.67$ Nm/rad and $k_{\text{tsph}} = 24.46$ Nm/rad.

3 Experimental Validation

An initial prototype in aluminum 7075T6 has been built, see Fig. 5. The three actuators are based on a linear belt drive Igus ZLW-1040-02-S-100 coupled to a DC RE-40 Maxon servomotor with a GP-32 14:1 reduction. A NI-PXIe 1062 has been used to control the prototype motion in real time with a cascaded joint space control of position, velocity and current. The position loop cycle time is 5 ms.

Currently, intensive testing is being made to validate the results from the dynamic model. The main sources of uncertainty are the friction in the linear guides, the FEA estimated stiffness, the simplified kinematics of the joints, the bolted joints not modelled in FEA and the manufacturing and assembly errors on the prototype. Nevertheless, in Figs. 6 and 7 it can be seen the torque measured at the motors, which is an indirect measurement of the force, for two motions, first a sinusoidal motion in Z with an amplitude of 2 mm and 0,5 Hz, and then a

Fig. 5 Developed prototype

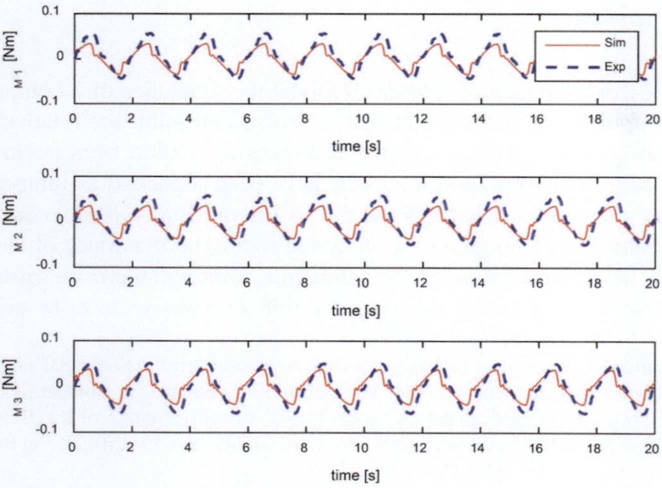

Fig. 6 Simulated motor torque (*red continuous line*) versus measured (*blue discontinuous line*)

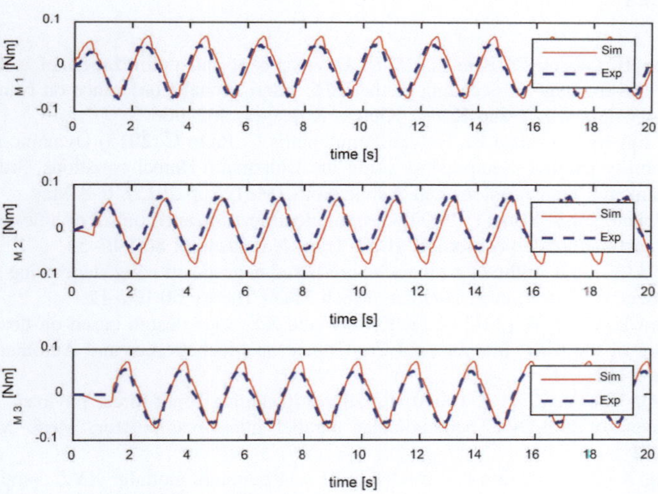

Fig. 7 Simulated motor torque (*red continuous line*) versus measured (*blue discontinuous line*)

sinusoidal motion of the three actuators with the same amplitude and frequency but a phase shift of 120°. Although the results are quite similar, there are deviations and even the trend is different, in the first test the simulated torque is lower and vice versa, so further testing must be made to adjust the model parameters to the real ones.

4 Conclusions

The present work shows a procedure to model the dynamics of a compliant 3PRS parallel manipulator. Given the fact that the compliant joints are relatively small in size comparing to the limbs and platform dimensions, it has been assumed that all the bodies behave as rigid, and the joints have been modelled as lumped torsional springs. The method used is the Principle of Energy Equivalence, combined with the Bolttzmann-Hamel equations to study the rotational dynamics of the platform. A prototype has been constructed for validation. Although there are several sources of discrepancy that are being analyzed, the initial results seem to be promising.

Acknowledgments The authors of this paper wish to acknowledge the financial support received from the Spanish Government via the Ministerio de Economía y Competitividad (Project DPI2015-64450-R), the ERDF of the European Union, the Government of the Basque Country (Project GIC07/78, IT445-10 and SAIOTEK 2013 SAI13/245, SPC13UN011), and the University of the Basque Country (Zabalduz-2012).

References

Ahn D, Chun B, Gweon D, Kim H (2010) Development and optimization of a novel 3-DOF precision flexure stage. Proceeding of the IEEE International Conference on Nanotechnology Joint Symposium with Nano Korea. Kintex, Korea, pp 903–906

Altuzarra O, Eggers P, Campa FJ, Roldan-Paraponiaris C, Pinto C (2015) Dynamic modelling of lower-mobility parallel manipulators using the Boltzmann-Hamel equations, 3rd Conference on Mechanisms, Transmissions and Applications (MeTrApp 2015), 6–8 May

Aphale S, Moheimani S, Yong Y (2009) Design, identification, and control of a flexure-based XY stage for fast nanoscale positioning. IEEE Trans Nanotechnol 8(1):46–54

Choi Y, Kim J (2012) A millimeter-range flexure-based nano-positioning stage using a self-guided displacement amplification mechanism. Mech Mach Theory 50:109–120

Choi H, Han Ch, Wang W (2011) 2-DOF kinematic XY stage design based on flexure element. Proceeding of the IEEE International Conference on Mechatronics and Automation. Beijing, China, pp 1412–1417

Gao F, Ge Q, Yue Y, Zhao X (2010) Relationship among input-force, payload, stiffness and displacement of a 3-DOF perpendicular parallel micro-manipulator. Mech Mach Theory 45:756–771

Hao G, Kong X (2012) Design and modeling of a large-range modular XYZ compliant parallel manipulator using identical spatial modules. J Mech Rob 4:021009

Howell L (2001) Compliant mechanisms. Wiley, New York

Kenton B, Leang K (2012) Design and control of a three-axis serial-kinematic high-bandwidth nanopositioner. IEEE/ASME Trans Mechatron, 17(2)

Li Y, Xu Q (2007) Kinematic analysis of a 3-PRS parallel manipulator. Rob Comput-Integr Manuf 23:395–408

Merlet JP (2000) Parallel Robots. Kluwer Academic Publishers, London

Dynamic Analysis of a Fatigue Test Bench for High Precision Flexure Hinges

D. Schoenen, M. Hüsing and B. Corves

Abstract General technological development and better knowledge of compliant mechanisms increase the growth in their applications. In current science the focus is on developing new hinge shapes or improving existing hinge types. Even though it can be assumed that the manufacturing based influences have a crucial impact on the performance of flexure hinges, experimental studies of the fatigue of compliant joints are currently missing in research activities. Thus, a test bench for fatigue testing was built to study the manufacturing based influences on the performance of high precision flexure hinges. First investigations show high load cycles. Hence it might be useful to increase the test frequency to reduce the overall testing time. Nevertheless, this causes increased inertia forces and moments which could additionally affect the test specimens. Accordingly, this contribution analyzed the dynamic behavior of the test bench. The kinematic equations are derived. Afterwards the actual forces acting on the test specimens are calculated and compared to a quasi-static load.

Keywords Flexure hinge · Dynamic analysis · Test bench · Test frequency · Inertia forces

---

The original version of this chapter was revised: The initial of author's name was corrected: The erratum to this chapter is available at 10.1007/978-3-319-45387-3_23

D. Schoenen (✉) · M. Hüsing · B. Corves
Department of Mechanism Theory and Dynamics of Machines,
RWTH Aachen University, Aachen, Germany
e-mail: schoenen@igm.rwth-aachen.de

M. Hüsing
e-mail: huesing@igm.rwth-aachen.de

B. Corves
e-mail: corves@igm.rwth-aachen.de

L. Zentner et al. (eds.), *Microactuators and Micromechanisms*,
Mechanisms and Machine Science 45, DOI 10.1007/978-3-319-45387-3_15

1 Introduction

Compliant mechanisms have many advantages like the monolithic structure, which induces a reduced assembly effort, less wear, reduced need for lubrication and weight reduction. This facilitates not only close to zero backlash and frictionless motion, but also significant potential for miniaturization. The disadvantages are deviation in motion, undesired rotational and/or translational parasitic effects, difficulties in analyzing and designing compliant mechanisms and the limited range of motion (Howell 2001).

However, experimental studies on the fatigue of precision flexure hinges are currently missing in research activities. Thus, in previous research a fatigue test bench for high precision flexure hinges was presented which is shown in Fig. 1 (Schoenen et al. 2015). The test bench allows simultaneous testing of six test specimens under the same conditions. Nevertheless, a clear detection of each test specimen is guaranteed through the design and the electronic control. With a combination of a double eccentric crank and a slotted guide system on the guide carriage, the swivel angles as well as the load case for each set of test specimens are adjustable. During adjustment a dial indicator is used to measure the variables.

The test bench is driven by a non-shifted slider crank mechanism which is powered by an electric drive. For reasons of simplification the schematic representation of the mechanism in Fig. 2 is illustrated two-dimensionally. However, the slider-crank mechanism moves in the horizontal plane, which is perpendicular to the plane used for the two-dimensional illustration.

Within this paper the dynamic load of the test specimens is analyzed. Initially, the moving masses have to be detected. Of course the main moving mass is the guide carriage. However, the guide carriage is positively driven by a slider crank mechanism. Therefore its inertia forces and moments are acting directly onto the linear guiding as well as the crank shaft. Hence the pendulum which is shown in Fig. 3 represents the relevant moving masses of the test bench. It is an inverted shifted slider rocker mechanism. The mechanism is spring loaded to separate the

Fig. 1 Test bench

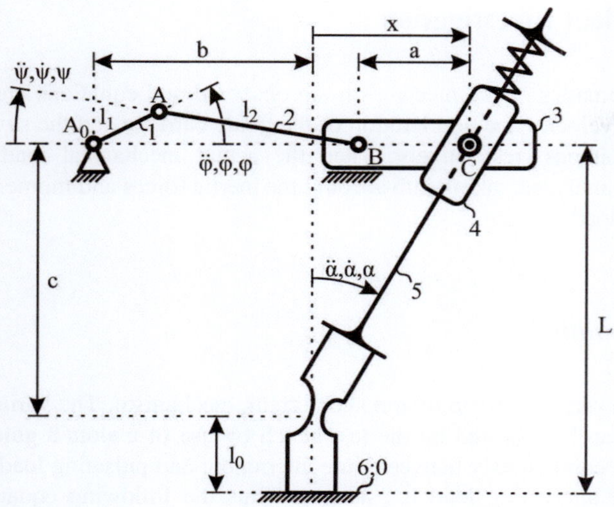

Fig. 2 Mechanism

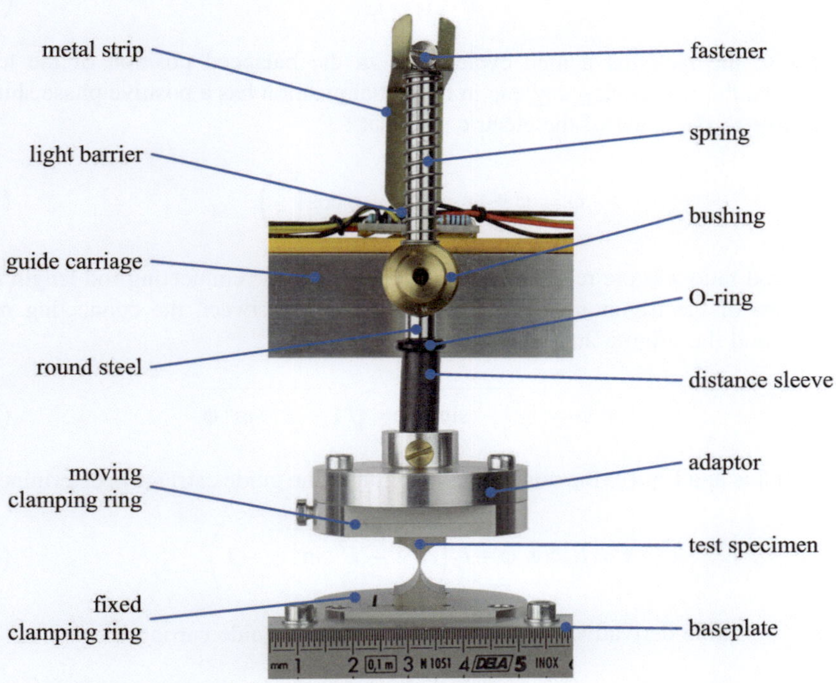

Fig. 3 Pendulum

hinge halves when fracture occurs. With this lifting motion of the spring a metal strip is also raised and the fracture is clearly detected by a light barrier. A distance sleeve and an O-ring hold the pendulum in a defined position after breaking.

2 Analytical Investigation

First the kinematics of the mechanism are observed and equations are derived for the position, velocity and acceleration of the guide carriage and the swivel angle of the test specimens, respectively. Then the actual mechanical load of the test specimens is analyzed, taking into account the inertia forces and moments as well as the external load.

2.1 Kinematics

The guide carriage is driven by the slider crank mechanism. The variables a and b (cf. Fig. 2) can be adjusted on the test bench by use of a slotted guide system in order to vary continuously between pure alternating and pulsating load. Within this study a pure alternating load is assumed. Thus the following equation must be verified.

$$b - a = l_2 \tag{1}$$

Due to the fact that a load cycle starts in the balanced position of the test specimens, the real driving angle φ in the initial position has a positive phase shift φ_i relative to the angle of the electric drive $\bar{\varphi}$:

$$\varphi = \bar{\varphi} + \varphi_i = \bar{\varphi} + \arccos\left(\frac{\lambda}{2}\right) \tag{2}$$

The rod ratio λ is the relation of crank length l_1 to the connecting rod length l_2. By means of the trigonometric functions coherence between the connecting rod angle ψ and the driving angle φ is given:

$$\cos\psi = \sqrt{1 - \sin^2\psi} = \sqrt{1 - \lambda^2 \sin^2\varphi} \tag{3}$$

With this and Eq. (2) the nonuniform motion of the guide carriage is determined:

$$x = l_1 \cos\varphi + l_2\left(\sqrt{1 - \lambda^2 \sin^2\varphi} - 1\right) \tag{4}$$

The first time derivative of x is the velocity of the guide carriage:

$$\dot{x} = -\dot{\varphi} l_1 \sin\varphi\left(1 + \lambda\frac{\cos\varphi}{\sqrt{1 - \lambda^2 \sin^2\varphi}}\right) \tag{5}$$

The slider crank mechanism is powered by an electric drive with a controlled constant angular velocity $\dot{\varphi}$. Accordingly the second time derivative of the driving angle cφ must be equal to zero. This simplifies the acceleration of the guide carriage according to the second time derivative of x:

$$\ddot{x} = -\dot{\varphi}^2 l_1 \left(\cos\varphi + \lambda \frac{\cos 2\varphi}{\sqrt{1-\lambda^2 \sin^2\varphi}} + \frac{\lambda^3}{4} \frac{\sin^2 2\varphi}{\sqrt{1-\lambda^2 \sin^2\varphi}^3} \right) \tag{6}$$

Compared to conventional hinges, flexure hinges do not have a stationary axis due to their flexibility. This is already taken into account during the design process by not forcing a fixed axis upon the test specimens (Schoenen et al. 2015). As a consequence additional tensions are avoided. The anticipated shift of the stationary axis is so low that for the calculation of the swivel angle α of the test specimen an idealized rotation axis is assumed. The swivel angle is defined positively in clockwise direction:

$$\tan\alpha = \frac{x}{c} \tag{7}$$

If necessary, the length c is also gradually adjustable on the test bench to facilitate testing of extra-long test specimens. Then two pillars on which the linear guiding of the guide carriage is mounted have to be replaced. The first time derivative of α is the angular velocity of the swivel angle:

$$\dot{\alpha} = \frac{c\dot{x}}{c^2 + x^2} \tag{8}$$

The second derivative of α with respect to time is the angular acceleration of the swivel angle:

$$\ddot{\alpha} = \frac{c}{c^2 + x^2} \left(\ddot{x} - \frac{2x\dot{x}^2}{c^2 + x^2} \right) \tag{9}$$

With the previously derived Eqs. (1)–(9) the kinematic variables are defined as functions of the driving angle φ and its angular velocity $\dot{\varphi}$. However, the angular velocity of the driving angle can be expressed as a function of the test frequency f:

$$\dot{\varphi} = 2\pi f \tag{10}$$

2.2 Kinetics

Below the inertia forces and moments of the pendulum are determined. For flexure hinges the characteristic load is calculated for a force application point (FAP) which

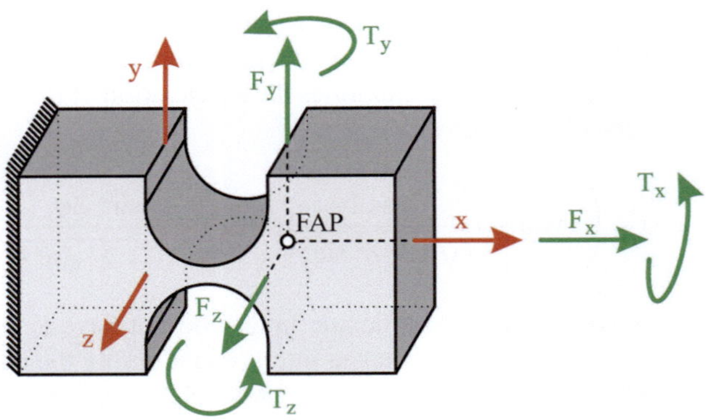

Fig. 4 General loading of a flexure hinge

is located, according to Fig. 4, at the intersection of the flexible part of the hinge and the part which is assumed to be ridged.

Contrary to the general loading of a flexure hinge, the design of the test bench avoids parasitic loading (Schoenen et al. 2015). Therefore, the loading of the pendulum can be simply considered as a two-dimensional problem. Nevertheless, the test specimens are tested upright on the test bench. Thus, acceleration due to gravity must be considered.

Accordingly, the free-body diagram of the pendulum is shown in Fig. 5. It consists of three parts. The first part is represented by the accumulated masses of the rocker (R) m_R which considers the mass of the moving clamping ring, the adaptor, the round steel, the O-ring, the distance sleeve, the spring, the metal strip, the fastener and the rigid part of the test specimen itself (cf. Fig. 3). The center of gravity (CG) is located at l_{CG_R} from the idealized rotation axis. The concentrated mass moment of inertia about the CG of the rocker is J_R.

The second part considers one half of the flexible part of the hinge (H). The center of gravity (CG) has a distance of l_{CG_H} from the rotation axis. Even though the mass m_H and the moment of inertia of the hinge J_H are very low at this point they are not neglected in order to avoid fundamental mistakes. Thus, the given equations are also valid for larger hinges. The length of the hinge l_H defines the distance of the FAP to the center of rotation.

The bushing (B) represents the third part. It is pivoted to the guide carriage. Hence, the bushing is linear guided and the guide carriage is loaded with the mass force of the bushing directly. Consequently, the mass of the bushing must not be taken into account when analyzing the mass forces of the pendulum. Nevertheless, the moment of inertia of the bushing J_B still acts onto the rocker.

Furthermore, there are two external spring loads acting onto the pendulum, a torsion spring and a compression spring. The torsion spring represents the spring stiffness of the test specimen itself. It is located at the rotation axis of the test specimen. The spring stiffness k_α is averaged for the motion range and is acting counterclockwise for a positive swivel angle:

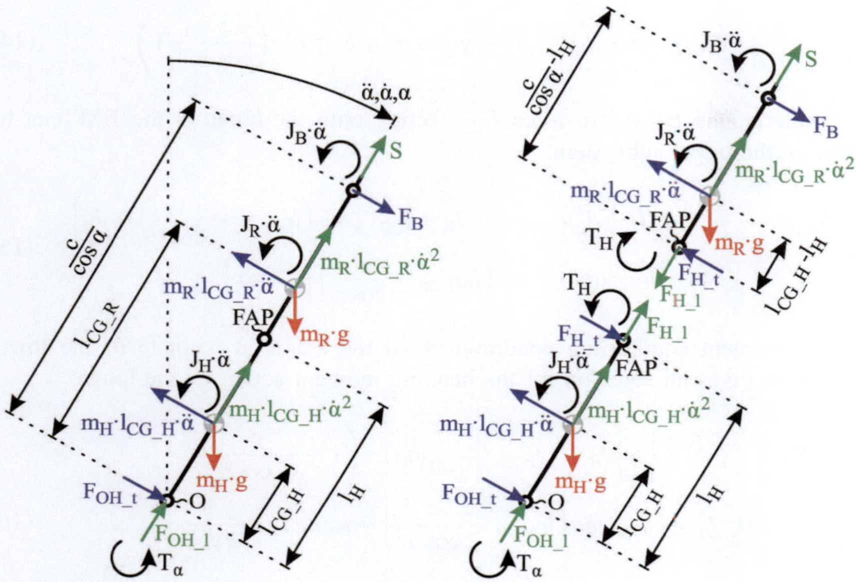

Fig. 5 Free-body diagram of complete pendulum and subdivision in subsystems (not to scale)

$$T_\alpha = k_\alpha \alpha \tag{11}$$

The second spring is a compression spring which tensions the flexure hinge. It supports itself against the bushing and the rocker. A low spring rate k_S was chosen, to minimize the change in the spring force, caused by elongation of the rocker. Moreover, the preload force S_i was set so low, that the weight force of the rocker and half the hinge can be raised safely. Hence, the spring force is as follows:

$$S = S_i + k_S c\left(\frac{1}{\cos\alpha} - 1\right) \tag{12}$$

The moment equilibrium condition about the z-axis at point O of the complete pendulum gives an equation for the force acting from the bushing onto the rocker:

$$\begin{aligned} F_{\mathrm{B}} = \frac{\cos\alpha}{c}\,[&\ddot{\alpha}(J_{\mathrm{H}} + l_{\mathrm{CG\_H}}^2 m_{\mathrm{H}} + J_{\mathrm{R}} + l_{\mathrm{CG\_R}}^2 m_{\mathrm{R}} + J_{\mathrm{B}}) \\ &-g\,\sin\alpha(l_{\mathrm{CG\_H}} m_{\mathrm{H}} + l_{\mathrm{CG\_R}} m_{\mathrm{R}}) + k_\alpha \alpha] \end{aligned} \tag{13}$$

Considering the force equilibrium condition of the upper subsystem yields the longitudinal force $F_{\mathrm{H\_l}}$ acting onto the hinge at the FAP.

$$F_{\mathrm{H\_1}} = m_{\mathrm{R}}\left(l_{\mathrm{CG\_R}}\dot{\alpha}^2 - g\,\cos\alpha\right) + S_i + k_S c\left(\frac{1}{\cos\alpha} - 1\right) \tag{14}$$

Similarly, the transverse force $F_{\mathrm{H\_t}}$ acting onto the hinge at the FAP can be found in the upper subsystem:

$$\begin{aligned} F_{\mathrm{H\_t}} = \frac{\cos\alpha}{c}\Big\{&\ddot{\alpha}\Big[J_{\mathrm{H}} + l_{\mathrm{CG\_H}}^2 m_{\mathrm{H}} + J_{\mathrm{R}} + l_{\mathrm{CG\_R}} m_{\mathrm{R}}\Big(l_{\mathrm{CG\_R}} - \frac{c}{\cos\alpha}\Big) + J_{\mathrm{B}}\Big] \\ &-g\,\sin\alpha\Big[l_{\mathrm{CG\_H}} m_{\mathrm{H}} + m_{\mathrm{R}}\Big(l_{\mathrm{CG\_R}} - \frac{c}{\cos\alpha}\Big)\Big] + k_\alpha \alpha\Big\} \end{aligned} \tag{15}$$

The moment equilibrium condition about the z-axis at point O of the lower subsystem gives an equation for the bending moment acting on the hinge:

$$\begin{aligned} T_{\mathrm{H}} = \frac{l_{\mathrm{H}}\cos\alpha}{c}\Big\{&\ddot{\alpha}\Big[J_{\mathrm{B}} + (J_{\mathrm{H}} + l_{\mathrm{CG\_H}}^2 m_{\mathrm{H}})\cdot\Big(1 - \frac{c}{l_{\mathrm{H}}\cos\alpha}\Big) \\ &+ J_{\mathrm{R}} + l_{\mathrm{CG\_R}} m_{\mathrm{R}}\Big(l_{\mathrm{CG\_R}} - \frac{c}{\cos\alpha}\Big)\Big] + k_\alpha \alpha\Big(1 - \frac{c}{l_{\mathrm{H}}\cos\alpha}\Big) \\ &- g\,\sin\alpha\Big[l_{\mathrm{CG\_H}} m_{\mathrm{H}}\Big(1 - \frac{c}{l_{\mathrm{H}}\cos\alpha}\Big) + m_{\mathrm{R}}\Big(l_{\mathrm{CG\_R}} - \frac{c}{\cos\alpha}\Big)\Big]\Big\} \end{aligned} \tag{16}$$

3 Results

With the previously derived equations the actual load of the test specimen can be calculated. Below, this is carried out for two significant test frequencies, 3 and 5 Hz. With a frequency equal to zero the test specimens are expected to remain in their initial position, but in this case this means the driving angle is changed very slightly to avoid any inertia effects. Thus, as a comparative value the quasi-static load of the test specimens is shown and labeled with 0 Hz. Even though the calculation for the dynamic load is done for two selected frequencies the missing values can be interpolated qualitatively. The variables and their values used for calculation are listed in Table 1.

Considering Eq. (14) for the longitudinal force acting onto the hinge it is clear that the largest percentage results from the compression spring preload. Figure 6 shows the course of the longitudinal force for the mentioned frequencies as a function of the driving angle φ. In addition, the swivel angle α is also shown. Interestingly, if the swivel angle reaches its maximum, the longitudinal force is independent from the test frequency. This effect occurs due to the fact that at this point the angular velocity $\dot{\alpha}$ is equal to zero. According to Eq. (14) the inertia effect concerning the longitudinal force is also nonexistent. The maximum deviation of the force for 3 Hz, and 5 Hz respectively, against the quasi-static value is less than 1.5‰, 4.2‰ respectively.

Table 1 Variables

Variable	Value	Unit	Explanation
l_1	3	[mm]	Crank length
l_2	105	[mm]	Connecting rod length
λ	2.857e−3	[–]	Rod ratio
c	57.55	[mm]	Height of the test bench
l_H	4.98	[mm]	Hinge length
l_{CG_H}	3.5	[mm]	Position of the CG of the hinge
l_{CG_R}	30.36	[mm]	Position of the CG of the rocker
m_H	0.04	[g]	Mass of the hinge
m_R	74.5	[g]	mass of the rocker
J_H	0.1	[kg mm^2]	Mass moment of inertia of the hinge
J_R	70.98	[kg mm^2]	Mass moment of inertia of the rocker
J_B	1	[kg mm^2]	Mass moment of inertia of the bushing
k_α	245.91	[N mm/rad]	Torsion spring stiffness of the hinge
k_S	0.113	[N/mm]	Spring stiffness of the compression spring
S_i	2.2	[N]	Initial compression spring preload
φ_i	89.181	[°]	Positive phase of the driving angle

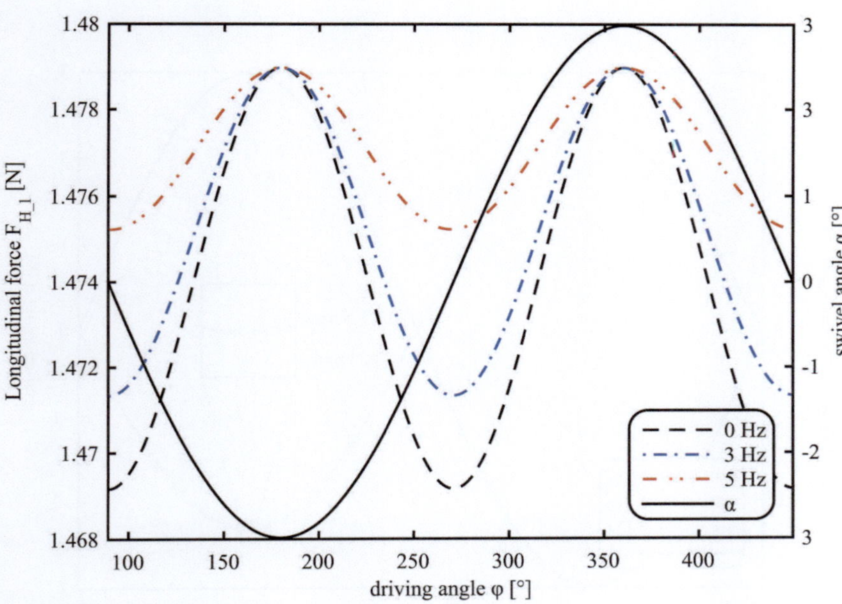

Fig. 6 Longitudinal force

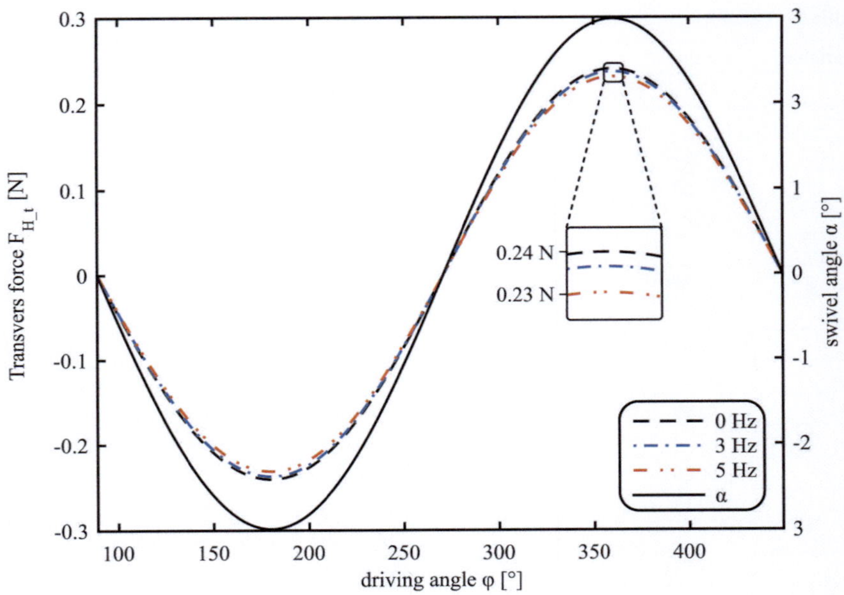

Fig. 7 Transverse force

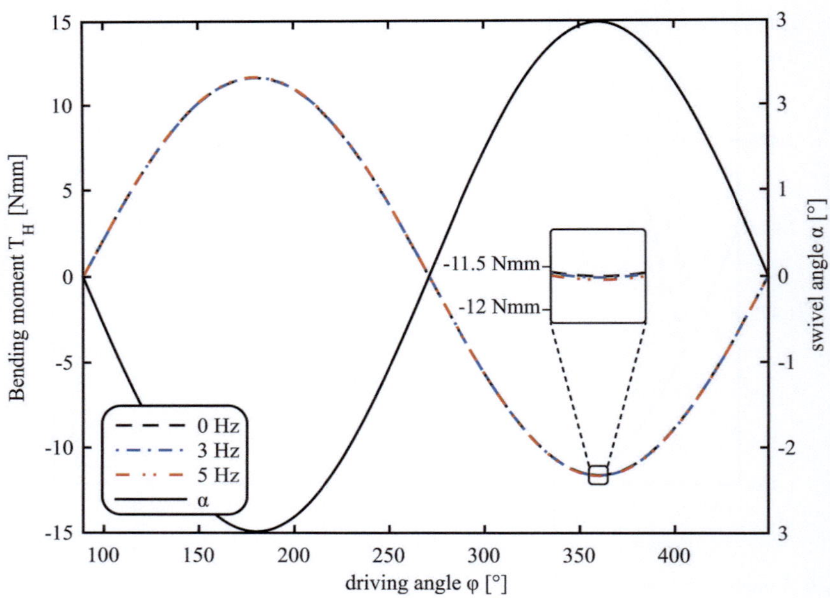

Fig. 8 Bending moment

Due to the fact that the absolute value of the transverse force compared to the longitudinal force is lower, the relative deviation increases. The maximum deviation of the transverse force for 3 Hz against the quasi-static value is less than −1.5 %. For 5 Hz it is less than −4.0 %. This time it can be observed that the maximum deviation in Fig. 7 occurs at the turning points of the swivel angle. This effect is because at this point the angular acceleration α is at its maximum value. Referring to Eq. (15) the change of $\ddot{\alpha}$ is responsible for the inertia effects of the transverse force.

Similar to the transverse force the bending moment is mostly influenced by $\ddot{\alpha}$ too. But the inertia effect increases the load. The deviation in the transverse force is hardly recognizable in Fig. 8. Nevertheless, the maximum deviation of the bending moment for 3 Hz, and 5 Hz respectively, is less than 1.3‰, 3.5‰ respectively.

4 Conclusions

Within this contribution an analytical calculation of the kinematics of a previous designed test bench for high precision flexure hinges is presented. For the purpose of increasing the test frequency due to high load cycles the equations for the dynamic loads are determined. For two selected frequencies the longitudinal force, the transverse force and the bending moment are calculated and the deviations against the quasi-static values are considered.

In consideration of the fact that a flexure hinge acts ideally as a pivot joint, the bending moment is the most significant load. However, the relative deviation of the bending moment against the quasi-static value with 1.3‰, 3.5‰ respectively, for 3 Hz, 5 Hz respectively, are negligible. Similarly, the relative deviation of the longitudinal force due to an increased test frequency is in the same range as mentioned before. Thus, change in longitudinal force due to inertia effects can be neglected in the intended frequency range too. The transverse force shows larger relative deviation with −1.5 %, and −4.0 % respectively, due to inertia effects. At this point it is important to mention that the inertia effect reduces the transverse force. Thus it has a positive effect. But this result has to be put into relation because the absolute value of the transverse force is very low. This means for the tested flexure hinges with a cross section of 15 mm times 40 μm for a quasi-static maximal value of the transverse force of 0.24 N, the resultant shear stress as an individual value is in the range of 0.4 MPa. For that reason even the change in transverse force due to inertia effects can be neglected too.

For the transferability of the results from the given flexure hinge to a different geometry only changes, for example, in the torsion spring stiffness, have to be considered. Of course the conclusions may alter. But the given equations remain valid.

Acknowledgments The research work reported here was supported by the German Research Foundation (Deutsche Forschungsgemeinschaft - DFG) within the Grant No. CO282/13-3.

References

Howell LL (2001) Compliant mechanisms. Wiley, New York

Schoenen D, Lersch S, Hüsing M, Corves B, Klocke F, Hensgen L, Klink A (2015) Entwicklung, Konstruktion und Anwendung eines Prüfstands zur Ermittlung der ertragbaren Lastzyklen filigraner hochgenauer stoffschlüssiger Gelenke, 11. Kolloquium Getriebetechnik, T. C. Lüth et al., Garching, Germany, pp 169–181

Self-setting Locks for Petal Type Deployable Space Reflector

V.I. Bujakas and A.A. Kamensky

Abstract The new design of petal type deployable space antenna is presented. To improve important characteristics of reflector—deployment repeatability, reflecting surfaces accuracy, rigidity and effective area of open mirror—the new kinematic scheme of petal mirror deployment is proposed. Within developed approach the deployment of large mirror is fulfilled in two stages. During the first stage the structure is transformed with low accuracy from folded position into a state close to the working one. High precision fixing of open mirror is performed during second stage of deployment. Self-setting locks are introduced into the structure and used for micromanipulations with the petals during second stage of mirror opening. The new kinematic scheme of deployment and the design of the locks are described. At the final stage of the reflectors deployment the self-setting locks operate as compliant mechanisms.

Keywords Petal-type reflectors · Self-setting locks · High precision deployment · Space mirror

1 Introduction

Large space antennas are developed and used in a variety of scientific and technical space projects. Technical requirements for reflecting surfaces accuracy of antenna depend from operating wavelength of the instrument. Large *mesh* designs are effectively used for antennas operating in the long-wavelength area of spectrum

V.I. Bujakas (✉)
P.N.Lebedev Physical Institute, Moscow, Russia
e-mail: bujakas@yandex.ru

A.A. Kamensky
ArtCor Company, Moscow, Russia
e-mail: kam@artcor.ru

L. Zentner et al. (eds.), *Microactuators and Micromechanisms*,
Mechanisms and Machine Science 45, DOI 10.1007/978-3-319-45387-3_16

(Thomson 2016). Large *solid* reflectors are developed for operation in short-wavelength instruments [1,2,3,4]. Petal type solid deployable space reflector was first suggested by Dornier Corporation within FIRST project development (Dornier 1987; Manfred Westphal (Dornier) 1990). The modified version of petal type reflector was used later for 10 m antenna of Radioastron space radio telescope (Kardashev et al. 2013; Arkhipov et al. 2008). This telescope operates at centimeter wavelengths. In order to use petal type design in the shortwave region of the spectrum it is necessary:

- increase the accuracy of reflecting surface of open antenna,
- enhance the rigidity of open design,
- ensure high repeatability of deployment.

To meet these requirements the new design of petal type reflector was proposed (Bujakas 2013, 2014). Within the design the deployment of large mirror is fulfilled in two stages. During the first stage the structure is transformed with low accuracy from folded position into a state close to the working one. High precision fixing of open mirror is performed during second stage of deployment. Self-setting locks are used for micromanipulations with the petals during second stage of deployment.

2 Classical Design of Petal Type Reflector

Classical design of petal type reflector contains a central mirror and a set of petals. *Each petal is connected to the frame of central mirror* via *cylindrical hinge*. In the folded state the petals are disposed above the central mirror. The deployment is realized by synchronous rotation of the petals around the axes of cylindrical hinges, Figs. 1 and 2. The actuators are located from the back side of the central mirror. It has been shown (Dornier 1987; Manfred Westphal (Dornier) 1990) if the directions of cylindrical hinges are correctly chosen, the deployment of the petals is realized without collisions.

However the classic petal type design has a serious shortcoming—in the open state the petals are not linked. Closed parabolic shell has a significant geometric rigidity. Being cut into petals the design loses this property. Moreover if petals are large it is very difficult to ensure high accuracy for outer part of open reflector. But this part of the reflecting surface introduces a decisive contribution to the effective area of open antenna.

We consider the transformable structure where after deployment the neighbor petals are linked and use the self-setting locks to ensure high repeatability of deployment and increase the rigidity of open reflector.

---

[1]http://lisa.nasa.gov/

[2]http://www.jwst.nasa.gov

[3]http://safir.jpl.nasa.gov

[4]http://www.asc.rssi.ru/millimetron

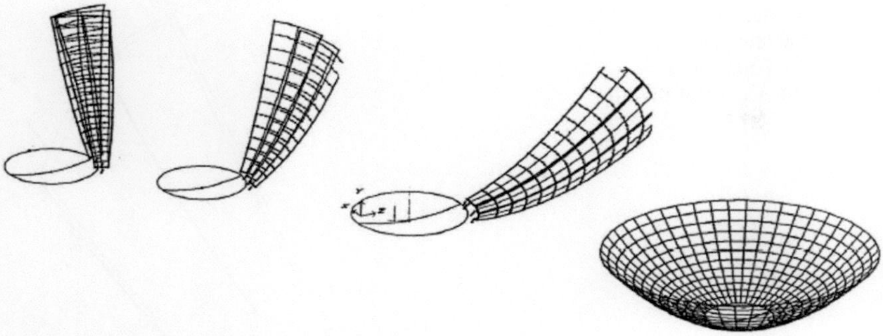

Fig. 1 Geometry of parabolic petal-type reflector deployment

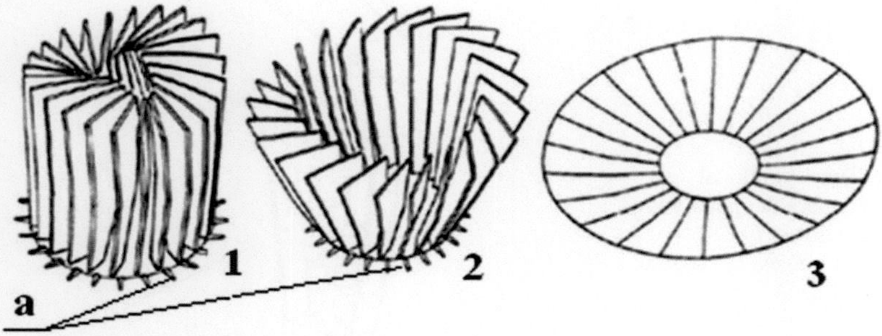

Fig. 2 Stages of classical petal type mirror deployment, *1*—folded mirror, *2*—partly deployed mirror, *3*—deployed mirror, *a*—cylindrical hinges that connect the petals with the central mirror

3 New Design of Petal Type Reflector

Within the new design the deployment of the mirror is fulfilled by moving the vertex of one petal along the outer edge of the adjacent petal (Fig. 3), so the links between the petals are kept during deployment.

Small linear actuators are used to implement the opening of the mirror. The actuator moves along the edge of the petal, and leads the vertex of the adjacent petal. Such placements of the actuators allow performing the precise alignment of the vertices of neighboring petals of open reflector.

To implement the desirable kinematics it is necessary:

- firstly, to weaken the connections of the petals with the central mirror; for this purpose, the cylindrical hinge that connects the petal with the central mirror; for this purpose, the cylindrical hinge that connects the petal with the central mirror in the classical design is replaced on the two-stage hinge (Fig. 4),
- secondly, to connect the holder of the actuator with the top of adjacent petal in such way to keep the structure stress free and geometrically unchangeable in

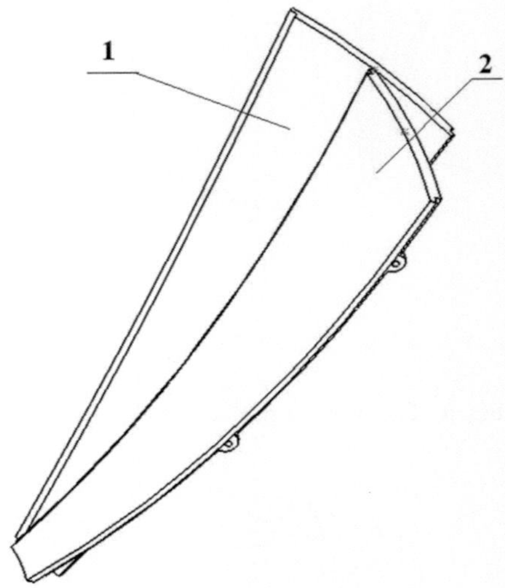

Fig. 3 The mutual movement of the petals. The vertex of the petal 2 moves along the outer edge of the petal 1

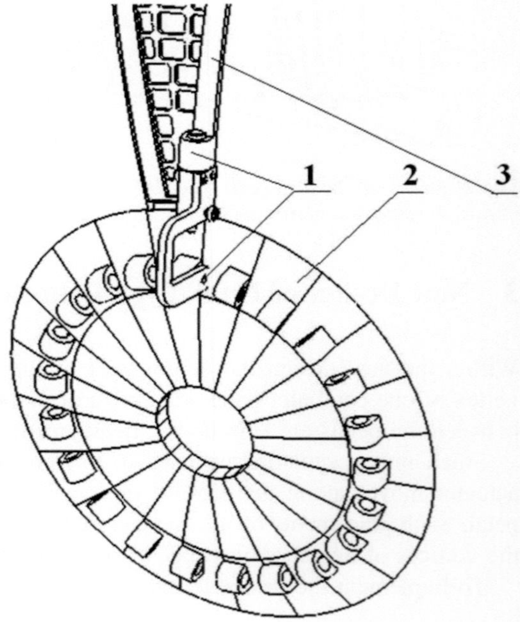

Fig. 4 Two cylindrical hinges connect the petal with a central mirror; *1*—cylindrical hinges, *2*—central mirror, *3*—petal

each moment of deployment; to meet these requirements the link between the holder of linear actuator and vertex of neighbor petal used in the new design contains the spherical and cylindrical hinges (Fig. 5).

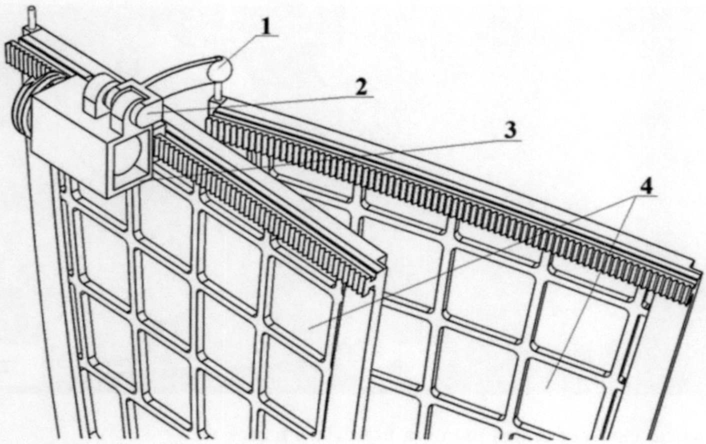

Fig. 5 New deployment system, *1*—spherical hinge, *2*—cylindrical hinge, *3*—linear actuator, *4*—petals

The new structure consists of n petals and a central mirror. Let N—total number of solid elements in the structure,

$$N = n + 1,$$

S_1 the number of kinematic constraints introduced by links between petals

$$S_1 = 2n,$$

S_2 the number of kinematic constraints introduced by the links between the petals and central mirror,

$$S_2 = 4n,$$

S the total number of kinematic constraints in the structure

$$S = S_1 + S_2 = 6n.$$

Thus Maxwell condition (necessary condition of statically determinacy)

$$S = 6N - 6$$

is fulfilled.

Direct calculations and computer model examination shows that the structure remains statically determinate *indeed* in each moment of deployment, if axes of

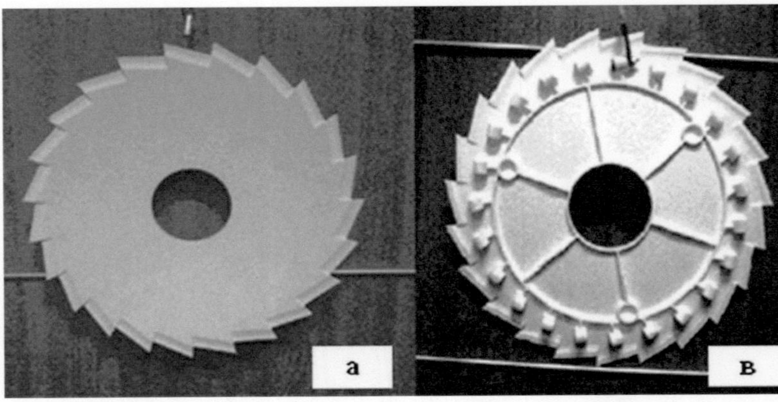

Fig. 6 Physical model of central mirror, **a** front view, **b** back view

hinges are correctly chosen. Therefore the structure remains stress free in each moment of deployment and is not requires precise synchronization for actuators operation during opening.

The physical model of new petal—type reflector is now under manufacturing. The petals and the central mirror of the layout have been manufactured by ArtCor Company [5] on CNC milling-machine. Elements of a model and process of fabrication are presented in Figs. 6, 7 and 8.

4 Self-setting Locks and Deployment Repeatability

To ensure high accuracy of large space structures, deployment repeatability and rigidity of open mirror the self—setting locks were proposed and developed (Bujakas and Rybakova 1998; Lake and Hachkowski 2000). To explain self-setting locks design and operation let us consider the device presented in Figs. 9 and 10. It consist the foundation equipped with three V-shaped grooves presented in Fig. 9a and heavy tripod with spherical supports (three polished balls) presented in Fig. 9b.

The position, at which the balls of tripod are located at the bottom of the corresponding grooves, is the only equilibrium state of device Fig. 10. Therefore the device is self-setting. If the walls of the grooves and spherical balls are perfectly polished the optical quality of repeatability for tripod installation on the foundation may be achieved. This wonderful and simple device is used in experimental optic during many years.

[5]www.artcor.ru/

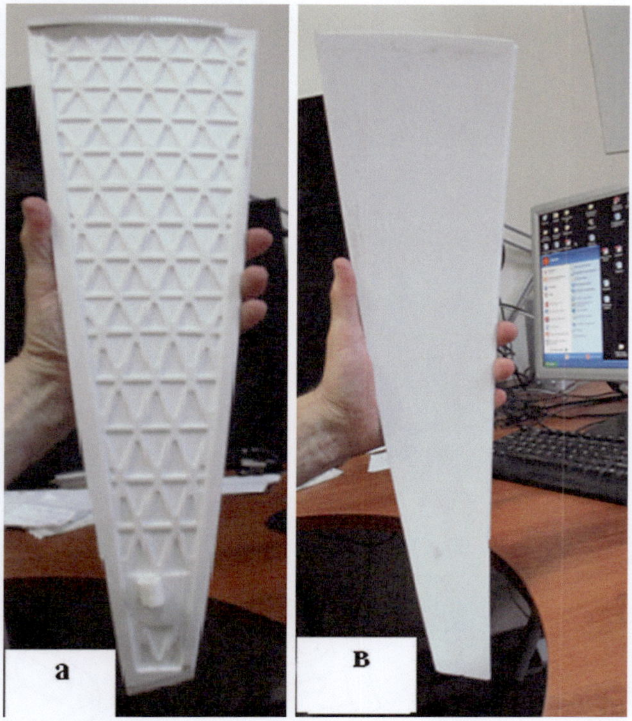

Fig. 7 Physical model of the petal, **a** back view, **b** front view

Fig. 8 Petal fabrication on CNC milling-machine

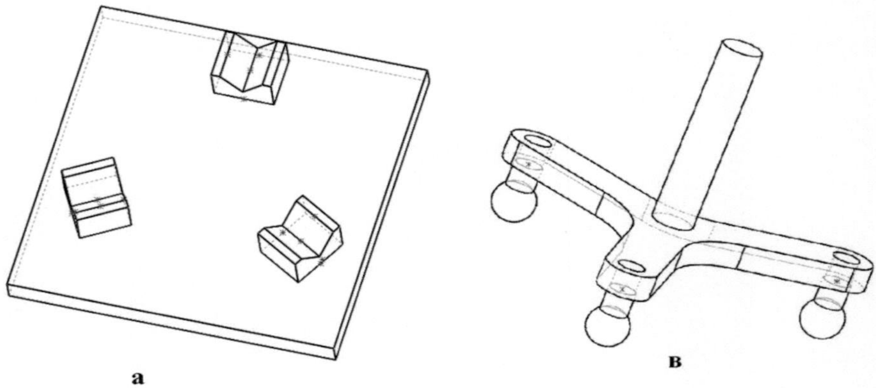

Fig. 9 The elements of self-setting optical device. **a** The foundation equipped with three V-shaped grooves, **b** the tripod with spherical supports

Fig. 10 Operation of self-setting optical device

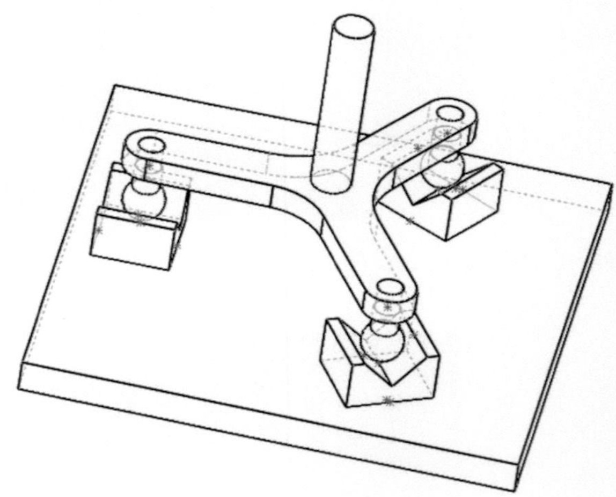

We modify the device as follows. Three V-shaped grooves and three spherical supports are placed on the lateral sides of petals according Fig. 11.

In the equilibrium state (in the open state of the reflector) the spherical supports of the petal are on the bottom of the grooves of neighbor petal and are held in this state by the elastic forces arising due to small deformation of the petals. Stages of latching of the lock are presented in Fig. 12. During final stage of the latching or initial stage of the unbuckling the deformations of the petals arise and the locks operate as compliant mechanisms.

If the spherical support displaced on the side wall of the V-shaped groove the forces component, returning the system to the equilibrium state at the bottom of the groove arise. Thus, self-setting effect is achieved.

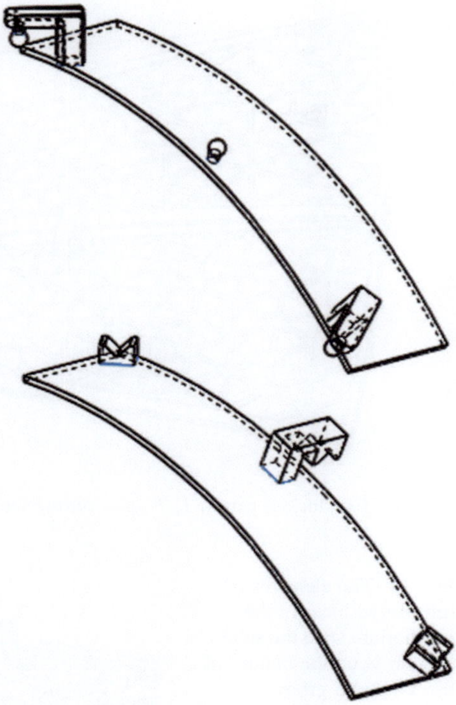

Fig. 11 V-shaped grooves and spherical balls on the lateral sides of the petals

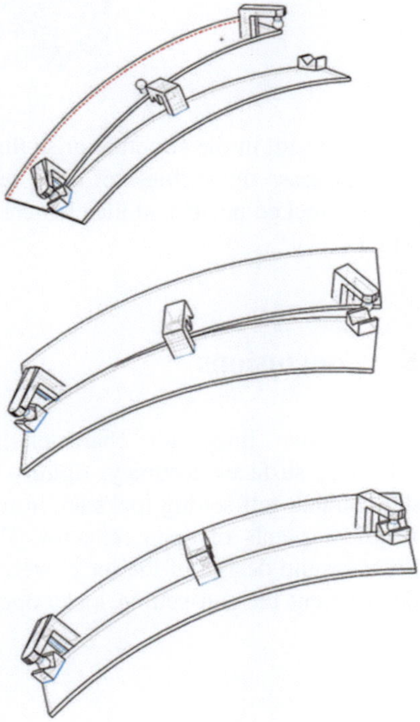

Fig. 12 Stages of self-setting lock latching

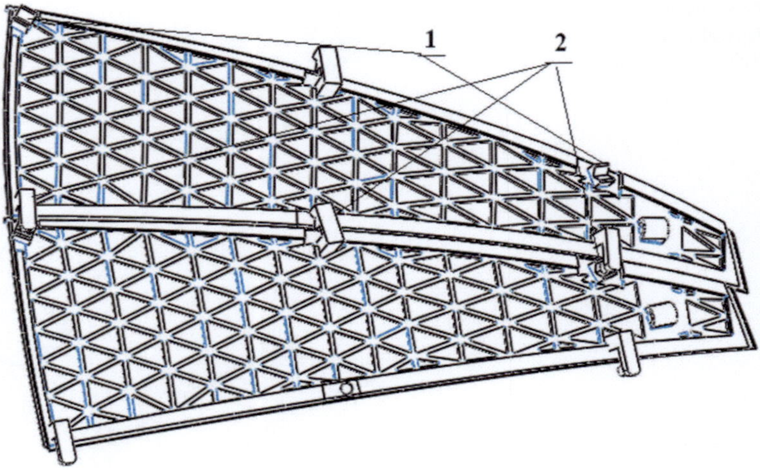

Fig. 13 Two locked petals, *1*, *2*—self setting locks

Fig. 14 The elements of self-setting lock. In the equilibrium state the spherical support is on the bottom of the V-shaped groove

As a result in the structure arise three closed loop of connections between petals what increase the stiffness of open reflector.

Two locked petals and the elements of self-setting lock are presented in Figs. 13 and 14.

5 Conclusions

To improve important characteristics of reflector—deployment repeatability, reflecting surfaces accuracy, rigidity and effective area of open mirror—statically determinate self-setting locks are introduced in the structure. The locks connect the neighbor petals of open reflector. The kinematic scheme of the new petal type reflector and design of the locks were described. At the final stage of the reflectors deployment the self-setting locks operate as compliant mechanisms.

References

Arkhipov MY, Baryshev AM, Kardashov NS et al (2008) Deployable antennas for space radio telescope: radioastron and millimetron missions. In: Proceedings of 30th ESA antenna workshop, ESTEC, Noordwijk, Netherlands

Bujakas VI (2013) Deployable antenna, Patent of Russian Federation, No 126199, 2013

Bujakas VI (2014) New design of petal type deployable space mirror, multibody mechatronic systems. In: Proceedings of the MUSME conference, 2014

Bujakas VI, Rybakova AG (1998) High precision deployment and shape correction of multi-mirror space designs. In: Proceedings of IUTAM-IASS symposium on deployable structures: theory and application, pp. 51–57, Cambridge, UK

Dornier, FIRST Technology study: multi surface control mechanism for a deployable antenna. Final report. Dornier Report RP-FA-D003, 1987

Kardashev NS et al (2013) “RadioAstron”-A telescope with a size of 300 000 km: main parameters and first observational results, Astronomy Reports, 2013, т. 90, No 3

Lake MS, Hachkowski MR (2000) Mechanism design principle for optical-precision, deployable instruments. In: Proceedings of 41st AIAA/ASME/ASCE/AHS/ASC structures, structural dynamics, and materials conference AIAA, 2000

Manfred Westphal (Dornier), United States Patent 4,899,167, 1990

Thomson MW (2016) The AstroMesh deployable reflector. www.northropgrumman.com/. 2016

Monolithic and Statically Balanced Rotational Power Transmission Coupling for Parallel Axes

D. Farhadi Machekposhti, N. Tolou and J.L. Herder

Abstract A new fully compliant rotational power transmission mechanism is presented. The design is based on the Pseudo-Rigid-Body Model (PRBM) of the Oldham constant velocity coupling. It can be fabricated as a single piece device with planar materials which make it suitable for micro scale applications. The internal stiffness of the proposed structure is eliminated by static balancing technique. Therefore, the compliance and zero stiffness behavior compensate for the structural error and poor efficiency inherent in rigid-body Oldham coupling, resulting in high mechanical efficiency power transmission system. The device is designed and its motion, torsional stiffness, and torque-angular displacement relations are predicted by the PRBM and finite element modeling. A large/macro scale prototype was manufactured and measured to evaluate the concept. This high efficient power transmission system can be applied in different applications in precision engineering and the relevant field such as micro power transmission system.

Keywords Compliant mechanism · Static balancing · Power transmission · Coupling · MEMS

1 Introduction

In many cases such as microtransmissions and microengines, a constant velocity power transmission system is required to transfer the rotational motion from one rotational axis to another rotational axis with a parallel offset (Sniegowski et al.

D. Farhadi Machekposhti (✉) · N. Tolou · J.L. Herder
Department of Precision and Micro System Engineering, Mechatronic System Design Group, Delft University of Technology, Delft, The Netherlands
e-mail: d.farhadimachekposhti@tudelft.nl

N. Tolou
e-mail: n.tolou@tudelft.nl

J.L. Herder
e-mail: j.l.herder@tudelft.nl

L. Zentner et al. (eds.), *Microactuators and Micromechanisms*,
Mechanisms and Machine Science 45, DOI 10.1007/978-3-319-45387-3_17

1996; Garcia et al. 1995). In the field of precision engineering, this problem was simply solved by using a couple of gears with the same number of tooth. However, this solution has a poor mechanical efficiency due to the gearing contact. Moreover, a precise micro assembling process is required to have a proper gearing system (Miller et al. 1996).

In prior art, different types of constant velocity couplings have been offered as a solution to accommodate with alignment errors for power transmission lines. However, they are rigid-body linkages, and like gearing systems, they have many disadvantages such as friction, backlash, wear, need for assembly, lubrication, and maintenance.

Compliant mechanisms have shown a great deal of potential in providing better solutions for transmission mechanisms (Farhadi Machekposhti et al. 2015a). These mechanisms drive their motion from elastic deformation of flexible segments rather than using conventional kinematic pairs (Howell 2001). This articulation became popular in the field of precision engineering due to the elimination of rigid-body kinematic joints which also eliminates the effect of backlash, wear, friction, and need for lubrication and assembly. A compliant coupling for all different and small alignment errors, angular and translational misalignment, was presented based on rigid-body Oldham coupling (Goodknight 1999; Soemers 2011). Wire flexures were used in those designs in order to provide an extra degrees of freedom for angular misalignment as well as translational misalignment between input and output axes. However, this cannot grantee a constant velocity transmission between input and output shafts due to the asymmetric configuration in respect to the homokinetic plane (Hunt 1973). Besides, they can only be applicable for low torque transmission due to the low torsional stiffness of their compliant design. Moreover, the design has a poor mechanical efficiency due to the overall compliancy of the designed structure.

Traditional kinematic pairs offer near zero stiffness around a desired axis and large stiffness along the other axes. However, the monolithic nature of compliant systems gives rise to a drawback: the elastic deformation of the monolithic structure requires significant force and energy which is considered a ‘necessary evil’ in compliant mechanism designs. In other words, the mechanical efficiency is poor, and it takes continuous force to hold the mechanism in position (Herder et al. 2000). This fact potentially results in problems such as rotation transmission with non-constant torque, overheating of the actuators, high energy consumption, insufficient travel range, low speed, and the need for larger actuators. However, the deformation energy is not lost; it is stored. It is a conservative force, and therefore it can be statically balanced (Herder et al. 2000; Tolou et al. 2010). As a consequence, these systems can be operated with much less effort as compared to the unbalanced situation.

The purpose of this work is to design a statically balanced and compliant constant velocity power transmission coupling which can efficiently transfer the rotational motion between two parallel axes.

The design of the mechanism is discussed in the following section. The conceptual design is demonstrated and the theoretical model, Pseudo rigid-body model and static balancing conditions, are studied. The model is then verified with experimental results and Finite Element Modeling (FEM).

2 Theoretical Model

The proposed compliant power transmission coupling is designed by applying conventional kinematic synthesis and using a Pseudo-Rigid-Body Model (PRBM) to create an equivalent compliant design. Besides, Static balancing technique is considered to provide static balancing conditions for the proposed design in order to have a high efficiency power transmission system.

The next sections describe the design in terms of rigid-body kinematic and then the PRBM is used to study the torque-angular displacement relation and static balancing conditions.

2.1 Rigid-Body Oldham Coupling

The rigid-body Oldham coupling is a specific type of four-bar linkage (Freudenstein et al. 1984). It has three movable parts, one coupled to the input shaft, one coupled to the output shaft, and a middle component which is connected to the first two parts by prismatic joints. These two sliders, prismatic joints, are two frictional joints which lose lot of energy during the motion. The direction of the prismatic joint on one side is perpendicular to the direction of another prismatic joint on the other side. The schematic of the mechanism is shown in Fig. 1.

It has two fixed and fully rotating frame joints which act as bearings for two parallel axes, crank 2 and 4. If the input link, 2, rotates through some angle, then the coupler link, 3, rotates through the same angle. This in turn rotates the output shaft, 4, through the same angle. Therefore, the two links, 2 and 4, rotate together and the

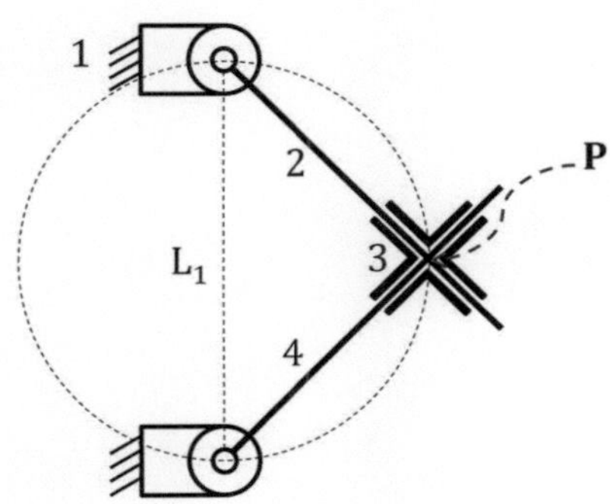

Fig. 1 Rigid-body Oldham coupling

velocity ratio remains constant at 1:1 throughout the rotation. Link 1 is ground link with length L_1 which actually is the offset between input link 2 and output link 4. The direction of two prismatic joints intersect each other at point P, the center of the coupler link. For a full cycle motion of the input link, the point P moves along a circular path with the diameter L_1. Therefore, the range of motion in each prismatic joint in each cycle is exactly the distance between the two parallel axes or the offset.

2.2 *Pseudo Rigid-Body Model and Static Balancing Conditions*

The pseudo-rigid-body model is used to convert the rigid-body Oldham coupling into a compliant design. This is done by rigid-body replacement synthesis (Howell 2001). The full cycle revolute joints of the input and output axes remain as a conventional rigid-body revolute joints since the mechanism will be connected to the input and output shafts, Fig. 2. The two prismatic joint can be replaced by translational compliant suspensions. Therefore, two translational springs K_1, and K_2 are considered as a replacement of the compliant translational suspension to simplify the analysis based on the pseudo-rigid-body model.

The input torque required to deal with stiffness of the system can be found by means of the virtual work principle. The analysis is based on static equilibrium and the energy equation can be written as

$$T_2 \delta\theta_2 - \delta V_e = 0. \tag{1}$$

where, T_2 and θ_2 are the input torque and angle of input shaft, respectively. V_e is the elastic potential energy which can be given by

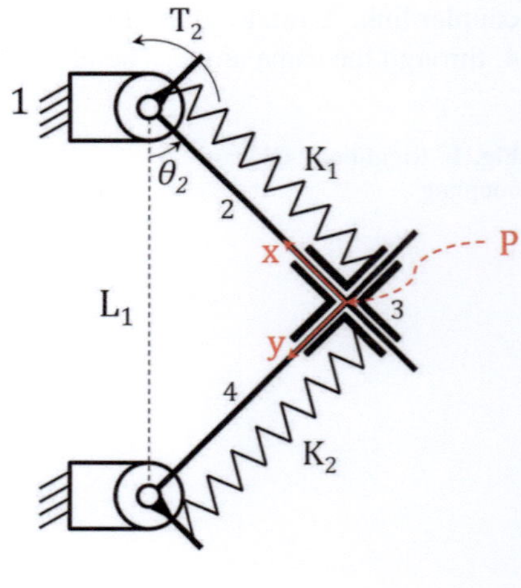

Fig. 2 The Pseudo-rigid-body model for an Oldham coupling

$$V_e = \frac{1}{2}\left(K_1 x^2 + K_2 y^2\right) \tag{2}$$

The mechanism can be statically balanced, the required input T_2 can be equal to zero, if the total elastic potential energy of the mechanism be a constant value over different input angle. This can be achieved by two constant and zero length springs K_1 and K_2 which have equal stiffness and they are in rest position at point P, input and output axes are collinear when the springs are in rest position. Therefore, these conditions can simplify the total elastic potential energy of the system as

$$V_e = \frac{1}{2}K_1\left(x^2 + y^2\right) = \frac{1}{2}K_1 L_1^2 = const. \tag{3}$$

Therefore, the mechanism should be designed such that the parallel offset of the two rotational axes provides the preloading to compensate the internal stiffness of the mechanism.

2.3 Compliant Design

The compliant counterpart of the monolithic and statically balanced power rotational transmission coupling for parallel axes is illustrated in Fig. 3.

The design comprises two translational compliant suspensions which are sharing the main frame. The main shuttles are decoupled and connected to the input and output axes.

The desired range of motion, high off-axis stiffness, and compact size are considered as design criteria to select the desired compliant translational suspension among the several different types of the compliant translational joints which were found and studied in prior art (Farhadi Machekposhti et al. 2012, 2015b).

3 Model Validation

This section describes prototyping, static torque, and actuation torque measurements, the measurement protocols and the results obtained from the measurements and finite element modeling.

3.1 Fabrication

A prototype with flexures out of stainless steel AISI316L was made for the experimental evaluation, Fig. 4. The set of parameters that defined the compliant structure are presented in Table 1.

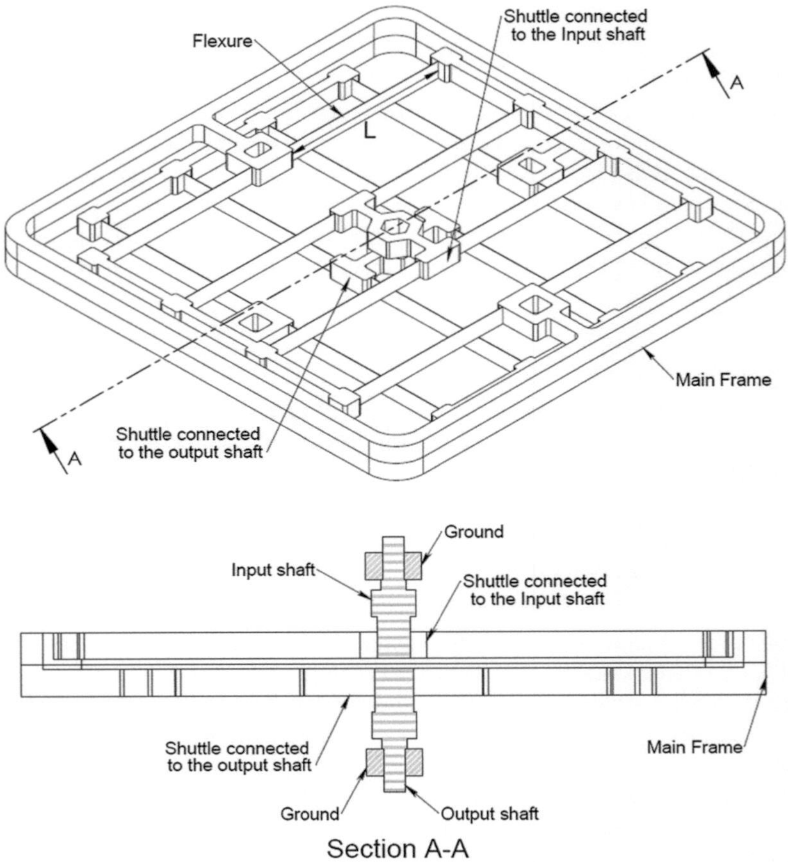

Fig. 3 Compliant counterpart of the monolithic and statically balanced power rotational transmission coupling

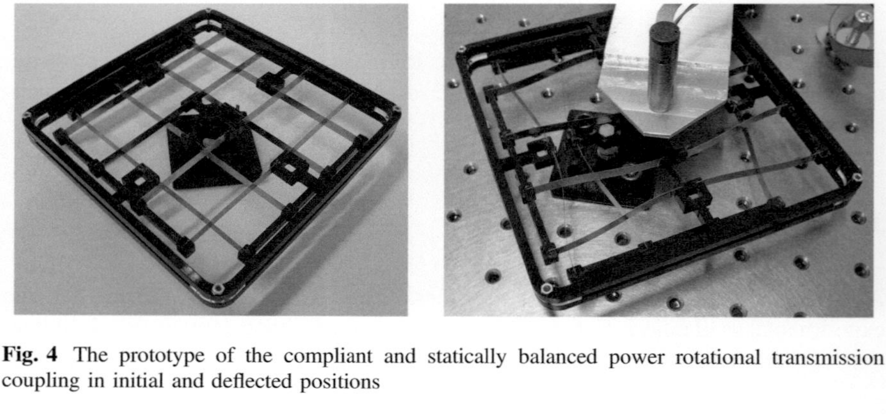

Fig. 4 The prototype of the compliant and statically balanced power rotational transmission coupling in initial and deflected positions

Table 1 Set of parameters of the flexures for the statically balanced compliant power rotational transmission coupling

Parameters	Values (mm)
Flexure length	50
Flexure thickness	0.2
Flexure out of plane thickness	5

The prototype is designed and fabricated for the maximum offset. This was limited by the yield strength of austenitic stainless steel AISI316L rated at 500 MPa.

3.2 *Experimental Evaluation*

An experimental setup was designed and built in order to measure both the torsional stiffness and minimum actuation torque of the designed statically balanced compliant power transmission coupling, Fig. 5.

The input axis is connected to the drive shaft. In order to find the torsional stiffness through static torque measurement, the output shaft is directly connected to

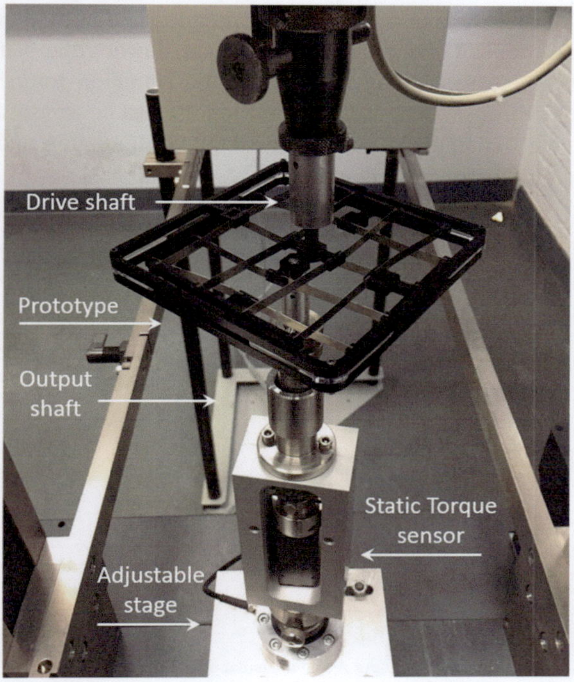

Fig. 5 Experimental setup for torsional stiffness and minimum actuation torque measurements on compliant and statically balanced power rotational transmission coupling. In this figure the setup is adjusted for 10 mm offset

the static torque sensor, HBM T20WN, which was mounted on an adjustable stage for different offset between input and output shafts. The input shaft was actuated for failure of the structure, and the reaction torque at the output shaft was measured by a static torque sensor which was linked to the angular displacement of the input shaft.

The torque-angular displacement characteristic, torsional stiffness, of the proposed compliant rotational power transmission coupling for different offset for both static torque experiment and finite element modeling are shown in Fig. 6. This shows the torque capacity of the designed compliant power transmission coupling. As can be seen, by increasing the offset between two axes the torsional stiffness will decrease and as result the maximum torque capacity of the proposed power transmission coupling will decrease as well.

Approximately 4 % error is observed between experimental result and finite element modeling. However, this can be explained by inaccuracy of the adjustable stage and the flexibility of the intermediate stages.

In order to find the minimum actuation torque, the output shaft is disconnected from the static torque sensor and it is free to rotate, quasi static torque measurement. The results for the minimum actuation torque of the proposed power transmission system for one full cycle rotation are shown in Fig. 7.

As can be seen, the experimental result and finite element modeling shows the proposed compliant rotational power transmission coupling is nearly statically balanced as this feature was also predicted by pseudo-rigid-body model. Therefore, The torsional stiffness of the proposed compliant power rotational transmission coupling can be increased without increasing the actuation torque. This can be increased by different approaches such as design for exact constraint folded flexures using slave flexures, change the compliant design configuration and beam combinations, increasing the dimensions, etc.

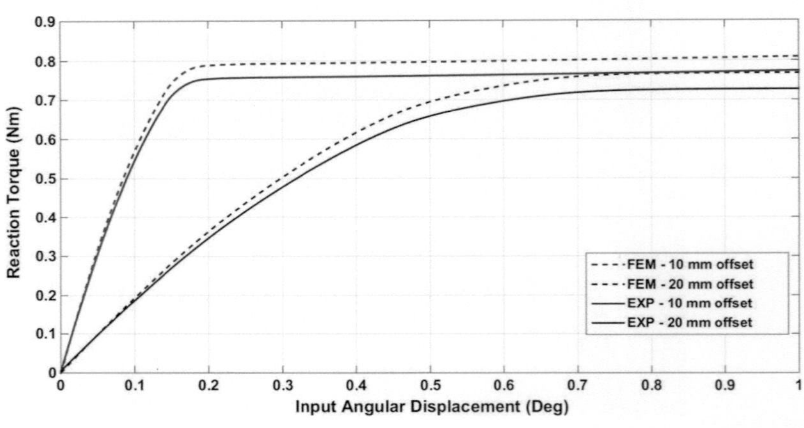

Fig. 6 Torque-angular displacement relation, torsional stiffness, for the compliant and statically balanced power rotational transmission coupling, for experiment (*solid line*) and finite element modeling (*dash line*) in two different parallel offsets

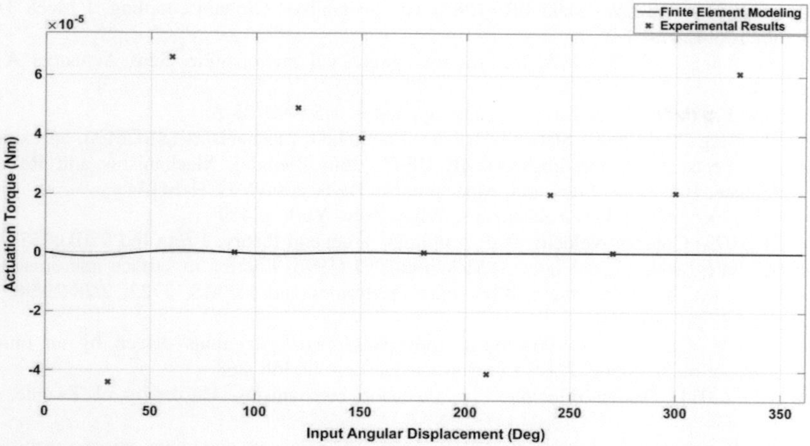

Fig. 7 The experimental results and finite element modeling of the minimum actuation torque of the compliant and statically balanced power rotational transmission coupling for 20 mm parallel offset between input and output axes

4 Conclusions

A new fully compliant and statically balanced rotational power transmission coupling is presented and its principle is described. The design is from the PRBM of the rigid-body Oldham coupling. The planar nature of the design makes it suitable for a manufacturing process on both macro and micro scales. The internal positive stiffness of the proposed compliant system is cancelled out by static balancing technique in order to have a power transmission coupling with high mechanical efficiency. This feature makes the design capable to be used for high torque applications. Prototypes were manufactured and evaluated by experimental data and finite element modeling. Results support the design and the theoretical analysis, PRBM. Moreover, It was shown, this compliant power transmission system has a great potential for high torque applications on both macro and micro scales.

References

Farhadi Machekposhti D, Tolou N, Herder JL (2012) The scope for a compliant homokinetic coupling based on review of compliant joints and rigid-body constant velocity universal joints. DETC2012-71514, Chicago, USA

Farhadi Machekposhti D, Tolou N, Herder JL (2015a) A review on compliant joints and rigid-body constant velocity universal joints towards the design of compliant homokinetic couplings. J Mech Des. doi:10.1115/1.4029318

Farhadi Machekposhti D, Tolou N, Herder JL (2015b) A fully compliant constant velocity universal joint. DETC2015-46813, Boston, USA

Freudenstein FF, Tsai LW, Maki ER (1984) The generalized Oldham coupling. J Mech Trans Autom 106(4):475–481

Garcia EJ, Sniegowski JJ (1995) Surface micromachined microengine. Sens Actuators A 48: 203–214

Goodknight FA (1999) Tape drive coupling apparatus. US5992782 A

Herder JL, Berg FPA (2000) Statically balanced compliant mechanisms (SBCM's), an example and prospects. In: Proceedings ASME DETC 26th Biennial, Mechanisms and Robotics Conference, Baltimore, Maryland, paper number DETC2000/MECH-14144

Howell LL (2001) Compliant mechanisms. Wiley, New York, p 459

Hunt KH (1973) Constant-velocity shaft couplings: a general theory. J Eng Ind 95(B):455–464

Miller SL, Sniegowski JJ, LaVigne G, McWhorter PJ (1996) Friction in surface micromachined microengines. In: Proceedings SPIE smart electronics and MEMS, 2722, 2/28-29/96, San Diego, CA, pp 197–204

Sniegowski JJ, Garcia EJ (1996) Surface-micromachined gear trains driven by an on-chip electrostatic microengine. IEEE Electron Device Lett 17:366–368

Soemers H (2011) Design principles: for precision mechanisms. University of Twente, The Netherlands

Tolou N, Henneken VA, Herder JL (2010) Statically balanced compliant micro mechanisms (SB-MEMS): concepts and simulation. DETC2010-28406, Montreal, Canada

Investigation of the Novelty Brackets "Gold-S"

F. Pollok, C. von Mandach, S. Griebel, V. Böhm and L. Zentner

Abstract The paper presents the new alternative orthodontic bracket Gold-S. The current applied force of wire on the compliant mechanism and thus its stiffness, will be determined and decreased. Furthermore, the investigation aims on decreasing stiffness by 25 %. Firstly, the geometric CAD model is simulated using the Finite Element Method (FEM), specific parameters are then altered. Accordingly boundary conditions, assumed simplifications and meshing is explained. Secondly, an experimental investigation is conducted, using an additive 3D printing process for the original as well as the improved model. Finally, the occurring forces in both models of the compliant mechanism are displayed in a graph for comparison of simulation and experimental investigation. It turns out that both results are quite similar, even though friction problems were encountered during investigation due to the used model material.

Keywords Compliant mechanisms · Modeling · Brackets · 3D-Printing

F. Pollok (✉) · S. Griebel · L. Zentner
Department of Mechanical Engineering,
Mechanism Technology Group, Technische Universität Ilmenau,
Ilmenau, Germany
e-mail: fabian.pollok@tu-ilmenau.de

S. Griebel
e-mail: stefan.griebel@tu-ilmenau.de

L. Zentner
e-mail: lena.zentner@tu-ilmenau.de

V. Böhm
Department of Mechanical Engineering, Technical Mechanics Group,
Technische Universität Ilmenau, Ilmenau, Germany
e-mail: valter.boehm@tu-ilmenau.de

C. von Mandach
Company Gold-S, Brugg, Switzerland
e-mail: cvmandach@bluewin.ch

L. Zentner et al. (eds.), *Microactuators and Micromechanisms*,
Mechanisms and Machine Science 45, DOI 10.1007/978-3-319-45387-3_18

1 Introduction

Brackets are the most common devices in orthodontics. There are various types and systems, but nowadays only usage of the edgewise bracket has spread throughout the world. An example of a bracket system is shown in Fig. 1. Since its invention in 1925 it maintained its shape and dimensions and consequently its problems. Too strong forces, friction between wire and slot, difficulties with torque control, difficulties with vertical height adjustment and generally with space closure. The reason is:

The high profile causes a lot of functional troubles. Unfortunately, the principle of the invention makes it impossible to reduce the size. There are neither advantages in scaling down the slot nor in reducing the wings. Even self-ligating brackets need additional wings with common dimensions.

Inventor Edward Angle's intention was to reach a precise positioning of the teeth regarding all three planes in space. But already within the first months of development he became aware of the fact, that precise positioning remained complicated. In retrospect several shortcomings of the back then advanced bracket have been revealed. Firstly, the bracket wings in occlusal direction are responsible for both masticatory disturbances and attrition of the upper dentition. They also obstruct tooth movements in the upper jaw. In addition, its design causes a relatively wide gap between wire and tooth surface. Therefore, the wire is required to withstand mastication forces that are many times greater than the forces it has to apply to the teeth. The wire straightens the teeth along its axis mainly in relation to its cross section. The more cross sections of the wire and the slot resemble, the higher are applied forces. The relation is not at all linear, in fact an exponential growth of forces takes place. Hence, wires that correspond to the slot dimensions do not simplify the positioning, but rather complicate it. This seems to be the reason for numerous different and sometimes contradictory treatment techniques, as well as for the amount of contradictory literature on forces, the use of lever arms and moments of force (Burstone 1995; Fiorelli 2016; Hanson 1980; Kesling 2006; Bennett 2014; Antoszewska et al. 2009).

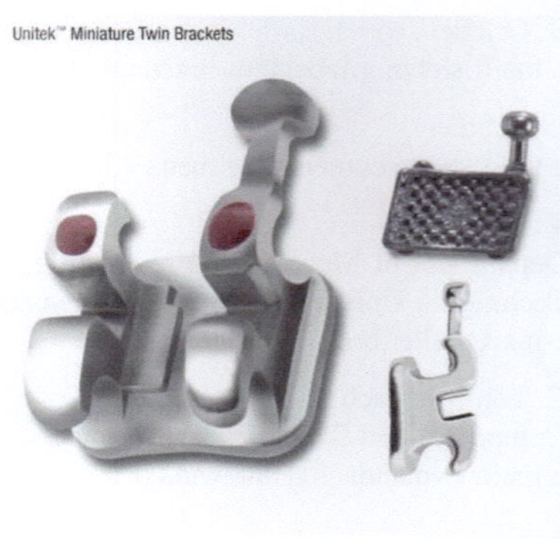

Fig. 1 Bracket from 3 M-Unitek (Unitek 2016)

However, no agreement has been reached regarding the ideal magnitude of forces and moments of force for moving teeth. Good results based on experiences have been published (Burstone 1995; Fiorelli 2016; Hanson 1980; Kesling 2006; Bennett 2014; Antoszewska et al. 2009), but they cover a large range and are usually combined with special treatment methods. Even the highly sophisticated industrial as well as the fully individualised production of brackets do not guarantee a perfect or an easy going treatment. This is because the bracket's slots are only about the same dimension, also the wires have only about the same cross section lengthwise. The faults further increase through bends those are common for precise adjustments. These small invisible deviations will then be 30-fold greater compared to the tooth. So many unpredictable forces exist between wire and slot, varying largely both in direction and range of magnitude. These generate a number of different biological reactions in the periodontium, causing temporally unpredictable side effects and leading to unwanted positions of the teeth. So a considerable part of the effort, made by the precision mechanics industry, crackles without any positive therapeutic influence.

2 Function of the Bracket "Gold-S"

The S-bracket of Gold-S proposes to counter the named disadvantages. The absence of a bonding base and a new wire fixation with a system of different retaining springs (compliant mechanisms) substantially reduce the horizontal contour. The bracket body is no longer milled or casted, but formed by bending a thin foil of metal (cobalt-chrome-alloy), see Fig. 2 for a possible geometry of the base. To produce a functional bracket out of it, several utilities of plastic are part of the invention. An additional protector prevents any soiling of functional areas around the bonding material and a bracket holder (base plate) fixes the tiny spring, which is otherwise difficult to hold within tweezers. Moreover, the bracket holder forms the outer contour of the bonding material and consequently a part of the final bracket body. In addition it fixes a pointer, which is designed to assist the precise positioning of the bracket on the tooth. With this the occlusal contour of the bracket remains a distance less than 1.3 mm away from the tooth surface and the gap

Fig. 2 Side (*left*) and isometric view (*right*) of the bracket system (base plate and compliant mechanism)

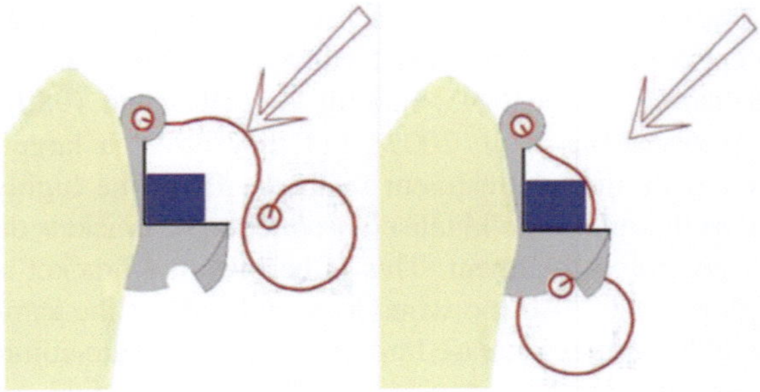

Fig. 3 Open (*left*) and close position (*right*) of the compliant mechanism of Gold-S

between the wire and the tooth surface remains below 0.3 mm. The common slot is replaced by just only a pair of rectangular planes and the fixation of the wire is done by an exchangeable retaining spring, as shown in Fig. 3.

Firstly, the nature of the invention leads even very thin wires within a corresponding small tube, in order to enable the wires great gentle therapeutic potential. Secondly, the spring's elasticity will prevent friction from jamming the wire. Further on it is possible to change the spring, so the rotation of teeth is better controllable. The advantage of a small width lies in an increase of inter-bracket distance and a cubic increase of the wire's elastic range.

This study discusses only the retaining spring with its resilient mechanism and its allotted resilience. A model of this resilient mechanism will be constructed with the finite element method (FEM) to reveal its stiffness. Further on the stiffness will be reduced till the point, where the function as a retaining mechanism ends.

3 Modeling of the Brackets "Gold-S"

Initially, the geometry of the compliant mechanism, given by the patent (VonMandach 2014), is implemented in CAD. The task of this analysis is to identify the applied force of a displaced wire on the compliant mechanism and improve the model, by decreasing its stiffness and hence its force. For this purpose the original version of Gold-S is altered by selected parameters. It aims on decreasing the stiffness of the improved model by 25 % in relation to the original model. This simulation is not a quantitative but a qualitative analysis.

3.1 Geometry and Material of the Brackets "Gold-S"

The flat pattern of the sheet metal with a thickness of 0.08 mm is displayed in Fig. 5. The model is divided into three segments, see Table 1 and corresponding

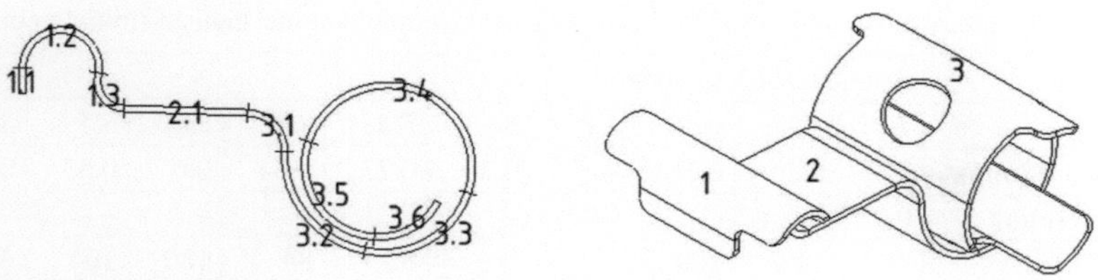

Fig. 4 Sub-segments and ISO-view of the compliant mechanism

Fig. 5 Dimensions of the flat pattern, bending lines *top* and geometrical edges *bottom* (Pollok 2015)

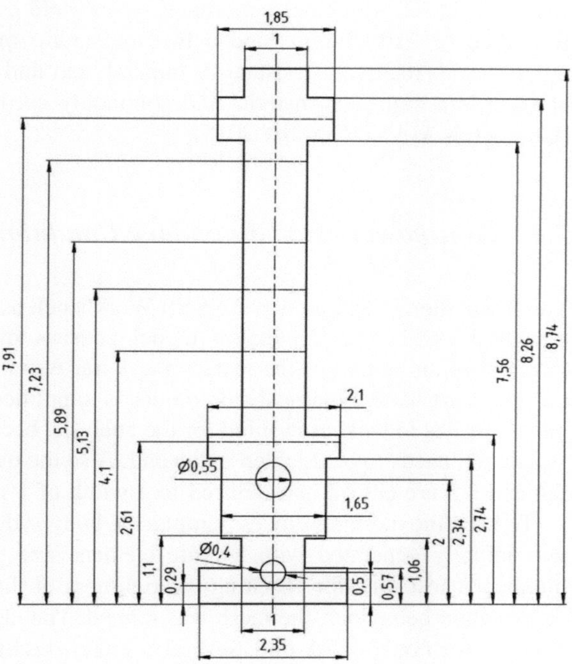

Fig. 4. Segment 1 is the first support A, Segment 2 is the contact zone between compliant mechanism and wire and Segment 3 is support B. Part 1.2 is the contact surface of the compliant mechanism to the base and allows rotation only. At the subsegment 2.1 is the main contact between the compliant mechanism and the wire. This is the sub-segment through which force is transmitted. Segment 3 is a roll spring with one and a half winding and has the purpose, to compensate the wire's displacement against the compliant mechanism. Its last sub-segment 3.6 is in contact with the base on one edge and can be slightly rotated on this fixed axis (Pollok 2015). For further information see Chap. 3.2.

The wire has a rectangular shape. Its dimensions are $0.4 \times 0.4\ mm^2$. The used material is a steel laminate of CoCr20Ni16Mo7 and is called "Phynox". The sheet is provided by the enterprise Lamineries MATTHEY SA. The exact material

Table 1 Geometrical parameter for the beam axis of the compliant mechanism (initial geometry)

Parameters	Segment. Sub-segment									
	1.1	*1.2*	*1.3*	*2.1*	*3.1*	*3.2*	*3.3*	*3.4*	*3.5*	*3.6*
Curvature (mm)	–	0.26	0.18	–	0.26	0.73	0.64	0.6	0.53	0.47
Length (mm)	0.2	–	–	0.92	–	–	–	–	–	–
Arc (°)	–	193	100	–	90	80	84	180	104	67
Width (mm)	1	1.85	1	1	1	1	1	2.1	1.65	2.35

chosen is R1500 which has a minimum offset yield point of 1500 N/mm^2. It has an E-module of 210 kN/mm^2 and a Poisson's ratio of 0.3. Phynox is immune to organic acid, relatively resistant to mineral acid and well accepted by the human body. It is a high tech material and commonly used for bio-medical applications (Lamineries MATTHEY SA 2015).

3.2 Groundwork and Boundary Conditions

The Simulation is executed in ANSYS Workbench using geometric nonlinear static structural FEM. The simulation model consists of two bodies, the compliant mechanism and a part of the rectangular wire. As mentioned in Chap. 3 and since the geometry is symmetrical, the model is simplified by dividing both bodies in half. This has to be kept in mind for the analysis, because the resulting force of the simulation needs to be doubled in the end. Also the outer parts in sub-segments 1.2, 3.4 and 3.5 are cut off and reduced to a width of 1 mm.

To keep the meshing process simple, the body is divided into sections. The mesh was initially generated with a finite element size of 0.15 mm. After passing a number of test runs for checking the rightness of the boundary condition and the deformation behaviour, the mesh was refined. The elements of the final refinement have a size of only 0.045 mm. Several comparisons of the averaged and unaveraged stress on surfaces of random samples with a stress gradient reached a difference smaller than 1 %. There are two elements in a row to analyse the stress gradient within the thickness. Used element types are Solid186 and Solid187 with a quadratic approach function (Pollok 2015).

Supports of the model are displayed in Fig. 6. The cut surface is a boundary condition as a *Displacement*, allowing movement in x- and y-direction, but blocking z-movement. The surface of sub-segment 1.2 is a boundary condition as *Remote Displacement* (support A), blocking all movement and rotation except for the rotation around the z-axis in the centre of the radius. The same applies to the support B in sub-segment 3.5, except for this time it is not an area but an edge. The simulation is started by moving the wire 0.35 mm in positive y direction. There are two contact zones. One is between the compliant mechanism 2.1 and the wire, the second one is in segment 3 between the two sides 3.2/3.3 and 3.5/3.6 (Pollok 2015). See Fig. 4 for the selected sub-segments.

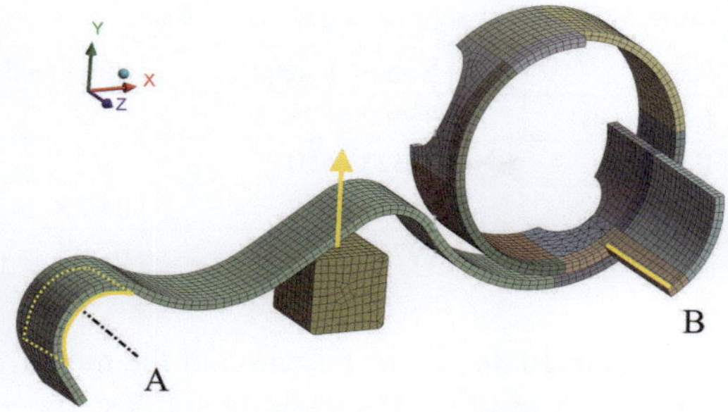

Fig. 6 Supports A and B of the compliant mechanism, allowing rotation around the z-axis. The wire is moved by 0.35 mm along the y-axis (*yellow arrow*). The back of the model, the cut surface, is blocked from moving along the z axis

Table 2 Cases for parameter combination

Cases	h [mm]	b [mm]	r in s-s 1.3 [mm]
I (shape I)	0.05	1	0.18
II	0.05	1	0.73
III	0.05	0.8	0.18
IV (shape II)	0.05	0.8	0.73

3.3 Improvement of Compliant Mechanism Geometry

As monotone behaviour of the model is expected with the given boundary conditions, only the threshold is considered and specific parameters are decreased. The thickness has the most influence on the stiffness of the compliant mechanism. Although the model started with a thickness of 0.08 mm, this is lowered to the final thickness of 0.05 mm before even starting the simulation. Another possible parameter is the width. It is decreased in sub-segments 1.1 to 3.3 from 1 mm to 0.8 mm, so that it still remains in the allowable stress range. Finally the rather hard curvature in sub-segment 1.3 is eased from 0.18 mm to 0.73 mm. The new curvature is tangential to the sub-segments 1.2 and 2.1. The new sub-segment 2.1 can still cover bigger wires than the small used wire (Pollok 2015). Since only the original and the final improved model are interesting for comparison, those are called shape I and shape II. These two shapes are tested on a test setup in Chap. 5. Cases II and III reveal the influence of the adjustments on the model's stiffness. All the parameter changes are listed in Table 2.

4 Results of Modeling

Every alteration on the parameters influenced the stiffness of the simulation model. The enlargement of the curvature had a rather small influence but it is still not negligible. As expected, case IV had the lowest force output of all alternatives. The results of the forces are listed in Table 3. As mentioned in Chap. 3.2, the force has

Table 3 Resulting simulated and actual force

Cases	Nodes/elements	Simulated force [N]	Actual force [N]
I (shape I)	35562/6773	0.791	1.58
II	34323/6510	0.77	1.54
III	31575/5873	0.625	1.25
IV (shape II)	29756/5468	0.598	1.19

to be doubled due to the bisection of the model. A comparison between shape I and II reveals a resulting decrease on stiffness by $\sim 24\ \%$.

5 Experimental Investigations

The experimental investigations will be carried out to verify the modeling results for the compliant mechanism. It should be confirmed experimentally that the stiffness of the new shape II is reduced by about 25 %. Therefore, shape I and shape II are manufactured, measured and compared under the aspect of decreasing stiffness. Because the real dimensions are very small, enlarged demonstrators are investigated experimentally. In order to compare the measurement for these two demonstrators and the simulation model results for both shapes, theoretical considerations should be made first.

Two pairs of systems are considered: (i) shape I and shape II with real dimensions and from material A, (ii) shape I and shape II, which are enlarged in a geometrically similar manner with coefficient k, from material B. The Force F depends on bending stiffness EI_z (E—Young's modulus, I_z—area moment of inertia), the initial curvature κ_0, the length L of the compliant mechanism and the given displacement u_0 of its contact point with the wire (Zentner 2014). To reach the displacement u_0, we require following forces for the first pair (i):

$$F_I = F_I(E_A I_{zI}, L, \kappa_{0I}, u_0), \quad F_{II} = F_{II}(E_A I_{zII}, L, \kappa_{0II}, u_0) \tag{1}$$

The stiffness ratio for pair (i) is F_I/u_0: $F_{II}/u_0 = F_I/F_{II} = \delta$. For the second pair (ii), they are needed forces $F_{k,I}$ and $F_{k,II}$:

$$F_{k,I} = F_{k,I}\left(E_B k^4 I_{zI}, kL, \frac{\kappa_{0I}}{k}, ku_0\right), \quad F_{k,II} = F_{k,II}\left(E_B k^4 I_{zII}, kL, \frac{\kappa_{0II}}{k}, ku_0\right) \tag{2}$$

The stiffness ratio for this pair (ii) is $F_{k,I}/ku_0$: $F_{k,II}/ku_0 = F_{k,I}/F_{k,II} = \delta k$. According to expressions (1) and (2), ratio δ is equal to ratio δk. Therefore, model results for the shapes I and II and the measurement results for the demonstrators with scale 20:1 can be compared with one another.

5.1 Materials and Methods

For the experimental investigation of the characteristic of the two compliant mechanisms with shape I and shape II, the demonstrators were manufactured in a scale of 20:1.

Due to the complex spatial geometry of the compliant mechanism an additive manufacturing process is chosen. The two mechanisms, the base plate and a stirrup (models the wire) are printed by means of an SLA desktop 3D printer (1 + form, shape Labs Inc., Somerville, United States). In the further course, the stirrup and the base plate are clamped into the pneumatic clamping jaws of a material testing machine. For this reason adhesively-linked supports for clamping are modeled on those. To map the elastic behavior of the compliant mechanisms, a resin (clear resin, shape Labs Inc., Somerville, United States) is used having predominantly brittle and only slightly viscoelastic behavior as a material. In contrast, the stirrup and the base plate made of resin are printed with impact-resistant distinctive properties (tough resin, shape Labs Inc., Somerville, United States) whereby a brittle fracture during the subsequent clamping of the two components can be avoided. All components are printed with a layer thickness of 50 microns. It was ensured that the support structures of the print were at the compliant mechanisms on the front edge side and, in case of stirrup and base plate, at not functionally relevant surfaces. After removal of the support structures, the components are post-cured. Therefor the components are exposed in direct sunlight for several hours. Finally, the components are finished by using emery paper with descending grain size to straighten the end points of the removed support structures and to smoothen irregularities on the contact surfaces of the components to each other.

Figure 7a and b show the two compliant mechanisms of shapes I and II in a top and side view. While shape I has a width of b_I = 20 mm and a radius of curvature

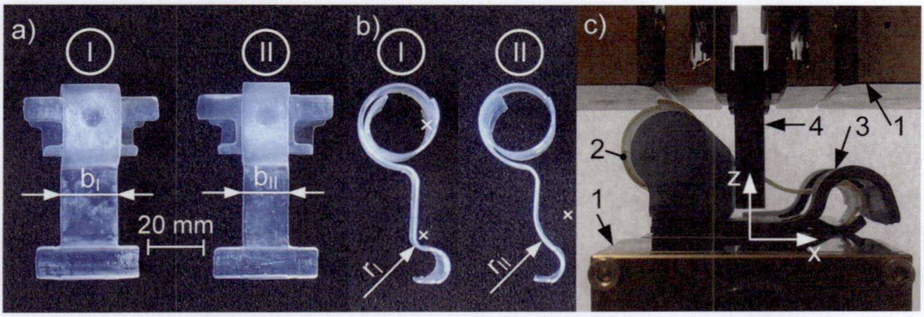

Fig. 7 **a** top and **b** side view of the two, additive manufactured compliant mechanisms with shape I and II at a scale of 20:1; **c** test setup for investigation the compliant mechanism characteristic: (*1*) pneumatic clamping jaws, (*2*) compliant mechanism, (*3*) base plate and (*4*) stirrup

of $r_I = 3.6$ mm, the improved shape II has a decreased width of $b_{II} = 16$ mm and a radius of $r_{II} = 14.4$ mm.

5.2 *Experimental Investigation of the Compliant Mechanisms*

A materials testing machine (ProLine table testing machine Z005, Zwick GmbH & Co. KG, Ulm, Germany) is used for the metrological examination of the two compliant mechanisms. For this purpose, the base plate is fixed with the bottom pneumatic clamping jaws. Accordingly the assembly of base plate and clamping jaws is immoveable. Then, one of the compliant mechanisms is threaded into the stirrup over an existing slot and the compliant mechanism is spanned on the base plate. After this, the second pneumatic clamping jaw which is fixed on the moveable traverse, is approached to the stirrup so that it can be fixed by the upper clamping jaw. The arrangement of the upper and bottom jaws is rotated by 90° against each other. Hence, the stirrup is perpendicular to the longitudinal axis of the compliant mechanism. Before each trial, the stirrup is moved in z direction until touching the compliant mechanism. Thereby, the zero point is set. A displacement s of 7 mm in positive z-direction is used by moving the stirrup fixed on the traverse. This procedure is done ten times for every compliant mechanism.

The displacement of the traverse is determined with accuracy class 0.5 (according to DIN EN 9513) and a positioning repeatability of the traverse and the associated displacement sensors of $\pm$ 2 μm. The corresponding tensile force F in displacement direction is determined by using a 100 N load cell with accuracy class 1 (ISO 7500-1). All measurement values are recorded in the loading phase of the compliant mechanism at a distance of 50 μm and with a 24 bit resolution. The speed of the traverse in this case is set to 15 mm/min. This is a compromise between total test time and influence of viscoelastic material properties on the measured characteristic of the compliant mechanism.

5.3 *Results of the Experimental Investigation*

Figure 8 (left) shows the raw data of the recorded measurement points of ten test series. It should be noted that taking a global view, the force values of the two compliant mechanisms increase along with the traverse displacement. In addition, the recorded force value F of shape I is higher compared to shape II at the same displacement value of the traverse. Furthermore, a non-continuous rise in the data is determined. Accordingly, the curves have sections with negative slope.

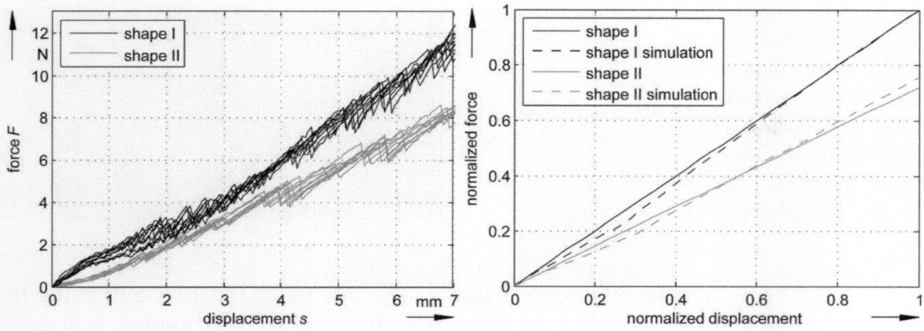

Fig. 8 *Left* Raw data of the n = 10 test series for determination of the compliant mechanism characteristic of shape I (*black line*) and shape II (*grey line*); *right* averaged linear characteristic lines of the compliant mechanisms (shape I and II), average of n = 10 test series, normalized on shape I and compared with the simulation

6 Discussion

The discontinuity of the data from all test series was a result of friction in the real assembly. This is caused by the stick-slip phenomenon whereby the value of sticking friction is significantly greater than the value of sliding friction. Accepting this data, the reason for this may be increased roughness of sliding surfaces (amongst others due to the presence of burrs) as well as a small fit of the components coupling surfaces to each other. Assuming that the compliant mechanism characteristic is linear, each test series was linearized using linear regression in such a way that the resulting straight line extends through the origin. (Hence, the stick-slip phenomenon is no longer visible and the friction is taken into account in the characteristic.) Subsequently, the ten straight lines were averaged for each compliant mechanism and normalized to the final value of the compliant mechanism with shape I, see Fig. 8 (right).

The slope of the straight lines can be equated to the Hooke's law with spring stiffness c. Thus, average spring stiffnesses c_I and c_{II} of 1.587 N/mm and 1.147 N/mm can be calculated for the compliant mechanisms with shape I and II. The ratio between the spring stiffness c_I to c_{II} is $k_E = 0.723$ (k_E – Experiment ratio). Compared with ratio $k_S = 0.757$ (k_S – Simulation ratio) at the end of simulation, a deviation of 4.5 % (0.723/0.757) can be determined.

The deviation is mainly caused by the effects resulting from the manufacturing process. At the anchoring points of support structures, for example, not exactly straight edge line can be achieved. Also, there are manufacture-related variations in thickness and small burrs are present. These are marked in Fig. 9. Through the abrasive finishing procedure of the printed compliant mechanisms, locally material is removed. This also creates deviations from the exact geometry. Further, results from the study of real structures show friction between the components, which was not considered in the simulation.

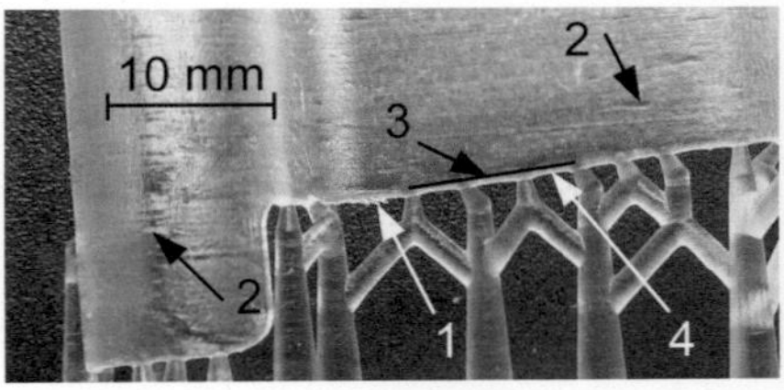

Fig. 9 Detail view of the compliant mechanism with support structures: geometric deviations from exact geometry pointed as (*1*) burr, (*2*) thickness variations, (*3*) and (*4*) not exactly straight running edge course

The simulation results of shape I and II do not have the problem with the stick-slip phenomenon, because of the simplification of the model and other material properties. Therefore there is a bend in the graph visible. This bend is the point of contact between the sub-segments 3.2/3.5 and 3.3/3.6, see Fig. 4. Of course this is not noticeable in the experimental investigation, due to the addressed necessary averaging and linearizing of ten test series.

7 Conclusion and Outlook

As the deviations in both investigations are very small, the requested decrease in stiffness of 25 % could be achieved. The results of the FEM simulations could be verified by the experimental investigations. For further investigations a plurality of compliant mechanisms should be examined to detect manufacture-related variations. Furthermore, measures such as lubrication or subsequent processing of the joining surfaces should be done to minimize the value of sticking friction. Also, measures changing the geometry of the joining surfaces are conceivable (changing the force flow through the joining surfaces). Thus, both can help to avoid stick-slip phenomenon.

References

Antoszewska J, Papadopoulos MA, Park HS, Ludwig B (2009) Five-year experience with orthodontic miniscrew implants: a retrospective investigation of factors influencing success rates. Am J Orthod Dentofacial Orthop 136(2):158.e1–158e10

Bennett JC, McLaughlin RP (2014) Fundamentals of orthodontic treatment mechanics. Hardcover LeGrande Publishing, UK

Burstone CJ, van Steenbergen E, Hanley KJ (1995) Modern edgewise mechanics and the segmented arch technique. Paperback, Ormco Corporation, Horsham

Fiorelli G, Melsen B (2016) Biomechanics in orthodontics, 4th edn. A multimedia interactive textbook. http://www.orthobiomechanics.com/en/

Hanson H (1980) The SPEED system: a report on the development of a new edgewise appliance. Am J Orthod Dentofacial Orthop 78:243–265

Kesling PC (2006) The Tip-Edge PLUS guide and the differential straight-arch technique, 6th edn. Hardcover TP Orthodontics, Westville
Lamineries MATTHEY SA (2015) Phynox, Product Catalogue
Pollok F (2015) Entwicklung eines Feder-Mechanismus für neuartige Brackets (Zahnspange). Bachelor Thesis, TU Ilmenau
Unitek M (2016) Orthodontic Solutions, Product Catalogue; 4.46:88
Von Mandach C (2014) Kit for an orthopedic bracket. Patent specification: WO2015140026 A1
Zentner L (2014) Nachgiebige Mechanismen. De Gruyter Oldenbourg, München

Dynamic Behavior of Active Lightweight Compliant Mechanisms with Integrated Piezoceramic Actuators by Under- and Overcritical Periodic Excitation

N. Modler, A. Winkler, A. Filippatos, Erwin-Christian Lovasz and D.-T. Mărgineanu

Abstract Integration of active elements as thermoplastic-compatible piezoceramic modules in lightweight compliant mechanisms offers the possibility to actively control its structural behavior by static, dynamic or vibro-acoustic loads. New active lightweight structures with material-integrated structural monitoring, energy-harvesting, or active vibration damping functionalities become possible. Previously, a demonstrator mechanism was designed and built, and its behavior under quasi-static excitation was simulated as a multi-body system, using two-layer cells with torsion and traction springs for compliance and linear motors for excitation. The mechanism was later tested to confirm the simulation results. This paper presents a study of the dynamic behavior of active compliant mechanisms obtained by integration of piezoceramic actuators into fiber-reinforced composite structures. The results of the dynamic mechanical simulation procedure were compared with experimental results for a given demonstrator mechanism. The comparison concluded that the simulation procedure describes fairly accurate the real behavior of the lightweight compliant mechanism, thus it can be used in the development of new active structures.

Keywords Compliant mechanisms · Piezoactuation · Dynamics · Oscillation

N. Modler · A. Winkler · A. Filippatos
Institute of Lightweight Engineering and Polymer Technology,
Technische Universität Dresden, Dresden, Germany
e-mail: niels.modler@ilk.mw.tu-dresden.de

A. Winkler
e-mail: anja.winkler@ilk.mw.tu-dresden.de

A. Filippatos
e-mail: angelos.filippatos@ilk.mw.tu-dresden.de

E.-C. Lovasz · D.-T. Mărgineanu (✉)
Department of Mechatronics, University Politehnica Timişoara,
Timişoara, Romania
e-mail: dan.margineanu@upt.ro

E.-C. Lovasz
e-mail: erwin.lovasz@upt.ro

L. Zentner et al. (eds.), *Microactuators and Micromechanisms*,
Mechanisms and Machine Science 45, DOI 10.1007/978-3-319-45387-3_19

1 Introduction

Compliant mechanisms, consisting of links with higher rigidity and/or with higher elasticity (compliant elements), revolute joints in hybrid structures or elastic joints in monolithic structures, can fulfill specified functions with a minimum number of elements and joints. Active compliant mechanisms (ACM) integrate sensors and actuators in the composite structure, further reducing complexity and costs. ACM movements depend both on rigid and elastic elements geometry and on load types which are influenced by elastic and inertial material properties (Modler 2008). Thus desired functions may be obtained only after consequently testing, re-design and adjustments, with resulting high time and cost effects.

A compliance comparison between multi-spring model and finite-element analysis was done (Gao 2010) for multidimensional accelerometers based on a fully decoupled compliant parallel mechanism. An under-actuated compliant variable stroke mechanism is analyzed (Tanık 2010) which employed Pseudo Rigid Body Model (PRBM) for prescribed output loading and constant input torque.

The development of FEM models requires dimensions, physical properties of used materials, etc. thus some dimensioning steps are previously required. Simulation models using cells composed of rigid links, revolute and prismatic joints, torsion and traction spring-dampers and linear actuators were presented in previous investigations (Modler 2009) as they were used for dynamic analysis of compliant mechanisms. The double layer model (Mărgineanu 2014) proved to be the most successful, as it has both compliance and actuation and the simulation parameters, e.g. the springs' rigidity and damping coefficients may be calculated according to the material properties and elements' dimensions. In this paper, the double layer model is used for dynamic analysis of a piezoelectric driven ACM by under- and overcritical periodic excitation.

2 Investigated Structure

The investigated mechanism presented in Fig. 1 under kinematic scheme is composed of an output link B_0B (4) connected to the frame (1) through the revolute joint B_0, with a counterweight for equilibration. The actuator A_0A (2) is rigidly

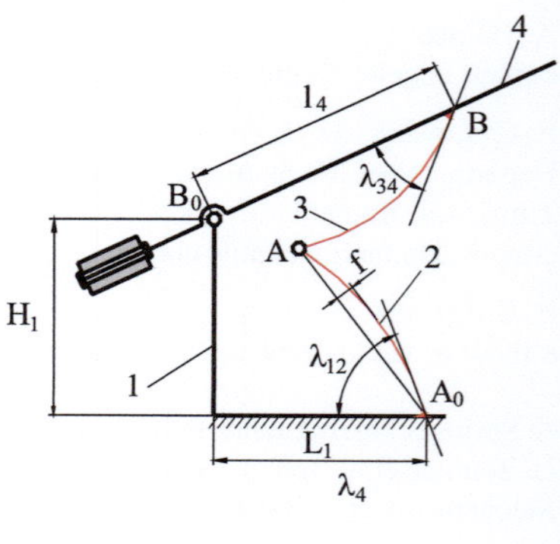

Fig. 1 Kinematic scheme of the active compliant mechanism

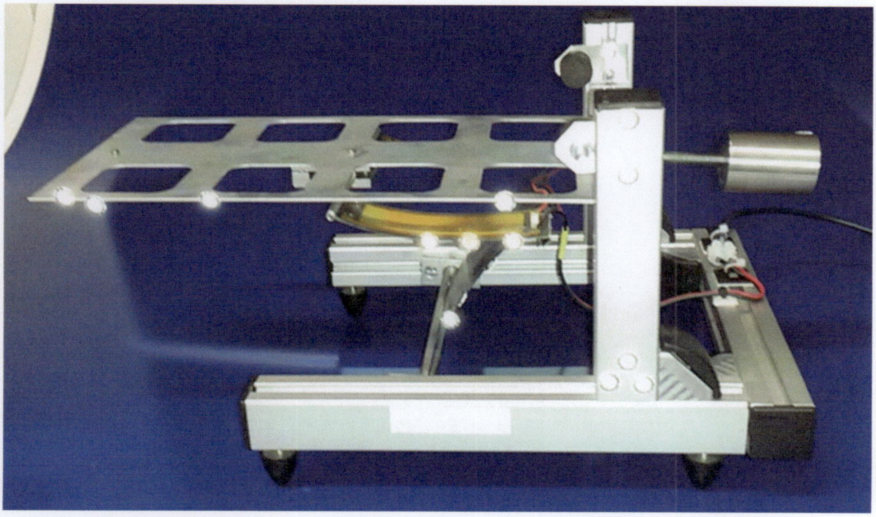

Fig. 2 The flap mechanism with two bending actuators demonstrator in the starting position

connected to the frame in A_0 at an adjustable angle λ_{12}, and the actuator AB (3) is also connected rigidly at an adjustable angle λ_{34} to the flap on B. The revolute joint A connects the two actuators. The two piezo-actuators have a free curvature f. The flap moves due to the actuators' bending as the curvature radius decreases. At first, the actuators bend under the flaps weight (see Fig. 2). The horizontal starting position is adjusted by the counterweight's position on the flap's shaft.

The first zone of the actuator is composed of a carbon fiber-reinforced composite (CFC) passive layer 0.25 mm thick. In the second actuator zone, on the CFC layer is bonded a 0.28 mm thick macro fiber composite (MFC) actuator layer.

3 Equivalent Two-Layered Pseudo Rigid Body Active Model Cell

The active element presented in Fig. 3 consists of a passive base layer and an actuator layer. The base layer function is to enable a large non-linear elastic deformation of the flexible active element and a large strain without damage initiation.

By the inverse piezoelectric effect used for actuation, the strain S depends on the load T and the electric field strength E. Considering the tensor of mechanical parameters on the crystal surface, the vector linear equation in Voigt's notation is:

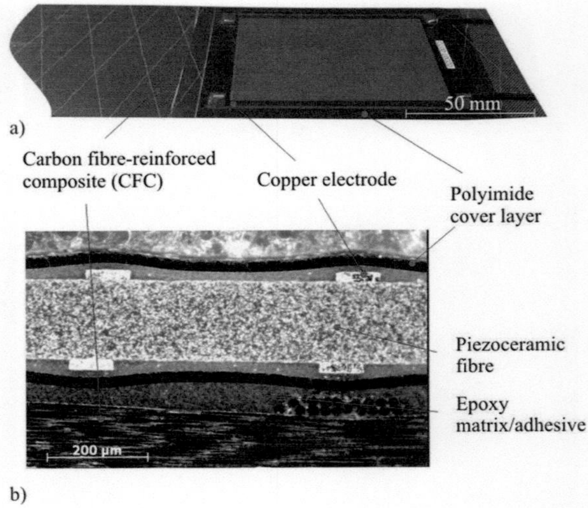

Fig. 3 Illustration of an active element

$$\{S\} = \left[s^E\right] \cdot \{T\} + \left[d^t\right] \cdot \{E\} \tag{1}$$

where $[d^t]$ is the matrix for the reverse piezoelectric effect and $[s^E]$ the compliance matrix for E = 0.

The two-layer cells simulation model (Mărgineanu 2014) for an ACM ensuring both compliance and an actuation is composed of a six-bar linkage as in Fig. 4. The passive layer is simulated by the rocker 6 $A_{p,j-1}A_{p,j}$ with the length L_0. The total deformation x_4 of the active layer 2, 3, 4, $A_{a,j-1}A_{a,j}$ rotates the interface link 5 around the revolute joint $A_{p,j}$ simulating the actuator's bending.

The bending compliance of the passive layer was simulated by torsion spring-dampers inserted in the revolute joints $A_{p,j}$. As two springs were connected on each side of the interface element to ensure equal angles with the current and the

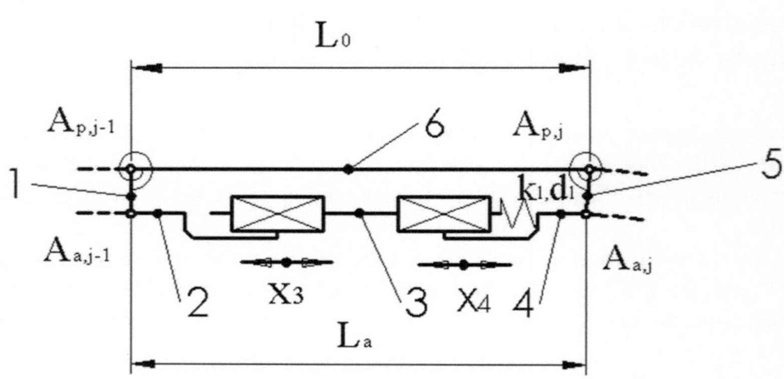

Fig. 4 Representation of a modelling cell

previous layer elements, the rigidity of these springs was double compared to the equivalent bending rigidity of the passive layer:

$$k_\theta = \frac{2M_j}{\theta} = 2E_p \cdot \frac{I_z}{L_0} = E_p \cdot \frac{B_p \cdot H_p^3}{6L_0} \quad (2)$$

with B_p, the width and H_p, the thickness of the passive layer, and the damping:

$$d_\theta = D_p \cdot \frac{B_p \cdot H_p^3}{6} \cdot \sqrt{E_p \cdot \rho_p} \quad (3)$$

where D_p is the damping grade, E_p is the Young's modulus, ρ_p is the density, and B_p, the width and H_p the thickness of the passive layer.

The deformation x_3 in the active prismatic joint under pulsating input Voltage signal with the amplitude U_{amp} and the frequency ω_e is:

$$x_3(t) = \frac{d_{33} \cdot U_{amp} \cdot L_0}{2 \cdot t_{IDE}} \cdot (1 - \cos(\omega_e \cdot t)) \quad (4)$$

where d_{33} is the piezoelectric constant and $t_{IDE,}$ the distance between electrodes.

The elastic deformation of the actuator layer under load is simulated by the spring-damper with the rigidity k_l:

$$k_l = E_a \cdot \frac{B_a \cdot H_a}{L_0} \quad (5)$$

and the damping coefficient:

$$d_l = 2D_a \cdot B_a \cdot H_a \cdot \sqrt{E_a \cdot \rho_a} \quad (6)$$

where D_a is the damping grade, E_a is the Young's modulus and ρ_a is the density of the piezoceramic's element's material, and B_a, the width and H_a the thickness of the active layer.

4 Analytic Dynamics Study of an Actuating Layer

The dynamic model for the active layer is shown in Fig. 5. The link 5 guides the exciter 3 moving with $x_3(t)$ according to Eq. 4. The driven link's 4 movement law $x_4(t)$ is determined by the dynamic equilibrium equation of forces acting on it:

$$\overrightarrow{F_i(t)} + \overrightarrow{F_d(t)} + \overrightarrow{F_{el}(t)} = \overrightarrow{F_{ext}(t)} \quad (7)$$

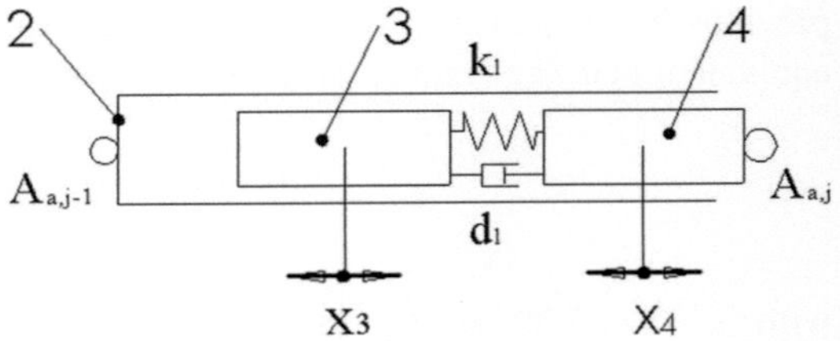

Fig. 5 Dynamic model of the piezo-active layer

where $F_i(t)$ is the inertial force:

$$F_i(t) = -m_{red} \cdot \ddot{x}_4 \tag{8}$$

with m_{red}, the ACM reduced mass on the cell's tip, $F_d(t)$ is the damping force:

$$F_d(t) = -d_l \cdot (\dot{x}_4 - \dot{x}_3) \tag{9}$$

and $F_{el}(t)$ is the spring force:

$$F_{el}(t) = -k_l \cdot (x_4 - x_3) \tag{10}$$

The differential equation may be written in scalar form, by substituting $x_3(t)$ as:

$$m_{red} \cdot \ddot{x}_4 + d_l \cdot \dot{x}_4 + k_l \cdot x_4 = \frac{k_l \cdot A_a}{2} - \frac{A_a}{2} \cdot (k_l \cos(\omega_e t) + \omega_e d_l \sin(\omega_e t)) + F_{ext}(t) \tag{11}$$

with the actuation amplitude A_a:

$$A_a = \frac{d_{33} \cdot U_{amp} \cdot L_0}{t_{IDE}} \tag{12}$$

The Eq. 11 shows that the piezo-actuation with a pulsating input Voltage signal is equivalent to an actuation with a step signal and an harmonic signal, both half excitation amplitude high.

Dividing by the reduced mass, the Eq. 11 becomes:

$$\ddot{x}_4 + 2\delta_l \omega_0 \dot{x}_4 + \omega_0^2 x_4 = \frac{\omega_0^2 A_a}{2} + \frac{\omega_0^2 A_a}{2}\left(2\frac{\omega_e}{\omega_0}\delta_l \sin(\omega_e t) - \cos(\omega_e t)\right) + \frac{F_{ext}(t)}{m_{red}} \tag{13}$$

where

$$\omega_o = \sqrt{\frac{k_l}{m_{red}}} \tag{14}$$

$$\delta_l = \frac{d_l}{2\sqrt{k_l m_{red}}} \tag{15}$$

is the natural frequency and the damping ratio of the system.

The solution for the Eq. 13 is obtained by adding the solutions of three linear differential equations, corresponding to constant force excitation, periodic excitation, and external load, respectively:

$$\ddot{x}_4 + 2\delta_l\omega_0\dot{x}_4 + \omega_0^2 x_4 = \frac{\omega_0^2 A_a}{2} \tag{16}$$

$$\ddot{x}_4 + 2\delta_l\omega_0\dot{x}_4 + \omega_0^2 x_4 = \frac{\omega_0^2 A_a}{2\cos(\phi)}\sin(\omega_e t - \phi) \tag{17}$$

where the excitation phase angle is:

$$\phi = \arctan\left(2\frac{\omega_e}{\omega_0}\delta_l\right) \tag{18}$$

$$\ddot{x}_4 + 2\delta_l\omega_0\dot{x}_4 + \omega_0^2 x_4 = \frac{F_{ext}(t)}{m_{red}} \tag{19}$$

The solutions for the Eqs. 16 and 17 are (Dresig 2014), for an under-damped system as is the case for most ACM, respectively:

$$x_{41}(t) = \frac{A_a}{2}(1 - e^{-\delta_l\omega_0 t})\sin\left(\sqrt{1-\delta_l^2}\omega_0 t\right) \tag{20}$$

$$x_{42}(t) = \frac{A_a}{2\omega_e\sqrt{\left((2\omega_0\delta_l)^2 + \left(\frac{\omega_0^2}{\omega_e^2} - 1\right)^2\right)}}\sin\left(\sqrt{1-\delta_l^2}\omega_0 t + \varphi\right) \tag{21}$$

The solution for the Eq. 19 depends on the external loads, and is difficult to be obtained analytically. For this, a dynamic Multi Body System (MBS) analysis using SolidWorks Motion was performed and is presented in the following part of the manuscript.

5 Dynamic Simulation of a Multi Body System

A simplified CAD model for the ACM (see Fig. 6) is used for the simulation. A wide bending actuator was used instead of the two narrow ones. All the passive elements which were not influencing the system's dynamic behaviour (axes,

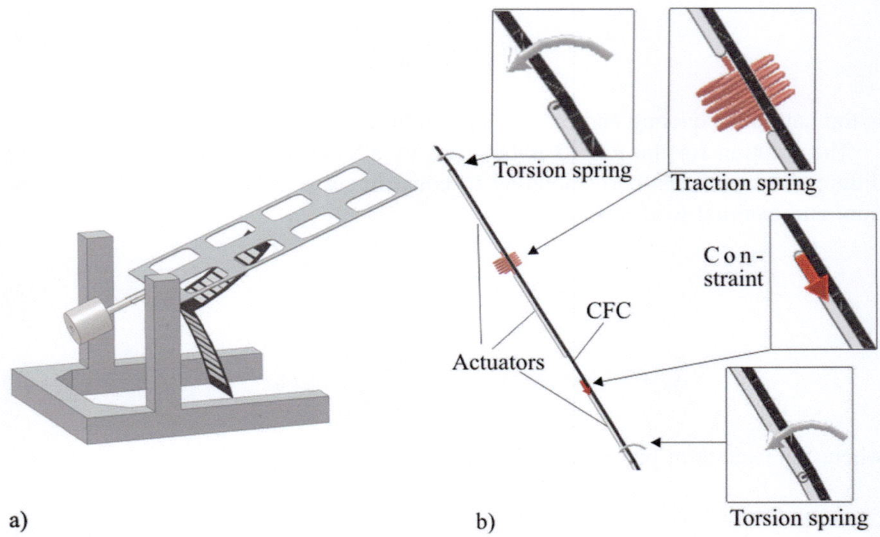

Fig. 6 ACM-Modell: **a** CAD model used for simulations; **b** simulation cell

bearings, housings, fixtures, bumpers, spacers, etc.) were substituted with constraints.

Using the simulation cell in Fig. 4, the bending actuators were simulated in the initial position with an angle constraint between the passive layers of the two subsequent simulating cells.

The simulating cell with two layers is shown in Fig. 6, with traction springs in red, torsion springs in grey and actuating movement shown by a red arrow. The revolute joints were simulated by coaxial constraints, and the prismatic joints, by co planarity constraints. The simulation parameters, calculated according to the Eq. 2–6 with the values given in Table 1 are summarized in Table 2.

Table 1 Dimensions and densities of the simulation elements for bending actuators

Parameter/element j	Thickness H [mm]		Length Lj [mm]	Width B [mm]		Density ρ [kg/m^3]	
1, 5, 6, 10,	0.25		13.5	75		1750	
2, 3, 4, 7, 8, 9	H_p	H_a	29	B_p	B_a	ρ_p	ρ_a
	0.25	0.28		75	57	1750	4750

Table 2 Simulation parameters for the equivalent spring dampers and linear motors

Torsion spring damper in the passive layer		Traction spring damper in the active layer		Linear motor
Rigidity [N mm/°]	Damping coefficient [N mm/(°/s)]	Rigidity [N/mm]	Damping coefficient [N/(mm/s)]	Stroke [mm]
2	1	16,700	100	0.02

The flap's movement was recorded as the relative angular displacement in the coaxial constraint simulating the revolute joint B_0. The data were exported and were used for the comparison with the experimental results.

6 Experimental Investigation on an ACM Demonstrator

The demonstrator ACM was tested with similar parameters as used in simulations and measured with acceptable accuracy by an optical contactless dynamic 3D analysis system PONTOS (type 5 M, GOM mbH), which avoids any mechanical influence of measuring forces on the ACM. The experimental setup for both actuating and the measuring devices are shown in Fig. 7.

The input voltage for the compliant piezoactuators was provided by a high voltage power amplifier (type PA05039, smart materials GmbH) controlled by an adapted LabVIEW program used to generate various inputs with adjustable amplitude, frequency and number of cycles. A pulsating 5 or 10 cycles with the frequencies of 0.5, 1, 2 and 5 Hz input signal from 0 to 750 V was applied to the MFC.

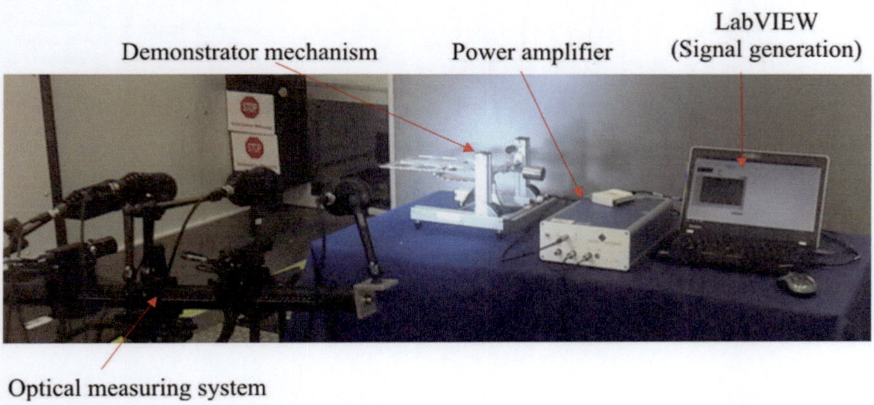

Fig. 7 Experimental set-up of the ACM demonstrator

7 Comparison Between Simulation and Experimental Results

A comparison between simulation and experimental results is presented in Fig. 8 up to Fig. 11.

For the overcritical case, above 1 Hz input excitation signal, the flap opened 10 degrees up. Then it oscillated around that position with ±4° in simulation and ±2° in experiment, as it is shown in Fig. 9. The phase difference was also obvious between the input signal and the excited oscillation.

At a higher 2 Hz frequency, shown in Fig. 10, the excited oscillation was visible also in the rising stage, accompanied by a free lower (0.75 Hz) frequency oscillation at the first 4 to 5 cycles. The oscillation amplitude decreased to ±1.5° in simulation and ±0.5° in experiment.

In Fig. 11, at 5 Hz, the excited oscillation was less visible, both at simulation and experiment, as predicted by Eq. 21, as the free oscillation were clearly noticeable.

After the end of excitation, the flap dropped and oscillated freely around the horizontal position. At all frequencies, the hysteresis effect was visible in tests, as the flap did not regained its horizontal position, as by the purely mechanical dynamic simulation.

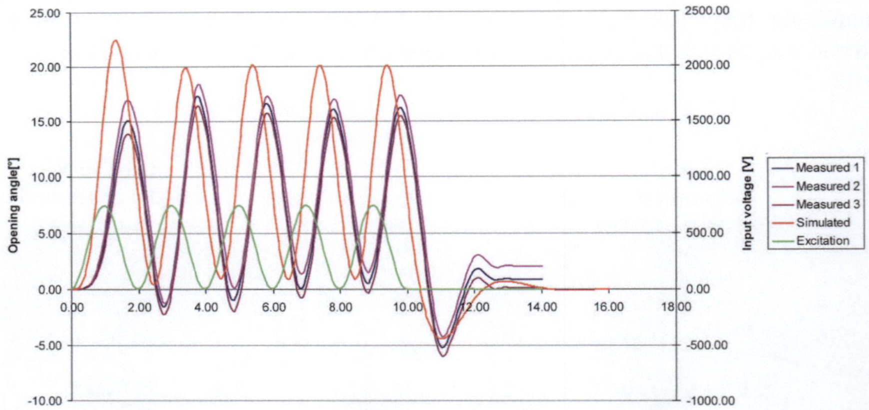

Fig. 8 Simulation and experimental results for an input voltage of 750 V at 0.5 Hz

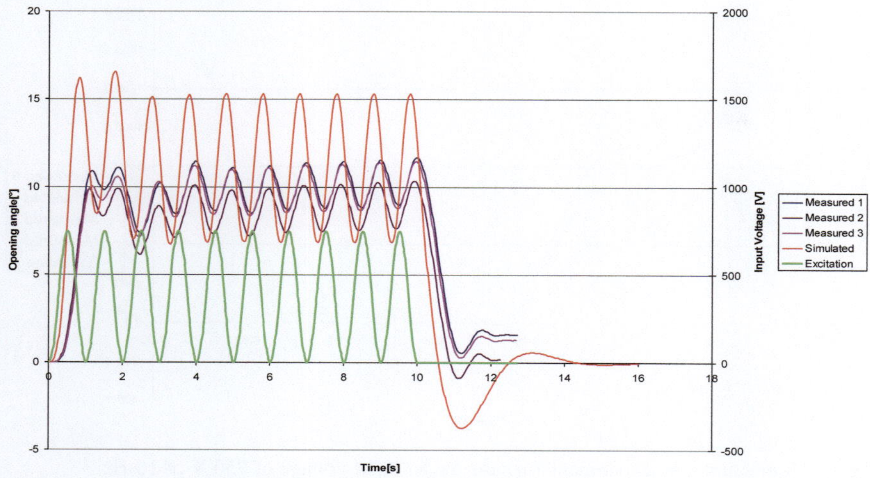

Fig. 9 Simulation and experimental results for an input voltage of 750 V at 1 Hz

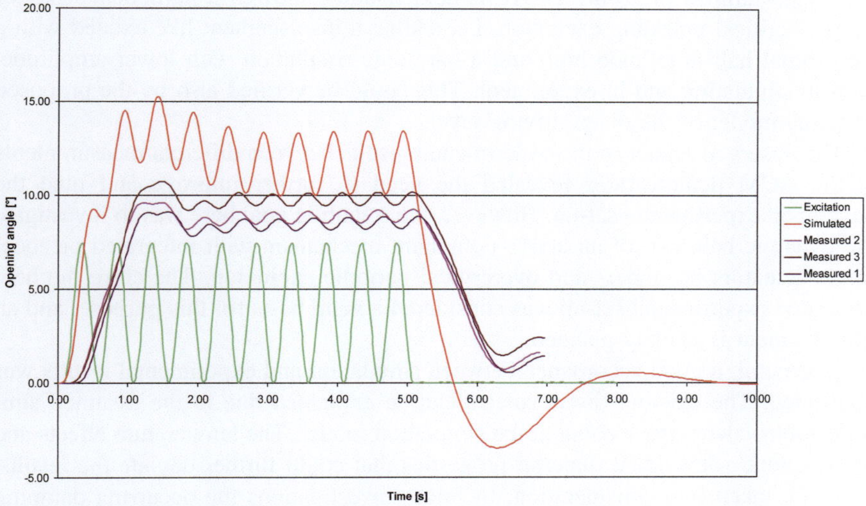

Fig. 10 Simulation and experimental results for an input voltage of 750 V at 2 Hz

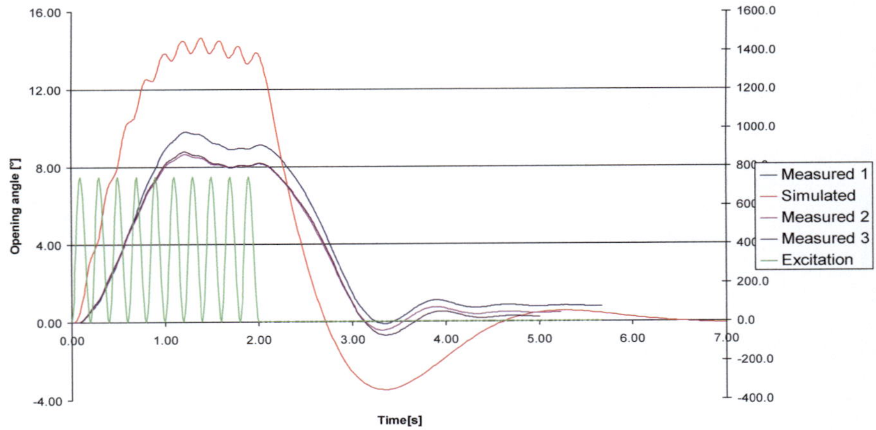

Fig. 11 Simulation and experimental results for an input voltage of 750 V at 10 Hz

8 Conclusions

An under-damped piezo-driven ACM demonstrator showed a particular behavior by over-critical pulsating excitation, i.e. adding a displacement like excited with a step signal half amplitude high and a harmonic oscillation with lower amplitude, both in simulation and in experiment. This could be verified also by the proposed dynamic model of the piezo-driving layer.

The observed scatter of the experimental results between different measurements at the ACM demonstrator revealed the need to further improve and tune the developed experimental set-up. However, the scientific goal here was to investigate the dynamic behavior of an active compliant mechanism with integrated piezoceramic actuators by under- and overcritical periodic excitation. Therefore the here presented experimental set-up was considered a valid basis for this purpose, and an improvement is already planned.

In general, a good accordance between simulation and experimental results was confirmed. The existing discrepancies can be explained due to the assumed simplifications taken into account at the simulation model. The temperature effects and the frequency-dependent material properties that could further deviate the results, were not taken into consideration. In future investigations the occurring damping effects and the required damping coefficients must be further modeled and a new comparison must be undertaken.

Acknowledgments This research is supported by the Deutsche Forschungsgemeinschaft (DFG) in context of the Collaborative Research Centre SFB/TR39, subproject B4 and the Collaborative Research Centre SFB 639, subproject D2.

References

Dresig H, Fidlin A (2014) Schwingungen und mechanische Antriebssysteme: modellbildung, Berechnung, Analyse, Synthese. 3. Auflage. Springer Vieweg Verlag

Gao ZH, Zhang, D (2010) Design, analysis and fabrication of a multidimensional acceleration sensor based on fully decoupled compliant parallel mechanism. Sens Actuators A 163

Mărgineanu D et al (2014) Multibody system simulation of hysteresis effect by micro textile-reinforced compliant mechanisms with piezo-electric actuators, MAMM Timişoara

Modler N (2008) Nachgiebigkeitsmechanismen aus Textilverbunden mit integrierten aktorischen Elementen. Dissertation Technische Universität Dresden

Modler N et al (2009) A design of compliant mechanisms with integrated actuators. SYROM - Braşov (Romania), pp 655–664

Tanık E, Söylemez E (2010) Analysis and design of a compliant variable stroke mechanism. Mech Mach Theory 45:1385–1394

Synthesis of Compliant Mechanisms With Defined Kinematics

A. Hasse, M. Franz and K. Mauser

Abstract A mechanism is designed to transform forces and/or displacements from an input to one or multiple outputs. This transformation is essentially ruled by the kinematics, i.e. the defined ratio between input and output displacements. Although the kinematics forms the basis for the design of conventional mechanisms, some common approaches for the topology and shape optimization of compliant mechanisms do not explicitly include the kinematics in their optimization formulation. The kinematics is more or less an outcome of the optimization process. A defined kinematics can only be realized by iteratively adjusting process-specific optimization parameters within the optimization formulation. This paper presents an optimization formulation that solves the aforementioned problem. It bases on one of the authors former publications on the design of compliant mechanisms with selective compliance. The formulation is derived by means of an intensive workup of the design problem of compliant mechanisms. The method is validated for a common design example: a force inverter.

Keywords Compliant mechanisms · Topology optimization · Selective compliance · Defined kinematics

1 Introduction

Over the course of more than two decades of ongoing research into the structural topology and shape optimization of compliant mechanisms, several optimization problem formulations have already emerged—but none of these are universally accepted (Deepak et al. 2008). The first contribution was presented in the beginning

A. Hasse (✉) · M. Franz · K. Mauser
Department of Mechanical Engineering, Engineering Design,
Friedrich-Alexander-University Erlangen-Nuremberg, Erlangen, Germany
e-mail: hasse@mfk.fau.de

M. Franz
e-mail: michael.mfr.franz@fau.de

L. Zentner et al. (eds.), *Microactuators and Micromechanisms*,
Mechanisms and Machine Science 45, DOI 10.1007/978-3-319-45387-3_20

of the 1990s by Ananthasuresh et al. (1994). Based on the common view that a compliant mechanism must be flexible enough to provide a certain motion and at the same time stiff enough to sustain external loads, they used a weighted sum of mutual potential energy (MPE) and strain energy (SE) as the objective function to minimize, where the MPE is a measure for the structure's flexibility and the SE is a commonly used measure of the structure's stiffness (Shield and Prager 1970). Since then, numerous researchers (Frecker et al. 1999; Kota et al. 2001; Chen et al. 2001) have followed with modified and extended MPE/SE formulation. A different approach was presented by Sigmund (1997). The objective of his approach was to maximize the so-called mechanical advantage (MA), the ratio between input and output force, subject to constraints on the input displacement and volume. Later, this formulation was reduced to maximize the output displacement actuated by a force at the mechanism's input (Bendsøe and Sigmund 2003). In the following years, various other objectives, for instance the energy efficiency formulation presented by Hetrick and Kota (1999) and the artificial I/O spring formulation presented by Rahamatalla and Swan (2005), have been introduced. However, a study performed by Deepak et al. (2008) showed that this formulation produces essentially similar designs. Furthermore, something that nearly all of the formulations presented so far have in common is that they apply additional springs at the output in order to support the optimization. It was observed that the stiffness values chosen for these springs have strong influence on the optimization result—it requires some experience to choose the right stiffness value in order to realize the desired kinematical behavior. Without these additional springs, the structural optimization could result in non-connected designs between input and output (Deepak et al. 2008; Sigmund 1997). Another drawback of the aforementioned formulations is the occurrence of point flexures—concentrated regions with extremely low stiffness compared to the remaining regions. With the desire to avoid the occurrence of point flexures, Chen and Wang (2007) proposed the characteristic stiffness formulation which requires no additional springs. This formulation can be applied to single-input, single output mechanisms. With a conventional sub-structuring approach they reduced the global stiffness matrix to the input and output degrees of freedom. They presented different possible objective functions based on the diagonal terms of the reduced stiffness matrix and the geometric advantage (the desired ratio between input and output displacement). An alternative approach (Hasse and Campanile 2009) presented by the author also relies on a reduced stiffness matrix and offers the same advantages as Wang's approach. Thereby, the objective function integrates modal quantities obtained from a specially synthesized eigenvalue problem in order to optimize the kinematical behavior. This paper presents a reformulation of this objective function based on the same idea in order to make the method more accessible. Therefore, we start with a discussion of the design problem of compliant mechanisms. Having understood what needs to be realized in terms of stiffness distribution makes it possible to derive a formal optimization problem under linear assumptions. This formulation is then implemented in a complete optimization procedure and tested on several design problems related to the well-known force inverter.

2 Design Problem

Before discussing the design of compliant mechanisms, we consider their conventional counterparts: classical mechanisms. They gain their mobility from relative movement of stiff members against each other. Seen as an elastic structure, a classical mechanism principally has an infinite number of structural degrees of freedom (*structural DoFs*). For practical reasons, the structure is mostly discretized to a finite number p of DoFs, and the general displacement field will be described by the vector

$$\mathbf{u} \in \mathbb{R}^p. \tag{1}$$

Due to the presence of conventional hinges, it is possible to model all members as rigid bodies by preserving the mechanism's capability to deform and perform its function. This allows for a simpler motion analysis with a reduced number m of *kinematic DoFs*.

If the motion is linearized about a given configuration, m expresses the number of linearly independent deformation modes

$$\boldsymbol{\chi}_i \in \mathbb{R}^p, \quad i = 1, \ldots, m \tag{2}$$

that can be generated when starting from this configuration. Hence, there will be a m-dimensional subspace $\mathbb{K} \subseteq \mathbb{R}^p$, which defines the possible motion of the mechanism

$$\mathbb{K} = \left\{ \sum_{i=1}^{m} \alpha_i \boldsymbol{\chi}_i | \alpha_i \in \mathbb{R} \right\}. \tag{3}$$

In the following, this subspace will be denoted as the *kinematics* of the mechanism. $\mathbf{u} \in \mathbb{K}$ holds for all the displacements that the mechanisms can produce—of course only under the assumption of rigid members.

The kinematics is normally the basis, i.e. the primary input parameter, of the synthesis of the mechanism, because it sets the ratio of displacement between the inputs and outputs of the mechanism and therefore determines the transformation of motion and/or force.

In principle, a compliant mechanism must fulfill the same task as its conventional counterpart: transforming motion and/or force from an input to an output. Thus, the kinematics should be the driving input parameter when designing compliant mechanisms as well. Nevertheless, there is a substantial difference in the functional principle: the motion is produced by elastic deformation. Therefore, the assumption of rigid members is not admissible—so the mechanism's displacements $\mathbf{u}$ due to an external load are principally not constrained to be $\mathbf{u} \in \mathbb{K}$. It is therefore useful to divide the displacements into desired $\mathbf{u}_d$ and undesired $\mathbf{u}_{ud}$ displacements, whereas $\mathbf{u}_d \in \mathbb{K}$. The complement

$$\overline{\mathbb{K}} = \mathbb{R}^p \backslash \mathbb{K} \tag{4}$$

is denoted as the subspace of the undesired deformations, so it holds:

$$\mathbf{u}_{ud} \in \overline{\mathbb{K}}. \tag{5}$$

Under linear considerations, this subspace $\overline{\mathbb{K}}$ is spanned by $p - m$ linear independent *undesired deformation modes* $\overline{\boldsymbol{\chi}}_i \in \mathbb{R}^p$ (analogously to Eq. (3)). The *desired deformation modes* are furthermore denoted with $\boldsymbol{\chi}_i \in \mathbb{R}^p$, as with the classical mechanisms.

The components of $\mathbf{u}$ in $\mathbb{K}$ and $\overline{\mathbb{K}}$ as a result of external loads $\mathbf{f}$ at the mechanism's inputs and outputs are determined by the stiffness of the mechanisms—which in turn is determined by the mechanisms topology, shape and material. The force-displacement-relationship of a general structure is normally described by a $p \times p$ stiffness matrix $\mathbf{k}$ so that holds

$$\mathbf{ku} = \mathbf{f}. \tag{6}$$

Now, we assume that the desired and undesired deformation modes are collected in a basis

$$\mathrm{X} = \begin{bmatrix} \boldsymbol{\chi}_1 & \cdots & \boldsymbol{\chi}_m & \overline{\boldsymbol{\chi}}_{m+1} & \cdots & \overline{\boldsymbol{\chi}}_p \end{bmatrix} \tag{7}$$

and are defined in an orthogonal fashion, so that holds

$$\frac{1}{2}\mathrm{X}^T\mathbf{k}\mathrm{X} = \frac{1}{2}\begin{bmatrix} k_1 & 0 & & \cdots & & 0 \\ 0 & \ddots & & & & \\ & & k_m & \ddots & & \vdots \\ \vdots & & \ddots & \bar{k}_{m+1} & & \\ & & & & \ddots & 0 \\ 0 & & \cdots & & 0 & \bar{k}_p \end{bmatrix}. \tag{8}$$

The terms

$$k_i \in \mathbb{R}, \quad i = 1, \ldots, m \tag{9}$$

related to the desired deformation modes are denoted as *primary stiffness* and the terms

$$\bar{k}_i \in \mathbb{R}, \quad i = m+1, \ldots, \mathrm{p} \tag{10}$$

related to the undesired deformation modes are denoted as *secondary stiffness*. Under normalization condition

$$\begin{aligned} \chi_i^T \chi_i &= 1, \qquad \forall i = 1 \ldots m \\ \overline{\chi}_i^T \overline{\chi}_i &= 1, \qquad \forall i = m+1 \ldots p \end{aligned} \tag{11}$$

Equation (8) gives insight into the physical meaning of the primary and secondary stiffness. If the structure is deformed according to a given deformation mode, it is

$$\mathbf{u}_a = \alpha \chi \tag{12}$$

where the scaling factor α has the dimension of a length. The corresponding deformation energy is

$$SE = \alpha^2 \frac{1}{2} \chi^T \overline{\mathbf{k}} \chi \tag{13}$$

and therefore

$$\frac{1}{2} \chi^T \overline{\mathbf{k}} \chi = \frac{SE}{\alpha^2} \tag{14}$$

i.e. the left side of Eq. (8) corresponds to the deformation energy divided by the square of the scaling factor. Its numerical value is equal to the energy stored in the system when the scaling factor is set to a unit length (1 m for SI units). Its dimension, however, equates to the energy per square length (Joule per square meter). A primary or secondary stiffness value is twice this energy per square length unit. For this reason, deformation modes with high stiffness values are stiffer, which means that more energy must be spent in order to reach the amplitude level defined by the condition $\alpha = 1$ m. Due to the low required energy, deformation with low stiffness values will dominate the static response of the system i.e. the system will preferentially deform according to a displacement distribution given by a linear combination of these deformation modes.

If a mechanism is required only to deform with $\mathbf{u} \in \mathbb{K}$ then an infinite secondary stiffness needs to be realized—this is impossible with real structures. However, it is typically for classical mechanisms that the stiffness of the members, and thus the secondary stiffness, is chosen so high that possible undesired deformation are negligible. Even if it is difficult to achieve, compliant mechanisms may have a similar behavior: the displacements are limited to the kinematics independent of the external loads. Then they are called "compliant mechanisms with selective compliance" (Hasse and Campanile 2009).

It can be stated that the higher the secondary stiffness compared to the primary stiffness is the more the compliant mechanism tends towards selective compliance behavior. In the following section the described design problem is translated into a formal optimization problem.

3 Optimization Formulation

Normally, the kinematics is not defined for all structural DoFs. In practice, the kinematics only refers to certain input and output DoFs (together from hereon named as *master DoFs*). It is therefore appropriate only to refer with the desired and undesired deformation mode to this reduced set of master DoFs. Therefore, we divide the p DoFs into q master DoFs and $p-q$ *slave DoFs*. The master DoFs are collected in the set a and the slave DoFs in c. Under the assumption that no external forces $\mathbf{f} \in \mathbb{R}^p$ act on the slave DoFs, the linear system equations

$$\begin{bmatrix} \mathbf{k}_{aa} & \mathbf{k}_{ac} \\ \mathbf{k}_{ca} & \mathbf{k}_{cc} \end{bmatrix} \begin{bmatrix} \mathbf{u}_a \\ \mathbf{u}_c \end{bmatrix} = \begin{bmatrix} \mathbf{f}_a \\ 0 \end{bmatrix} \tag{15}$$

can be reduced to

$$(\mathbf{k}_{aa} - \mathbf{k}_{ca}\mathbf{k}_{cc}^{-1}\mathbf{k}_{ac})\mathbf{u}_a = \hat{\mathbf{k}}\mathbf{u}_a = \mathbf{f}_a \tag{16}$$

where $\hat{\mathbf{k}}$ is a $q \times q$ matrix. Within this paper, we constrain our consideration to single-input single-output mechanisms with one kinematic DoF, i.e. $\hat{\mathbf{k}}$ is a 2×2 matrix and the input to the optimization formulation is the normalized desired deformation mode

$$\chi^T \chi = 1, \quad \chi \in \mathbb{R}^2. \tag{17}$$

A normalized undesired deformation mode

$$\overline{\chi}^T \overline{\chi} = 1, \quad \overline{\chi} \in \mathbb{R}^2 \tag{18}$$

exists for every $\hat{\mathbf{k}}$ and it holds

$$\chi^T \hat{\mathbf{k}} \overline{\chi} = 0. \tag{19}$$

The design problem derived in the last section could be directly translated in the following optimization

$$\begin{aligned} \max \quad & \bar{k} = \overline{\chi}^T \hat{\mathbf{k}}(\mathbf{x}) \overline{\chi} \\ s.t. \quad & k - k_{desired} = \frac{1}{2} \chi^T \hat{\mathbf{k}}(\mathbf{x}) \chi - k_{desired} \leq 0 \end{aligned} \tag{20}$$

where the vector $\mathbf{x} = (x_1, \ldots, x_n)^T$ contains the design variables and $k_{desired}$ is the upper bound on the primary stiffness. Problematic with this direct formulation is that the undesired deformation mode $\overline{\chi}$ changes with the design variables throughout the optimization. The undesired deformation mode has to be determined by a suitable procedure, such as the Gram-Schmidt-orthogonalization, within every iteration of the optimization procedure. A corresponding approach was already presented by the author in (Hasse and Campanile 2009). However, within this publication we want to present a more compact formulation. Therefore, we rewrite condition (19)

$$\boldsymbol{\chi}^T \hat{\mathbf{k}}(\mathbf{x}) \overline{\boldsymbol{\chi}} = \beta \boldsymbol{\chi}^T \boldsymbol{\delta}. \tag{21}$$

$\boldsymbol{\delta}$ can be defined prior to the optimization by using the normalization condition

$$\boldsymbol{\delta}^T \boldsymbol{\delta} = 1, \quad \boldsymbol{\delta} \in \mathbb{R}^2, \tag{22}$$

since $\boldsymbol{\delta}$ is independent of the design variables. Only the scaling factor β relates to the design variables. Thus, we can write the objective function in problem (20) as

$$f(\mathbf{x}) = \overline{\boldsymbol{\chi}}^T \hat{\mathbf{k}}(\mathbf{x}) \overline{\boldsymbol{\chi}} = \beta^2 \boldsymbol{\delta}^T \hat{\mathbf{k}}(\mathbf{x})^{-1} \boldsymbol{\delta}. \tag{23}$$

Through the normalization condition (18)

$$1 = \overline{\boldsymbol{\chi}}^T \overline{\boldsymbol{\chi}} = \beta^2 \boldsymbol{\delta}^T \hat{\mathbf{k}}(\mathbf{x})^{-1} \hat{\mathbf{k}}(\mathbf{x})^{-1} \boldsymbol{\delta}. \tag{24}$$

the scaling factor β can be determined

$$\beta^2 = \frac{1}{\boldsymbol{\delta}^T \hat{\mathbf{k}}(\mathbf{x})^{-1} \hat{\mathbf{k}}(\mathbf{x})^{-1} \boldsymbol{\delta}}. \tag{25}$$

Therefore, the objective function writes as follows

$$f(\mathbf{x}) = \frac{\boldsymbol{\delta}^T \hat{\mathbf{k}}(\mathbf{x})^{-1} \boldsymbol{\delta}}{\boldsymbol{\delta}^T \hat{\mathbf{k}}(\mathbf{x})^{-1} \hat{\mathbf{k}}(\mathbf{x})^{-1} \boldsymbol{\delta}}. \tag{26}$$

The optimization problem (20) is extended, additionally to the new objective function, with a further constraint limiting the distributable amount of material within the design domain and lower and upper bounds on the design variables

$$\begin{aligned}
\max \quad & f(\mathbf{x}) = \frac{\boldsymbol{\delta}^T \hat{\mathbf{k}}(\mathbf{x})^{-1} \boldsymbol{\delta}}{\boldsymbol{\delta}^T \hat{\mathbf{k}}(\mathbf{x})^{-1} \hat{\mathbf{k}}(\mathbf{x})^{-1} \boldsymbol{\delta}} \\
s.t. \quad & g(\mathbf{x}) = \frac{1}{2} \boldsymbol{\chi}^T \hat{\mathbf{k}}(\mathbf{x}) \boldsymbol{\chi} - k_{desired} \leq 0 \\
& h(\mathbf{x}) = \sum \mathbf{x} - v = 0 \\
& \mathbf{x}_{\min} \leq \mathbf{x} \leq \mathbf{x}_{\max}
\end{aligned} \tag{27}$$

where v is the limit to the amount of distributable material. Obviously, the only term in the optimization problem, which is now dependent on the design variables, is the stiffness matrix $\mathbf{k}$.

4 Implementation

The derived optimization formulation (27) is integrated in a complete optimization procedure. We have chosen a ground structure approach (Bendsøe and Sigmund 2003) as parametrization. Thereby, the design variables x_i scale the stiffness of the i-th ground-element:

$$\mathbf{k}(\mathbf{x}) = \sum_{i=1}^{n} x_i \mathbf{k}_i. \tag{28}$$

In our case, the ground elements are modelled by Bernoulli-Euler beam elements. The ground structure approach is exemplified in Fig. 2. The parametrization, the structural model and the derived optimization formulation need to be completed by a proper optimization algorithm in order to obtain a complete optimization procedure. Within this work, however, we have applied an interior-point method standardly implemented in MATLAB. This algorithm converged best.

5 Problem Statement of Design Examples

The validation example is a force inverter. The corresponding mechanism's input and the output are defined as master DoFs. The difference between the derived (27) and other formulations is that the derived formulation does not require explicit distinction between input and output. Because only two master DoFs are considered and the mechanism should only have one kinematic DoF, the kinematics is described by one desired deformation mode $\chi \in \mathbb{R}^2$.

The design domain, the support and the master DoFs are shown in Fig. 1. The design domain is described by the following geometric parameters

$$l = 100 \text{ mm}, \; h = 50 \text{ mm}, \; d = 10 \text{ mm} \tag{29}$$

where l is the reference, h is the height and d is the mechanism's width.

Concerning the parametrization, the design domain is meshed with ground elements as depicted in Fig. 2.

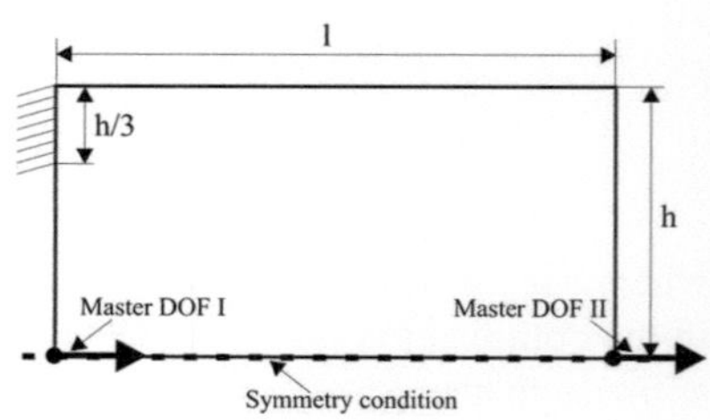

Fig. 1 Design domain with corresponding support and master DoFs

Fig. 2 Parameterized design domain

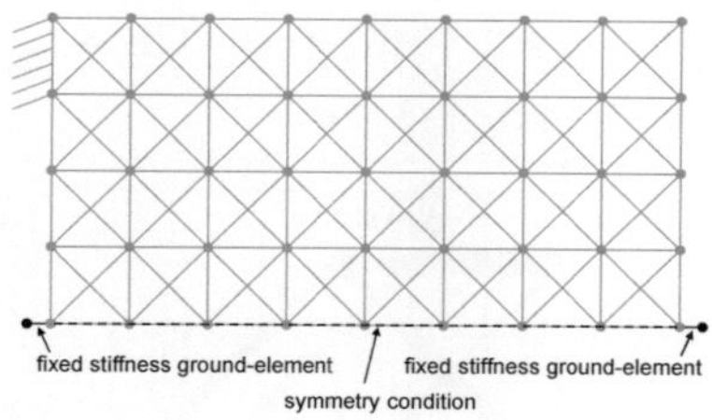

Every beam of the reference ground-structure is characterized by the following material and geometric parameters

$$E = 1500\ \text{MPa},\ I = 6.66\ \text{mm}^4,\ A = 20\ \text{mm}^2 \tag{30}$$

where E is the Young's modulus, I the moment of inertia and A is the cross-sectional area.

The lower and the upper bounds for the design variables are defined as

$$10^{-5} \leq x_i < 1, \quad i = 1 \ldots n. \tag{31}$$

As a starting point for the optimization, all design variables are set to the limit to the amount of distributable material v.

6 Results and Discussion

The optimization procedure was applied to force-inverter design tasks with different kinematics. The input parameters v and $k_{desired}$ to the optimization formulation (27) are

$$v = 0.06,\ k_{desired} = 8.5 \frac{\text{N}}{\text{mm}} \tag{32}$$

The results for different desired deformation modes are shown in Fig. 3. The beams are drawn with different line thicknesses. Each thickness is proportional to the corresponding design variable. Beams for which the design variable is smaller than 5×10^{-3} are omitted from the plot. The optimization procedure converged for every choice of kinematics. A clear mechanical structure has emerged for all cases. The values of the objective function are similar for all designs. The results show a dependency on the choice of the kinematics. The constraint function is almost zero for nearly all of the obtained results. Thus, the desired primary stiffness is realized.

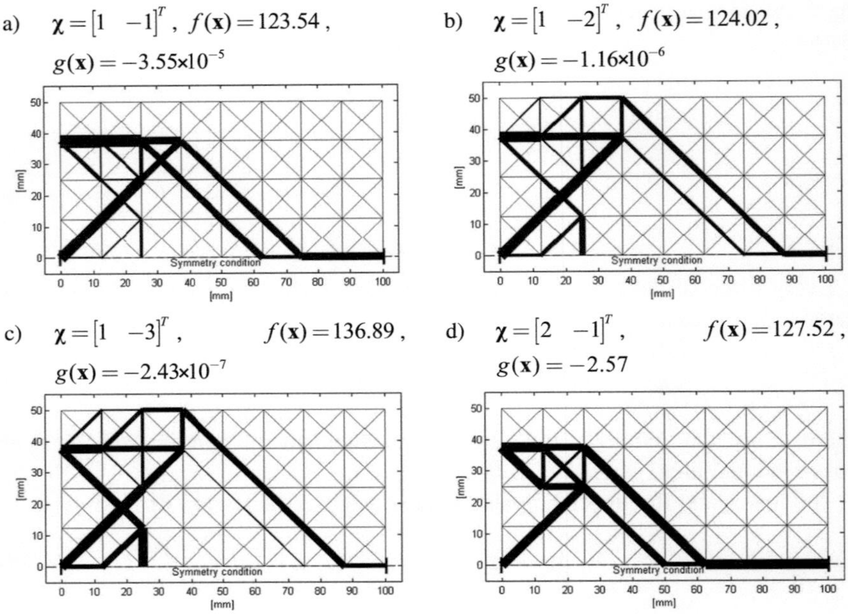

Fig. 3 Force inverter optimized for different desired deformation modes

The optimization can result in different local minima when starting from different positions within the search space. Therefore, the obtained results do not necessarily describe the global optimum. The possibility of finding a solution with higher objective function values is left open.

The following studies will examine whether or not selective compliance behavior is obtained for the different designs by applying different loads. The structure optimized for $\chi = [1 \;\; -1]^T$ (see Fig. 3) is chosen as representative. The other mechanisms behave similarly; therefore their representation has been left out. Load-case 1 is an applied force f_{in} = 300 N at master DoF I. The displacements u_{in} at master DoF I and u_{out} at master DoF II are calculated as the response. The deformed state is shown at the top of Fig. 4a. It can be observed, that the input and output displacements are not an exact scaling of the desired deformation mode, but they tend towards the desired kinematics.

Load-case 2 is an applied force f_{in} = 120 N at master DoF I plus an additional load f_{out} = −240 N at master DoF II (see Fig. 4b). The ratio between input and output displacement has now changed compared to Load-case 1. This indicates that the second eigenvalue is too small and both eigenvectors are therefore stimulated by the external loads.

Examinations have shown that by increasing the parameters ν and $\mathbf{x}_{max}$, there is a stronger implementation of the desired kinematics in the structure. Then the output and input displacement is more or less a scaling of the desired deformation mode.

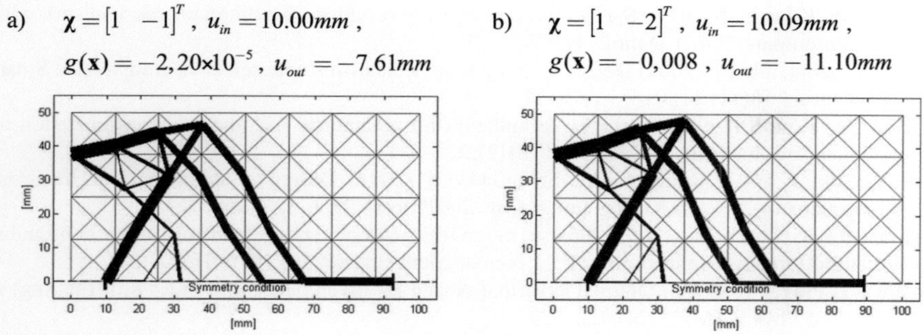

Fig. 4 Force inverter $\chi = [1 \quad -1]^T$ under different load-cases

Furthermore, the mechanism showed a higher independency of the applied external forces—the ratio of the input and output displacements do not significantly change with changing external loads.

7 Conclusion and Future Work

This paper presented a reformulation of the authors former optimization formulation for the design of compliant mechanisms with selective compliance. It is written in a compact manner and offers the advantages that no optimization supporting spring elements are needed. Furthermore the kinematics is directly considered.

Future examinations should consider stress constraints according to the applied material and large deformations. Furthermore, the design of mechanism with kinematic DoFs higher than one should be considered.

References

Ananthasuresh GK, Kota S, Kikuchi N (1994) Strategies for systematic synthesis of compliant MEMS. In: Proceedings of the 1994 ASME winter annual meeting, pp 677–686

Bendsøe MP, Sigmund O (2003) Topology optimization: theory, methods and applications. Springer-Verlag, New York

Chen BC, Silva EC, Kikuchi N (2001) Advances in computational design and optimization with application to MEMS. Int J Numer Meth Eng 52(1–2):23–62

Chen S, Wang MY (2007) Designing distributed compliant mechanisms with characteristic stiffness. In: ASME 2007 international design engineering technical conferences and computers and information in engineering conference, American Society of Mechanical Engineers, pp 33–45

Deepak SR, Dinesh M, Sahu D, Jalan S, Ananthasuresh GK (2008) A comparative study of the formulations for topology optimization of compliant mechanisms. In: ASME 2008 international design engineering technical conferences and computers and information in engineering conference, American Society of Mechanical Engineers, pp 205–216

Frecker M, Kikuchi N, Kota S (1999) Topology optimization of compliant mechanisms with multiple outputs. Struct Optim 17(4):269–278
Hasse A, Campanile LF (2009) Design of compliant mechanisms with selective compliance. Smart Mater Struct 18(11):115016
Hetrick JA, Kota S (1999) An energy formulation for parametric size and shape optimization of compliant mechanisms. J Mech Des 121(2):229–234
Kota S, Joo J, Li Z, Rodgers SM, Sniegowski J (2001) Design of compliant mechanisms: applications to MEMS. Analog Integr Circ Sig Process 29(1–2):7–15
Rahmatalla S, Swan CC (2005) Sparse monolithic compliant mechanisms using continuum structural topology optimization. Int J Numer Meth Eng 62(12):1579–1605
Shield RT, Prager W (1970) Optimal structural design for given deflection. Zeitschrift für Angew Math und Phys ZAMP 21(4):513–523
Sigmund O (1997) On the design of compliant mechanisms using topology optimization. J Struct Mech 25(4):493–524

A Concept of Adaptive Two Finger Gripper with Embedded Actuators

A. Milojević, N.D. Pavlović, S. Linß, M. Tomić, N.T. Pavlović and H. Handroos

Abstract Today in many industries there is a great need for grasping different shaped and soft objects. For safe grasping of such objects and specially a fragile one adaptability is required. Developing a gripper that can adapt the shape of its grasping surface to different shaped objects and achieve safe and reliable manipulations of that objects, represent a challenging task. Many micro domain applications would benefit from adaptive gripper and adaptive grasping, such as in medicine or biomedicine where there is a need for manipulations of human tissue or individual cells. This paper presents concept of a new adaptive two finger gripper with embedded actuators. By using compliant systems—compliant mechanisms with embedded actuators it is possible to develop an adaptive compliant gripper. By embedding the actuators gripper could morph or change the shape of its grasping surface and achieve different grasping patterns i.e. gripper would have structural adaptability. Synthesis methodology for the adaptive gripper that includes simultaneous topology optimization and actuators placement, is also presented. It will be

A. Milojević (✉) · N.D. Pavlović · M. Tomić · N.T. Pavlović
Department of Mechatronics and Control, Faculty of Mechanical Engineering in Niš, Niš, Serbia
e-mail: andrijamilojevic87@gmail.com; Andrija.Milojevic@lut.fi

N.D. Pavlović
e-mail: pavlovic@masfak.ni.ac.rs

M. Tomić
e-mail: mishatomic@gmail.com

N.T. Pavlović
e-mail: pnenad@masfak.ni.ac.rs

A. Milojević · H. Handroos
Laboratory of Intelligent Machines, Lappeenranta University of Technology, Lappeenranta, Finland
e-mail: heikki.handroos@lut.fi

S. Linß
Department of Mechanical Engineering, Mechanism Technology Group, Technische Universität Ilmenau, Ilmenau, Germany
e-mail: sebastian.linss@tu-ilmenau.de

L. Zentner et al. (eds.), *Microactuators and Micromechanisms*,
Mechanisms and Machine Science 45, DOI 10.1007/978-3-319-45387-3_21

shown that the developed adaptive gripper can achieve multiple shapes of its grasping surface when different contracting or extending actuators are active, whereby they realize different stroke.

Keywords Gripper · Adaptability · Compliant mechanisms · Embedded actuators · Synthesis

1 Introduction

Today in many industries there is a great need for grasping different shaped and soft objects. For safe grasping of such objects and specially a fragile one adaptability is required. Thus, developing adaptive gripper that can achieve safe and reliable manipulations of different objects, represent a challenging task. Such gripper would be able to morph the shape of its grasping surface and adapt to different shaped objects. Moreover, many micro domain applications would benefit from adaptive gripper and adaptive grasping, such as in medicine or biomedicine where there is a need for manipulations of human tissue or individual cells.

Many in the field developed different kind of soft grippers that can in some way achieve adaptability (Choi and Koc 2006; Jain et al. 2013; Jung et al. 2013; Giannaccini et al. 2014; Martinez et al. 2013; Ilievski et al. 2011; Araromi et al. 2014; Amend et al. 2012). But most of the developed grippers require assembling (Choi and Koc 2006), external drive (electrical motors, compressed air) (Choi and Koc 2006; Giannaccini et al. 2014; Martinez et al. 2013; Ilievski et al. 2011; Amend et al. 2012), high operating voltage (Araromi et al. 2014), they cannot be scaled to micro domain, and there is no unique syntheses methodology how to obtain an adaptive gripper.

In this paper different approach for achieving adaptability is presented. One way for realization of adaptive gripper is to use compliant systems (Milojevic and Pavlovic 2015). Compliant system represent compliant mechanism with embedded actuators and sensors. A compliant mechanism can be defined as single monolithic flexible structure that achieve force and motion transmission via elastic deformation of its individual elastic segments (Howell et al. 2013). These monolithic structures deform smoothly as a whole and can offer many benefits for realization of adaptability and adaptive structures, such as: no need for assembly, reduced complexity, no wear, no friction, no lubrication, ease of manufacture, better scalability, better accuracy, built in restoring force, low cost. By embedding sensors and actuators within the compliant structure it is possible to develop adaptive gripper that can both sense the object i.e. detect contact with objects and adapt its grasping surface to the shape of the object. By embedding the actuators gripper could morph or change the shape of its grasping surface and achieve different grasping patterns i.e. gripper would have structural adaptability. By embedding sensors gripper could detect contact with objects and via controller (that need to be developed as well) can decide when and which actuator to activate in order to adapt to the shapes of objects. In this paper more attention is paid to the problem of embedding actuators since the goal is first to develop a gripper that has structural adaptability (for gripper application sensors can be added after the adaptive gripper is obtained, as gripper need only to detect contact with objects at the grasping surface).

Compliant systems are biologically inspired as both sensors and actuators are embedded in one system and they deform with the whole structure, whereby the actuators serve both as structural elements and force generators (similar like in biological systems). Thus, to develop adaptive gripper actuators need to be included in the synthesis process as well i.e. structural topology of compliant structure (mechanism) and actuator placement must be simultaneously synthesize; the whole compliant system is obtained in one synthesis process. In this approach forming of compliant structure topology affects the placement of the actuators and vice versa placement of the actuators affect how the topology of compliant structure (mechanism) is formed.

Much attention has been paid to development of different synthesis methodologies for compliant mechanisms (Joo et al. 2000; Saxena and Ananthasuresh 2000), but little attention has been focused on synthesis of compliant mechanisms for shape morphing applications (Lu and Kota 2003, 2005; Zhao and Schmiedeler 2013). The methods presented in these papers focus on developing a compliant mechanisms that can achieve only one predefine shape (target shape). But these methods cannot be applied for development of adaptive gripper that need to achieve multiple shapes of its grasping surface i.e. multiple grasping patterns. Moreover, in the synthesis of classical compliant mechanisms as well as compliant mechanism for shape morphing application, actuators are added after the topology of compliant mechanism is formed. Synthesis methodology for compliant systems—compliant mechanisms with embedded actuators/sensors has been first introduced in (Trease and Kota 2009), but the presented method often give solutions of compliant systems with lot of intersections between structural elements as well as elements and actuators, which is very difficult to manufacture. Beside this, intersections often increase stiffness and complexity of the structure which can significantly lower the system functionality (Milojević and Pavlović 2013). In our previous publication (Milojevic and Pavlovic 2015) this method was improved so that solutions without intersecting elements as well as elements and actuators was obtained, whereby the obtained solutions are eases to manufacture. This method (Milojevic and Pavlovic 2015) was applied to develop adaptive compliant gripper with embedded actuators.

In this paper the concept of adaptive two finger gripper with embedded actuators is presented. It will be shown that the adaptive gripper can achieve multiple shapes of its grasping surface when different actuators are active. Such gripper could grasp different shaped objects. The concept of adaptive finger gripper was introduced in (Milojević and Pavlović 2014), and some of the results of the behavior of such gripper finger when only contracting, only extending as well as combination of contracting and extending actuators were presented. In this paper further analysis is preformed, new results and the concept of grasping the objects with two fingers is presented.

2 Synthesis Methodology for Adaptive Compliant Gripper with Embedded Actuators

The synthesis of adaptive compliant gripper finger with embedded actuators was presented in (Milojević and Pavlović 2014), here only the short overview is shown.

To develop adaptive compliant gripper compliant mechanisms with distributed compliance are used (Joo et al. 2000); distributed compliance provide a smooth deformation filed, which reduces the stress concentration. For synthesis of compliant mechanisms with distributed compliance topology optimization approach is usually used (Joo et al. 2000; Saxena and Ananthasuresh 2000; Milojević et al. 2013), and similar approach is adopted here; except now placement of the actuators is included in the synthesis process as well. In our previous publication (Milojevic and Pavlovic 2015) the computer-coded algorithm that preforms simultaneous synthesis of topology and actuator placement, was developed. The program utilizes Genetic Algorithms (Reeves and Rowe 2002) and linear Finite Element Method (FEM) to find the optimal solution of adaptive gripper.

The outline of the methodology that is used to develop adaptive compliant gripper with embedded actuators is shown in Fig. 1. Only one finger of the gripper is optimized, as based on one finger, two, three or multi fingered gripper could be developed. In the first step the problem specifications are defined (Fig. 1a and Table 1), which includes defining: size of the design domain (allowable space where the compliant structure should fit), grasping surface (left boundary of the design domain), number of output points that are controlled on the grasping surface (three output points are chosen to represent the output region where horizontal

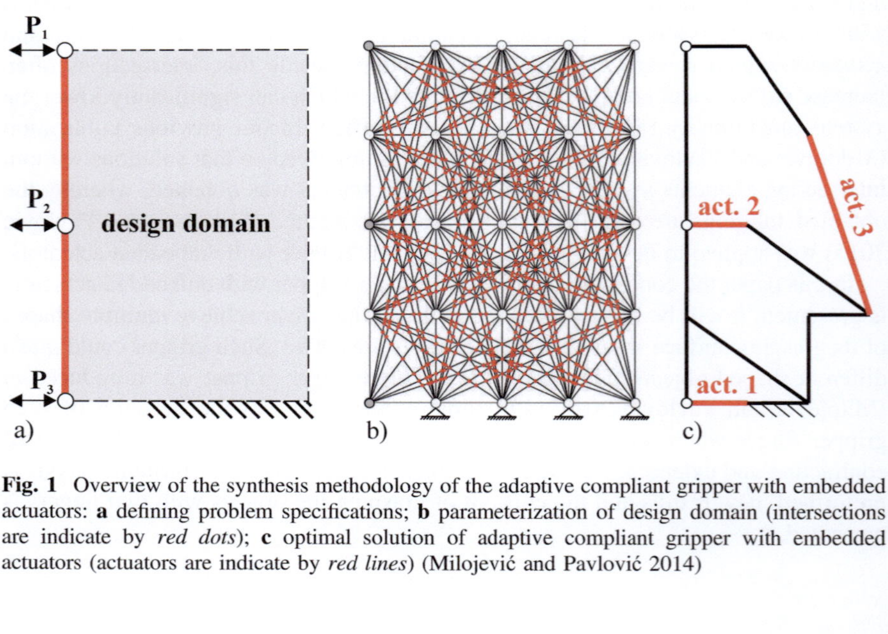

Fig. 1 Overview of the synthesis methodology of the adaptive compliant gripper with embedded actuators: **a** defining problem specifications; **b** parameterization of design domain (intersections are indicate by *red dots*); **c** optimal solution of adaptive compliant gripper with embedded actuators (actuators are indicate by *red lines*) (Milojević and Pavlović 2014)

direction is set as desired direction of the output deflection), supports (one part of bottom design domain boundary), property of the material (Young modulus) from which the mechanism should be built and other constraints such as minimum value of the output deflection $d_{min,target}^{act}$ and total element length L_t which is equivalent to the volume constraint. Table 1 summarizes the parameters that are used for development of adaptive gripper finger.

In the next step design domain is parametrized. The Ground Structure Approach is used for the parameterization (Joo et al. 2000; Saxena and Ananthasuresh 2000), thus, design domain is divided by network of nodes and beam elements connecting the nodes (Fig. 1b). This ground structure (Fig. 1b) serves as an initial solution for the optimization i.e. optimal solution is searched within this ground structure. Variables for optimization are beam element thickness and variable that selects element to be an actuator. Every beam element can have one of four predefined thickness values (Table 1) and if element has value of thickness equally to 0, this means that the element is eliminated from the initial ground structure. Actuator variables have values between 1 and total number of beam elements in the structure (Table 1). The objective function that is used for the synthesis of the adaptive compliant gripper finger is given in (Milojević and Pavlović 2014).

After the parameterization search method is applied to find the optimal solution of adaptive gripper finger with embedded actuators. As an optimization method Genetic Algorithms are adopted here, as they can perform global search over large design space with large number of design variables. The genetic algorithm parameters used in the syntheses of adaptive gripper finger are given in Table 1.

Table 1 Design specification for developing adaptive compliant gripper with embedded actuators

Design parameters	
Design domain	120 mm × 80 mm
Grid size	5 × 5
Degree of nodal connectivity	4
Number of beam elements	168
Element modulus	E_{el} = 2.48 GPa
Actuator modulus	E_{akt} = 500 MPa
Actuator block force	90 N
External load	0.1 N
Element out-of-plane thickness	1.5 mm
Element thickness choice	0.5, 1, 1.5 mm
Thickness of shape morphing surface	0.5 mm
Genetic algorithm parameters	
Initial population	200
Generations	1000
Selection function	Roulette
Crossover	95 %
Mutation	9 %
Elite count	2

To obtain the optimal solution of adaptive compliant gripper finger with embedded actuators more than twenty Genetic Algorithms are run. The optimization starts with 168 beam variables and total number of 1664 intersection in the initial ground structure (Fig. 1b). During the optimization some of the beam elements are chosen to be eliminated from the structure while some are chosen to become an actuator, remaining elements with elements selected as actuators form the optimal topology. Figure 1c, shows the results of optimization—optimal topology of adaptive compliant gripper finger with embedded actuators (in the obtained solution three embedded actuators).

3 Analysis of Behavior of the Adaptive Compliant Gripper with Embedded Actuators

Based on the obtained optimal solution (Fig. 1c) 3D solid model of the adaptive compliant gripper with embedded actuators was designed. Figure 2 shows the 3D model with two fingers of the gripper. In the 3D design, actuators were modeled as thin elastic elements in form of a spring (Fig. 2). These thin elastic elements can be easily deformed and thus they can simulate actuation (contraction or extension) i.e. can simulate contracting or extending actuators. The idea here was to simulate general type of actuators that can realize contraction or extension and not to simulate some particular actuator.

To investigate behavior of the developed compliant gripper with embedded actuators nonlinear FEM analysis was preformed; ABAQUS software was used. For the analysis, to simulate contraction and extension of the actuators, input displacement of ±5 mm was introduced at both ends of the actuators in the direction

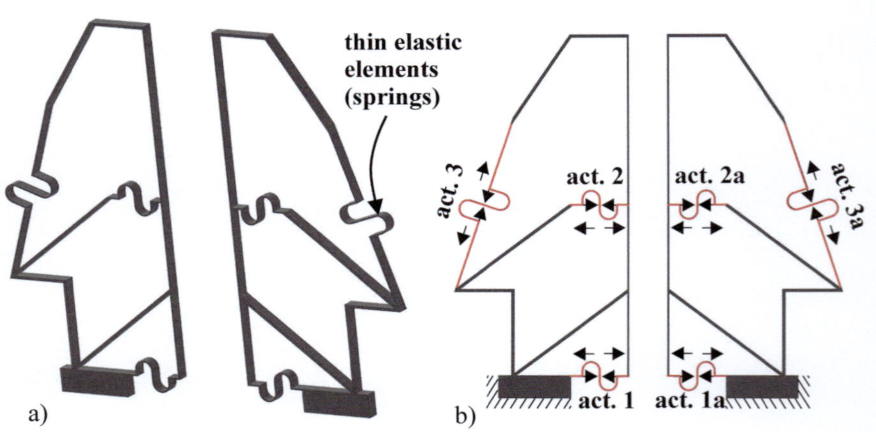

Fig. 2 Concept of adaptive two finger gripper with embedded actuators: **a** 3D design; **b** inputs for FEM analysis (actuators are indicate by *red lines*)

of actuator axis (Fig. 2b); this represents stroke of the actuators (total stroke is 10 mm). The results of simulation of gripper finger behavior with only contracting and only extending actuators as well as for some combination of active contracting and extending actuators, were presented in our previous publication (Milojević and Pavlović 2014). But in (Milojević and Pavlović 2014) only the grasping patters that gripper could achieve when actuators realize full stroke, were presented. Adaptive gripper could realize grasping patterns of different radii for different stroke of the actuators. This means that gripper could realize grasping of objects of a different size. Figure 3 shows the results of FEM simulation when contracting actuators realize different stroke (concept of two finger gripper grasping objects of a different shapes and size is shown).

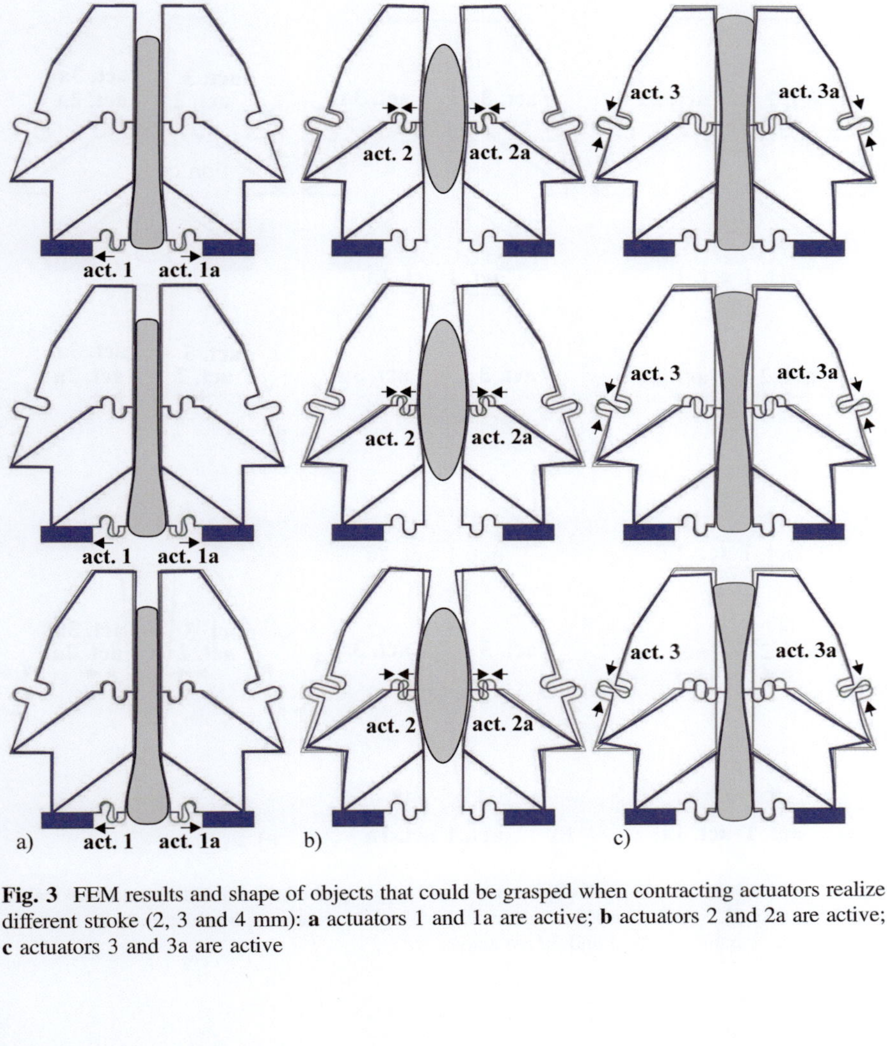

Fig. 3 FEM results and shape of objects that could be grasped when contracting actuators realize different stroke (2, 3 and 4 mm): **a** actuators 1 and 1a are active; **b** actuators 2 and 2a are active; **c** actuators 3 and 3a are active

The FEM results show that adaptive gripper could realize grasping of different shaped and sized objects when different actuators are active. Two common shaped objects could be grasped: convex objects (when actuator 2 is active, Fig. 3b) and concave objects (when actuator 3 is active, Fig. 3c). Moreover, for a different stroke of actuator 2, convex objects of a different size could be grasped, and for a different stroke of actuator 3, concave objects of a different size could be grasped. Also different shaped and sized objects could be grasped when actuator 1 is active (Fig. 3a). Many more different shaped and sized objects could be grasped when combination of different contracting actuators is used, whereby they realize different stroke (Fig. 4). For example convex-concave objects of a different size could be grasped when contracting actuators 1 and 2 are active and realize different stroke (Fig. 4a).

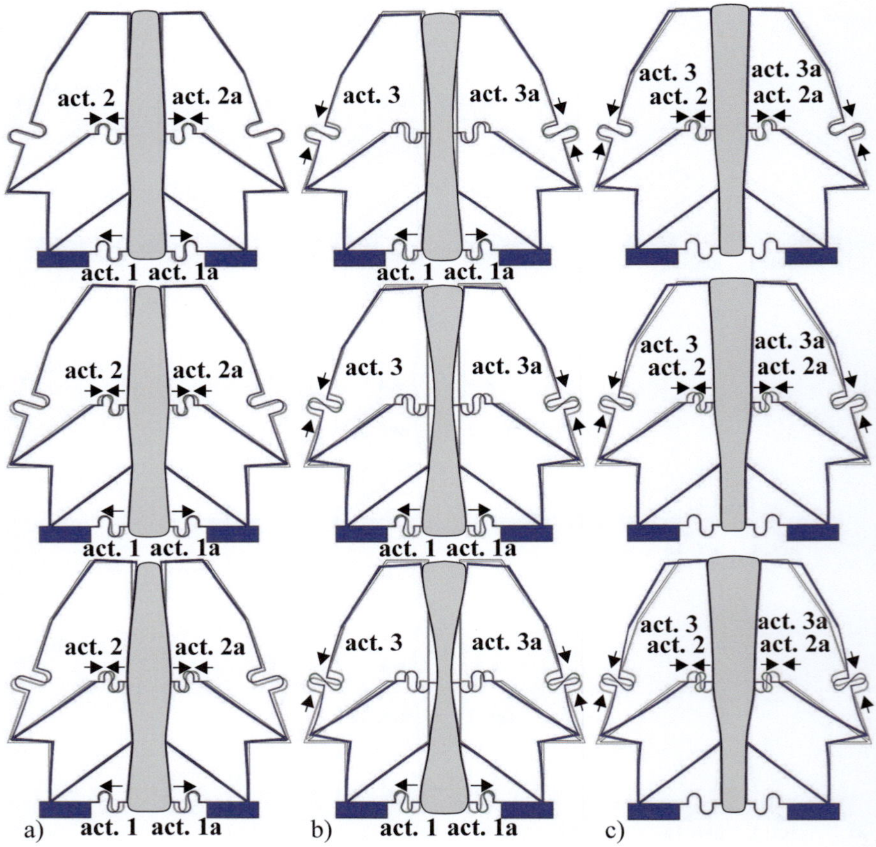

Fig. 4 FEM results and shape of objects that could be grasped when contracting actuators realize different stroke (2, 3 and 4 mm): **a** actuators 1, 1a, 2 and 2a are active; **b** actuators 1, 1a, 3 and 3a are active; **c** actuators 2, 2a, 3 and 3a are active

The adaptive compliant gripper could realize different grasping patterns when extending actuators are used. Figure 5 shows the concept of two finger gripper grasping objects of a different shapes and size when extending actuators realize different values of stroke.

Similar like in the case of contracting actuators, it could be seen that two common shaped objects could be grasped: convex (Fig. 5c) and concave shaped objects (Fig. 5b). But now for a different stroke of actuator 3 convex objects of a different size could be grasped (Fig. 5c) and for a different stroke of actuator 2 concave objects of a different size could be grasped (Fig. 5b). Also for a different stroke of extending actuator 1, objects of a different size could be grasped (Fig. 5a). The FEM results when different combinations of extending actuator are used, while

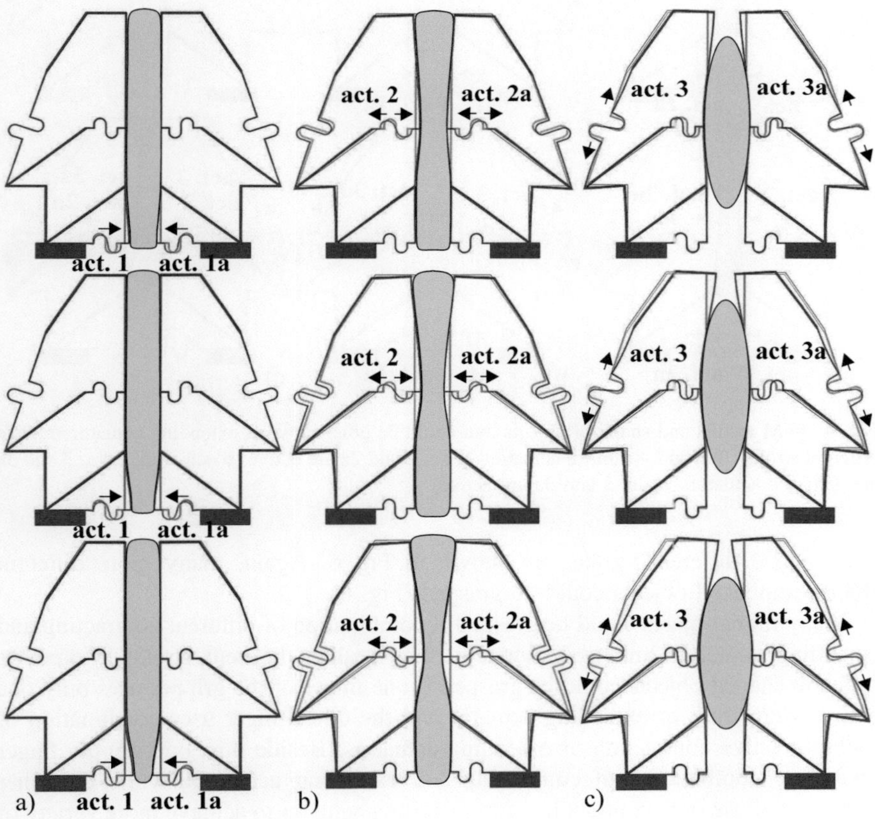

Fig. 5 FEM results and shape of objects that could be grasped when extending actuators realize different stroke (2, 3 and 4 mm): **a** actuators 1 and 1a are active; **b** actuator 2 and 2a are active; **c** actuators 3 and 3a are active

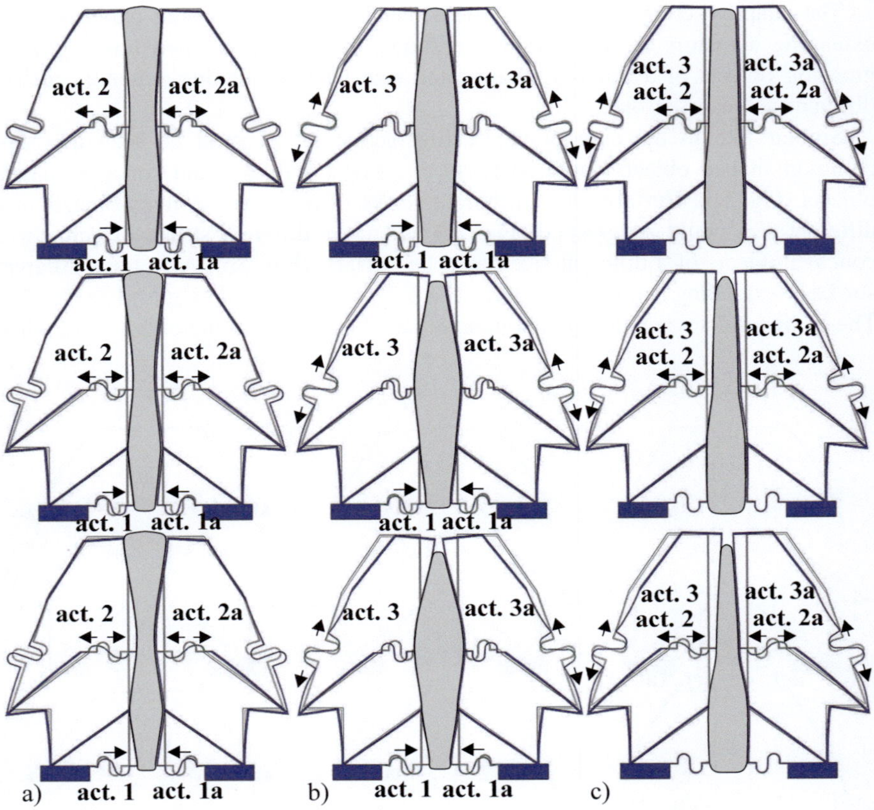

Fig. 6 FEM results and shape of objects that could be grasped when extending actuators realize different stroke (2, 3 and 4 mm): **a** actuators 1, 1a, 2 and 2a are active; **b** actuators 1, 1a, 3 and 3a are active; **c** actuators 2, 2a, 3 and 3a are active

they realize different stroke, are shown in Fig. 6. Again, many more different shaped and sized objects could be grasped (Fig. 6).

Many more objects could be grasped if combination of different contracting and extending actuators are used, whereby they realize different stroke. Moreover, different shaped objects could be grasped if one finger of the gripper uses only one active contracting or extending actuator and the other finger uses combination of multiple active contracting or extending actuators. Beside this, one gripper finger could use combination of contracting and extending actuators, while the other finger could use only contracting or only exiting actuator to achieve more variety of grasping patents and thus grasp more different shaped objects.

4 Conclusions

This paper presents a concept of adaptive two finger gripper with embedded actuators. By using compliant systems—compliant mechanisms with embedded actuators it is possible to develop an adaptive compliant gripper. By embedding the actuators within the compliant gripper structure, gripper could change the shape of its grasping surface and adapt to different shaped objects in order to achieve safe and reliable manipulation of that objects.

The design methodology for adaptive compliant gripper was also presented. Synthesis methodology includes simultaneous optimization of gripper structural topology and actuator placement (orientation, size and location). This design methodology represent one novel approach to synthesis of adaptive grippers and could be used to develop different grippers according to designer needs.

FEM analysis of the behavior of the developed adaptive compliant gripper was also presented. The FEM results demonstrated that by using only contracting or only extending actuators, gripper could realize different grasping patterns and thus could grasped different shaped objects. It was also shown that when actuators realize different stroke, objects of different size could be grasped; the shape of objects were also shown. Many more different shaped and sized objects could be grasped if combinations of contracting and extending actuators are used, whereby they realize different stroke. It is also possible that one finger of the gripper uses only contracting actuators or combination of multiple contracting actuators and other finger of the gripper to use only extending actuator or combination of multiple extending actuators to achieve more variety of different grasping patterns. These concepts will be further investigate in some of our future work.

The presented concept of an adaptive compliant gripper could be also scaled down to micro domain and could be applied for grasping and manipulation of individual cells or human tissue.

The results indicate that the adaptive two finger gripper with embedded actuators developed in this paper is a good starting point for further development and investigation.

Future work will cover designing the full two finger gripper i.e. designing a compliant mechanism that can provide rotation of the two fingers of the gripper.

Acknowledgments The research work reported here was supported by the DAAD and Ministry of Education, Science and Technological Development of the Republic of Serbia (PPP Serbia 2014, Project-ID 451-03-01858/2013-09/10). The financial support is gratefully acknowledged.

References

Amend JR Jr, Brown E, Rodenberg N, Jaeger H, Lipson H (2012) A positive pressure universal gripper based on the jamming of granular material. IEEE Trans Robot 28:341–350

Araromi OA, Gavrilovich I, Shintake J, Rosset S, Richard M, Gass V, Shea HR (2014) Roll-able multisegment dielectric elastomer minimum energy structures for a deployable microsatellite gripper. IEEE/ASME Trans Mech 20:438–446

Choi H, Koc M (2006) Design and feasibility tests of a flexible gripper based on inflatable rubber pockets. Int J Mach Tools Manu 46:1350–1361

Giannaccini ME, Georgilas I, Horsfield I, Peiris BHPM, Lenz A, Pipe AG, Dogramadzi S (2014) A variable compliance, soft gripper. Auton Robot 36:93–107

Howell LL, Magleby SP, Olsen BM (2013) Handbook of compliant mechanisms. Wiley, Somerset

Ilievski F, Mazzeo AD, Shepherd RF, Chen X, Whitesides GM (2011) Soft robotics for chemists. Angew Chem Int Ed 50:1890–1895

Jain RK, Majumder S, Dutta A (2013) SCARA based peg-in-hole assembly using compliant IPMC micro gripper. Robot Auton Syst 61:297–311

Joo J, Kota S, Kikuchi N (2000) Topological synthesis of compliant mechanisms using linear beam elements. Mech Struct Mach 28(4):245–280

Jung G-P, Koh J-S, Cho K-J (2013) Underactuated adaptive gripper using flexural buckling. IEEE Trans Robot 29(6):1396–1407

Lu K-J, Kota S (2003) Design of compliant mechanisms for morphing structural shapes. J Int Mater Syst Struct 14:379–391

Lu K-J, Kota S (2005) An effective method of synthesizing compliant adaptive structures using load path representation. J Int Mater Syst Struct 16:307–317

Martinez RV, Branch JL, Fish CR, Jin L, Shepherd RF, Nunes RMD, Suo Z, Whitesides GM (2013) Robotic tentacles with three-dimensional mobility based on flexible elastomers. Adv Mater 25:205–212

Milojević A, Pavlović ND (2013) Software for synthesis of compliant mechanisms without intersecting elements. Facta Univ Ser Mech Eng 11(2):153–168

Milojević A, Pavlović ND (2014) Development of adaptive compliant gripper finger with embedded actuators. In: Microactuators and micromechanisms, mechanisms and machine science, vol 30. Springer, New York, pp 33–49

Milojević A, Pavlović ND (2015) Development of a new adaptive shape morphing compliant structure with embedded actuators. J Int Mater Syst Struct (Accepted) doi:10.1177/1045389X15590270, first publish on June 16 (2015)

Milojević A, Pavlović ND, Milošević M, Tomić M (2013) New software for synthesis of compliant mechanisms. In: Proceedings of the 2nd international conference mechanical engineering in XXI century, Niš, pp 273–278

Reeves CR, Rowe JE (2002) Genetic algorithms-principles and perspectives: a guide to GA theory. Kluwer Academic Publishers, Dordrecht

Saxena A, Ananthasuresh GK (2000) On an optimal property of compliant topologies. Struct Multidisc Optim 19:36–49

Trease B, Kota S (2009) Design of adaptive and controllable compliant systems with embedded actuators and sensors. J Mech Design 131(111001):12

Zhao K, Schmiedeler JP (2013) Using rigid-body mechanism topologies to design shapechanging compliant mechanisms. In: Proceedings of the 2013 ASME international design engineering technical conferences and computers & information in engineering conference, Portland, 4–7 Aug. http://ame-robotics.nd.edu/LAB/Publications/DETC2013-12576-preprint.pdf

Implementation of Self Contact in Path Generating Compliant Mechanisms

Prabhat Kumar, Anupam Saxena and Roger A. Sauer

Abstract Path generating compliant mechanisms with self contact are synthesized using continuum discretization and negative circular masks. Hexagonal cells are employed to represent the design space and negative circular masks are used to remove material beneath them. Each mask is defined by three parameters which constitute the design vector, which is mutated using the Hill-climber search. Hexagonal cells circumvent various geometrical singularities related to single point connections however, many "V" notches get retained on the boundary edges. Self contact is permitted between various subregions of the continuum. To achieve convergence in contact analysis, it is essential to have smooth boundary surfaces for which a smoothing technique is employed. Consequently, many hexagonal cells get modified to six-noded generic polygons. Non-linear finite elements analysis is implemented with Mean Value Coordinates based shape functions. Self contact is modeled between two deformable bodies in association with augmented Lagrange multiplier method. To minimize the error in shape, size and orientation between the actual and specified paths, a Fourier Shape Descriptors based objective is employed. An example of path generating compliant mechanism which experiences self contact is presented and its performance is compared with a commercial software.

Keywords Complaint mechanisms · Self contact · Mean value coordinates · Boundary smoothing · Fourier shape descriptors

P. Kumar (✉) · A. Saxena
Indian Institute of Technology Kanpur, Kanpur, India
e-mail: kprabhat@iitk.ac.in

A. Saxena
e-mail: anupams@iitk.ac.in

R.A. Sauer
RWTH Aachen University, Aachen, Germany
e-mail: sauer@aices.rwth-aachen.de

L. Zentner et al. (eds.), *Microactuators and Micromechanisms*,
Mechanisms and Machine Science 45, DOI 10.1007/978-3-319-45387-3_22

1 Compliant Mechanisms

Compliant mechanisms (CMs) execute the desired work by deriving the mobility from the deflection of their deformable members. Such deflections may be large or small depending upon their applications. CMs assure various advantages e.g., ease in assembly and manufacturing, less wear and frictional losses, more precision and no backlash. Consequently, use of such mechanisms is increasing in the different engineering domains e.g., micro-electro-mechanical systems, aerospace, automotive, bio-medical devices and various other fields.

To synthesize compliant mechanisms, two well established approaches e.g., topology optimization (Bendsøe and Kikuchi 1988; Sigmund 1997) and Pseudo Rigid Body Models (Howell 2001) exist. Topology optimization entails systematic determination of the material layout via extremising a certain objective with/without some constraints. Methods that employed quadrilateral cells to discretize the design domain suffer from many geometrical singularities (Saxena 2011) mostly related to the point connection between two neighboring elements. Hexagonal cells have natural edge connectivity between two neighboring cells (Saxena 2011). To perform finite element analysis, each hexagonal cell is either divided into six triangular elements (Langelaar 2007), or two quadrilateral shapes (Saxena and Saxena 2007), or modeled as a single element (Saxena 2011).

For path generating CMs, to minimize the discrepancies in the specified and actual paths, objective based on least-square is used (Saxena and Ananthasuresh 2001). Least-square objective does not provide local control for discrepancies in shape, size and orientation of paths. Fourier shape descriptors (FSDs) based objective is employed by Ullah and Kota (1997) to compare individual discrepancies in the shape, size and orientation of paths.

Non smooth paths can be generated either using partial compliant (Rai et al. 2007) or contact-aided compliant mechanisms (Mankame and Ananthasuresh 2004). Prespecified contact surfaces were used in association with FSDs objective (Mankame and Ananthasuresh 2004). Pseudo hinges/sliding pairs (Kumar et al. 2016) get introduced at mating surfaces and consequently configurations (Midha et al. 1994) of CMs change. Reddy et al. (2012) used curve beams to discretize the design space and determined contact locations systematically rather than pre-specifying them to synthesize non-smooth path generating contact-aided compliant mechanism. Kumar et al. (2016) presented continuum discretization based approach to synthesize C^0 path generating contact-aided compliant mechanisms with mutual contact.

2 Motivation

Constituting members of CMs while undergoing large deformation may experience self contact, which is implemented herein. Hexagonal cells are used to represent the design space and negative circular masks are placed suitably over it to remove material within cells beneath the overlaying masks to impart the optimum material layout.

Contents of the paper are organized as follows: Sect. 3 briefs the continuum discretization and topology optimization with negative circular masks. Boundary smoothing and finite element formulation are discussed in Sect. 4. Section 5 outlines contact modeling and analysis. An example on path generating contact-aided compliant mechanism is presented in Sect. 6 followed by the concluding remarks in Sect. 7.

3 Continuum Discretization and Negative Circular Masks

A tessellation of hexagonal cells (Ω_H) represent the design domain and negative circular masks (Ω_M) are employed to assign the material state $\rho(\Omega_\mathrm{H})$ to each cell. Each mask is modeled with three parameters (x_i, y_i, r_i) which constitute the design vector, where r_i is the radius of the ith mask and (x_i, y_i) are the coordinates of its center. If N masks are employed, the design vector contains $3N$ parameters. $\rho_k(\Omega_\mathrm{H}) = 0$ implies cell is void, while $\rho_k(\Omega_\mathrm{H}) = 1$ represents solid cell. A compliant continuum therefore comprises of all unexposed cells. The boundary of such continua contains many V-notches which pose difficulty in manufacturing, generate stress-concentration regions, and importantly, pose difficulty in convergence of contact analysis (Kumar and Saxena 2015). Mathematically, the material states of any cell can be described as

$$\rho(\Omega_\mathrm{H}) = \begin{cases} 0, & \text{if } \Omega_\mathrm{H}^\mathrm{c} \subset \text{ any } \Omega_\mathrm{M}(d_\mathrm{ik} - r_\mathrm{i} \leq 0) \\ 1, & \text{if } \Omega_\mathrm{H}^\mathrm{c} \not\subset \text{ any } \Omega_\mathrm{M}(d_\mathrm{ik} - r_\mathrm{i} > 0) \end{cases} \tag{1}$$

where $\Omega_\mathrm{H}^\mathrm{c}$ represents the centroid of Ω_H, and d_ik is the distance between the centers of Ω_M and $\Omega_\mathrm{H}^\mathrm{c}$ (Fig. 1).

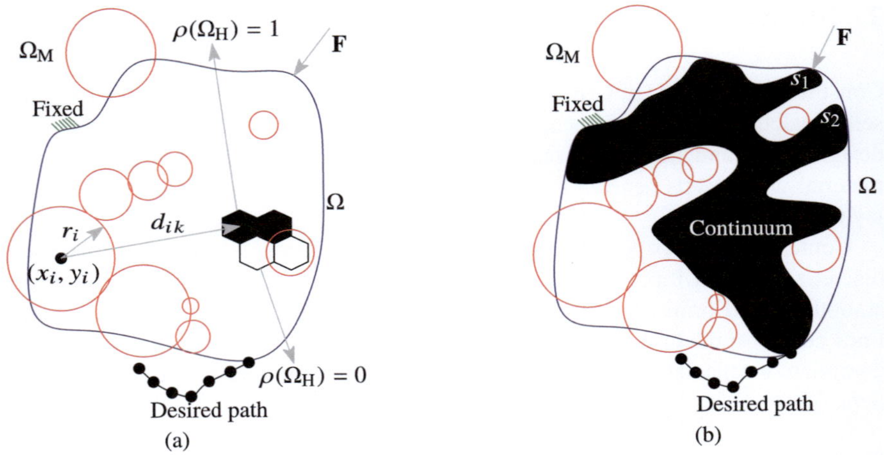

Fig. 1 Hexagonal cells Ω_H are used to discretize the design space Ω (*blue curve*) and negative masks Ω_M (*red circles*) are placed over them. Unexposed cells comprise the continuum (*black solid* in (**b**)). Input force (**F**), desired output path (*black curve* with *dots*) and fixed boundaries are shown. Surfaces s_1 and s_2 (shown in (**b**)), and many such surface pairs, may come in contact

4 Boundary Smoothing and Finite Element Formulation

The obtained final compliant continua by the use of hexagonal cells and negative circular masks contain serrated bounding surfaces (Kumar and Saxena 2015). However, to facilitate contact analysis, it is desirable to have smooth bounding surfaces (Corbett and Sauer 2014). Thus, boundary smoothing scheme is implemented herein to smoothen the bounding surfaces. In this section, we briefly present boundary smoothing scheme and finite element formulation.

4.1 Boundary Smoothing

Boundary smoothing (Kumar and Saxena 2015) is employed to alleviate V notches at the continuum boundary(ies) resulting due to hexagonal cells. Bounding edges of the continuum and boundary nodes are noted. Mid-points of these edges are joined via straight lines. Boundary nodes are projected to these lines along their shortest perpendiculars (Fig. 2a). Positions of boundary nodes are updated and used for finite element analysis. The connectivity matrix remain unchanged. The smoothing process can be implemented multiple times (Fig. 2b). The process may alter some of hexagonal cells into six-noded generic concave polygons (as shown in Fig. 2b). The continuum volume can be reduced further by removing regular cells, i.e., those do not get affected by boundary smoothing. This is equivalent to introducing more

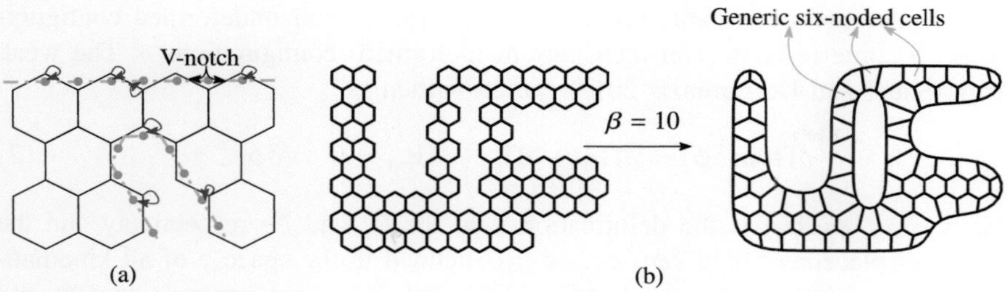

Fig. 2 Boundary smoothing **a** boundary nodes are projected along their shortest perpendiculars on *straight lines* joining mid-points of boundary edges **b** *left* with $\beta = 0$ and *right* with $\beta = 10$. β : number of boundary smoothing steps

negative masks, one for each regular cell, such that the centers of that respective cell and the mask coincide. Smoothing is performed again to eliminate the remnant notches.

4.2 *Finite Element Formulation*

The boundary smoothing process (Kumar and Saxena 2015) results in altering some regular hexagonal cells into six-noded generic polygonal cells. These cells are modeled with Mean Value Coordinates based shape functions (Floater 2003; Sukumar and Tabarraei 2004). Elemental stiffness matrix $\mathbf{K}_e$ is evaluated as follows: (i) Each cell is divided into six triangular regions with respect to the centroid of that cell. (ii) Integration over each triangle is performed with 25 Gauss-points[1] (Kumar and Saxena 2015). (iii) Summation of such six integrations is done and $\mathbf{K}_e$ is obtained. The global stiffness matrix $\mathbf{K}^{glob}$ is evaluated by assembling $\mathbf{K}_e$. Likewise the internal force vector $\mathbf{r}_e$ and global internal force vector $\mathbf{R}$ are obtained. When self contact arises the elemental contact force $\mathbf{f}_c^e$ and corresponding stiffness $\mathbf{K}_c^e$ are evaluated, as below, and placed at their global degree of freedom inside $\mathbf{R}$ and $\mathbf{K}^{glob}$.

5 Contact Modeling and Analysis

Self contact is modeled as quasi-static contact between two deformable bodies in association with augmented Lagrange multiplier method.

[1]for accuracy.

Consider two deformable bodies $\mathcal{B}_{01}$ and $\mathcal{B}_{02}$, in their undeformed configurations, to come in contact in their current (deformed) configurations[2] . The weak form (Sauer and De Lorenzis 2013) can be written as

$$\delta\Pi(\boldsymbol{\phi}, \delta\boldsymbol{\phi}) = \delta\Pi_{\text{int}} + \delta\Pi_{\text{c}} - \delta\Pi_{\text{ext}} = 0 \quad \forall\delta\boldsymbol{\phi} \subset \gamma \tag{2}$$

where $\boldsymbol{\phi} = (\boldsymbol{\phi_1}, \boldsymbol{\phi_2})$ is the deformation field for $\mathcal{B}_1$ and $\mathcal{B}_2$ respectively and the virtual displacement field $\delta\boldsymbol{\phi}(\delta\boldsymbol{\phi_1}, \delta\boldsymbol{\phi_2})$ is defined in the space γ of all kinematically permissible deformations. $\delta\Pi_{\text{int}}$, $\delta\Pi_{\text{c}}$ and $\delta\Pi_{\text{ext}}$ are internal, contact and external virtual works respectively. The virtual contact work $\{\delta\Pi_c = \delta\Pi_{c1} + \delta\Pi_{c2}\}$ (Sauer and De Lorenzis 2013) is defined as

$$\delta\Pi_{ck} = -\int\limits_{\delta_c\mathcal{B}_k} \mathbf{t}_k.\delta\boldsymbol{\phi}_k da_k, \tag{3}$$

where $\delta_c\mathcal{B}_k$ is the contact surface, $\mathbf{t}_k$ is the contact traction and a_k is the contact surface area. Further, indices c and k are used to represent contact and two deformable bodies respectively.

5.1 Normal Contact

Consider two points $\mathrm{P}_1(\mathbf{x}_1)$ and $\mathrm{P}_2(\mathbf{x}_2)$ on bodies $\mathcal{B}_1$ and $\mathcal{B}_2$ respectively[3] (Fig. 3b). The gap between them is defined as

$$\mathbf{g} = \mathbf{x}_2 - \mathbf{x}_1. \tag{4}$$

If $\mathbf{g}.\mathbf{n}_1 \geq 0$, bodies do not penetrate each other. Here $\mathbf{n}_1$ is the normal vector at $\mathrm{P}_1(\mathbf{x}_1)$. The distance d between points is $||\mathbf{x}_2 - \mathbf{x}_1||$. For a given $\mathbf{x}_2 \subset \partial\mathcal{B}_2$, we search for a point $\mathbf{x}_1 \subset \partial\mathcal{B}_1$ such that d becomes *minimum*. That is, we solve for

$$d_{min} = \min_{\mathbf{x}_1 \subset \mathcal{B}_1}(d) \tag{5}$$

We assume that the boundary surfaces are convex locally. Each material point on $\mathcal{B}_1$ is defined using a convective co-ordinates system $\boldsymbol{\xi} = (\xi^1, \xi^2)$, with $\frac{\partial\mathbf{x}_1}{\partial\xi^\alpha} = \mathbf{a}^1_\alpha; \alpha = 1, 2$. Here $\mathbf{a}^1_\alpha$ is the tangent vector. The solution of Eq. (5) is obtained by imposing $\frac{\partial d}{\partial\xi^\alpha} = 0$. Since $\frac{\partial d}{\partial\xi^\alpha} = \left\{\frac{\partial d}{\partial\mathbf{x}_1} \cdot \frac{\partial\mathbf{x}_1}{\partial\xi^\alpha}\right\}$, the following two conditions are obtained

---

[2]bodies are denoted as $\mathcal{B}_1$ and $\mathcal{B}_2$.

[3]$\mathrm{P}_1(\mathbf{X}_1)$ and $\mathrm{P}_2(\mathbf{X}_2)$ are the corresponding points in the undeformed configurations respectively.

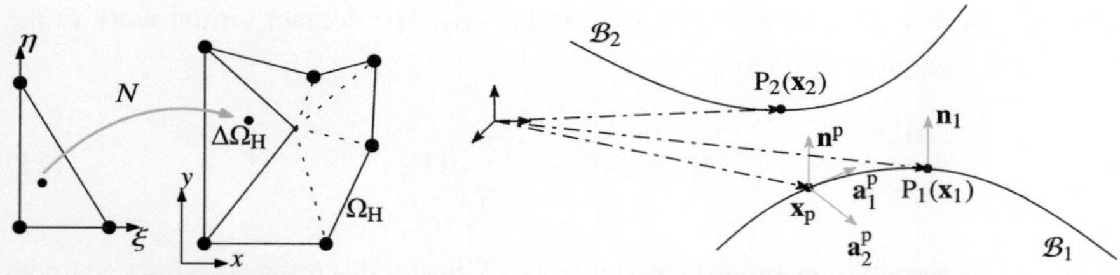

Fig. 3 **a** Each cell is divided into six Δ regions and integration is performed. **b** Points of surfaces coming in contact are determined by finding the minimum distance between them

$$\frac{\partial d}{\partial \xi^{\alpha}} = -\{\mathbf{x}_2 - \mathbf{x}_1(\xi)\}.\mathbf{a}^1_{\alpha} = 0 \tag{6}$$

Let Eq. (6) holds true for $\mathbf{x}_1(\xi)|_{\xi=\xi_p} = \mathbf{x}_p(\xi_p)$, where at $\mathbf{x}_p$, $\mathbf{a}^p_{\alpha}$ and $\mathbf{n}^p$ are the tangent and normal vectors respectively. The normal gap is evaluated as

$$g_n = (\mathbf{x}_2 - \mathbf{x}_p(\xi_p)).\mathbf{n}^p. \tag{7}$$

For discrete surfaces (C^0 continuous), surface normals vary at each nodal position. Therefore, for such surfaces, normal is equivalent to

$$\mathbf{n}^p = (\mathbf{x}_2 - \mathbf{x}_p(\xi_p))/g_n \tag{8}$$

where normal gap $g_n = \pm\sqrt{(\mathbf{x}_2 - \mathbf{x}_p).(\mathbf{x}_2 - \mathbf{x}_p)}$. In classical penalty method (Sauer and De Lorenzis 2013), traction $\mathbf{t}_{ck}$ is defined as

$$\mathbf{t}_{ck} = \begin{cases} -\varepsilon_n g_n \mathbf{n}^p & \text{for } g_n < 0 \\ \mathbf{0} & \text{for } g_n \geq 0 \end{cases} \tag{9}$$

where ϵ_n is a chosen penalty parameter.

5.2 *Contact Discretization*

We employ, the following 2D linear Lagrangian basis functions to represent contact segments.

$$N^1_{pk} = \frac{1}{2}(1 - \xi^k_P), \quad N^2_{pk} = \frac{1}{2}(1 + \xi^k_P) \tag{10}$$

where $\zeta_{\mathrm{p}}^{k}|$ $(k = 1, 2)$ is obtained by solving Eq. (6). The contact virtual work in the discretized system is defined as

$$\delta\Pi_{ck} \approxeq \delta\Pi_{ck}^{h} = \sum_{e=1}^{n_{sel}} \delta\Pi_{ck}^{e} \tag{11}$$

where n_{sel} is the number of surface elements Γ^{e} in the discretized contact segment $\partial_{c}\mathcal{B}_{k}^{h}$. We rewrite Eq. (3) in the discretized segments $\partial_{c}\mathcal{B}_{k}^{h}$ correspond to $\delta_{c}\mathcal{B}_{k}$ as

$$\delta\Pi_{ck}^{e} = -\int_{\Gamma^{e}} \delta\boldsymbol{\phi}_{ck}^{e}.\mathbf{t}_{ck} da_{k}; \quad k = 1, 2 \tag{12}$$

where $\delta\boldsymbol{\phi}_{ck}^{e} = \mathbf{N}_{ck}\mathrm{v}_{ck}^{e}$ and $\mathbf{N}_{ck}$ equals $[N_{\mathrm{p}k}^{1}\mathbf{I}_{2}, N_{\mathrm{p}k}^{2}\mathbf{I}_{2}]$ with $\mathbf{I}_{2}$ is 2×2 unit matrix and v_{ck}^{e} is the virtual displacement vector. Therefore, Eq. (12) is equivalent to

$$\delta\Pi_{ck}^{e} = -\mathrm{v}_{ck}^{e} \int_{\Gamma^{e}} \mathbf{N}_{ck}^{T}\mathbf{t}_{ck} da_{k}; \quad k = 1, 2 \tag{13}$$

From Eq. (13) the elemental contact force vector is evaluated as

$$\mathbf{f}_{ck}^{c} = -\int_{\Gamma^{e}} \mathbf{N}_{ck}^{T}\mathbf{t}_{ck} da_{k}; \quad k = 1, 2 \tag{14}$$

and thus, elemental contact stiffness matrix is calculated as

$$\mathbf{K}_{ck}^{e} = \frac{\partial \mathbf{f}_{ck}^{e}}{\partial \mathbf{u}_{k}^{e}} = -\int_{\Gamma^{e}} \mathbf{N}_{ck}^{T} \frac{\partial \mathbf{t}_{ck}}{\partial \mathbf{u}_{k}^{e}} da_{k} = -\int_{\Gamma^{e}} \mathbf{N}_{ck}^{T} \frac{\partial \mathbf{t}_{ck}}{\partial \mathbf{x}_{k}^{e}} \mathbf{N}_{ck} da_{k}, \tag{15}$$

where elemental displacement vector $\mathbf{u}_{k}^{e} = \mathbf{x}_{k}^{e} - \mathbf{X}_{k}^{e}$. Detailed expressions for contact force vector and elemental contact stiffness are mentioned in (Sauer and De Lorenzis, 2013). We have considered frictionless contact in our implementation.

6 Synthesis Example

We present an example where the compliant mechanism experiences self contact (Fig. 4b) while tracing the specified path. To attain this CM, the objective (Kumar et al. 2016) based on Fourier Shape Descriptors with the given volume constraint is minimized using the Random Mutation Hill Climber method (Kumar et al. 2016). Table 1 depicts the value of different parameters used in the optimization process. Figure 4a describes the design specifications for the compliant continuum. The CM

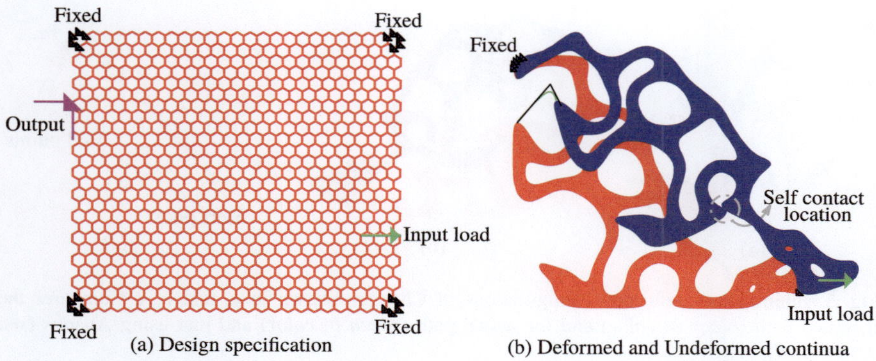

Fig. 4 **a** Design Specifications are shown. **b** The undeformed (*red in color*) and deformed (*blue in color*) configurations are shown. *Dotted-circle* is the zone of self contact. The specified path (*black*) and the actual path (*green*) is also shown

Table 1 Parameters used in the synthesis. *HCH* and *HCV* represent the number of hexagonal cells in the horizontal directions and in the vertical direction respectively

Parameter's name	Value	Parameter's name	Value
Design domain ($HCH \times HCV$)	25 × 25	Number of Ω_M in x-direction (N_x)	7
Number of Ω_M in y-direction (N_y)	7	Maximum radius of Ω_M	6.0 mm
Minimum radius of Ω_M	0.1 mm	Maximum number of iterations	40,000
Young's modulus	2100 MPa	Poisson's ratio	0.30
Permitted volume fraction ($\frac{V^*}{V}$)	0.30	Mutation probability	0.08
Contact surface radii factor	0.50	Maximum mutation size (m_{max})	6
Upper limit for input load	500 N	Lower limit for input load	−500 N
Weight for A_{err} (w_a)	50	Weight for B_{err} (w_b)	50
Weight for path length error (w_L)	1	Weight for path orientation error (w_q)	5
Boundary smoothing steps (β)	10	Penalty parameter (ε_n)	100

is also analyzed in Abaqus with neo-Hookean material and four-noded plain-strain elements (CPE4I) using the obtained optimal force, boundary conditions, and active self contact locations. The actual path traced by the CM using presented method closely follows the path obtained via Abaqus analysis (Fig. 5c).

Figure 4b depicts the deformed (in blue) and undeformed (in red) configurations of the final optimized compliant continuum. Two intermediate configurations of the CMs, one before contact and one after, are shown in 5a and 5b while tracing the path (green in color). Figure 5c compares the specified, actual and that obtained using Abaqus. The final compliant mechanism is obtained after 5741 search iterations with 104.61 N input force.

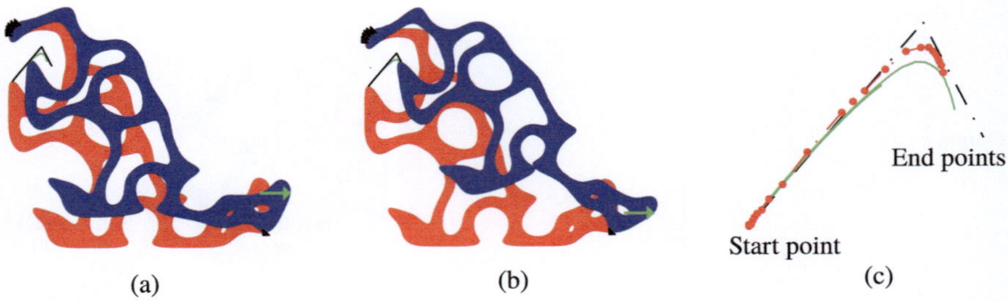

Fig. 5 Intermediate deformed configurations of CMs are shown in **a**, and **b**. **c** Compares the specified path (*black in color*) and the actual path (*green in color*) and that using Abaqus (*red dash-dotted*)

7 Conclusion

We present a continuum discretization based method to design the optimal topology of a compliant mechanism undergoing self contact. Hexagonal cells are employed for domain representation and negative masks are used for material assignment. Boundary smoothing is used to facilitate contact analysis. Mean Value Coordinates based shape functions are employed for Finite Element Analysis. Hill climber search process is used to optimize the Fourier Shape Descriptors objective to achieve the specified path. Contact-aided compliant mechanisms may find use in special purpose applications, say static balancing, negative stiffness etc. One can also implement both self contact and mutual (two different bodies) contact, a goal, planned for future work.

References

Bendsøe MP, Kikuchi N (1988) Generating optimal topologies in structural design using a homogenization method. Comput Methods Appl Mech Eng 71(2):197–224

Corbett CJ, Sauer RA (2014) Nurbs-enriched contact finite elements. Comput Methods Appl Mech Eng 275:55–75

Floater MS (2003) Mean value coordinates. Comput Aided Geom Des 20(1):19–27

Howell LL (2001) Compliant mechanisms. Wiley, New York

Kumar P, Saxena A (2015) On topology optimization with embedded boundary resolution and smoothing. Struct Multi Optim 52(6):1135–1159

Kumar P, Sauer RA, Saxena A (2016) Synthesis of c0 path-generating contact-aided compliant mechanisms using the material mask overlay method. J Mech Des 138(6):062301

Langelaar M (2007) The use of convex uniform honeycomb tessellations in structural topology optimization. In: 7th world congress on structural and multidisciplinary optimization, Seoul, South Korea, pp 21–25

Mankame ND, Ananthasuresh G (2004) Topology optimization for synthesis of contact-aided compliant mechanisms using regularized contact modeling. Comput Struct 82(15):1267–1290

Midha A, Norton TW, Howell LL (1994) On the nomenclature, classification, and abstractions of compliant mechanisms. J Mech Des 116(1):270–279

Layout: T1 Standard Book ID: 417395_1_En Book ISBN: 978-3-319-45386-6
Chapter No.: 23 Date: 3-12-2016 Time: 3:32 pm Page: 1/1

Erratum to: Dynamic Analysis of a Fatigue Test Bench for High Precision Flexure Hinges

D. Schoenen, M. Hüsing and B. Corves

Erratum to: Chapter "Dynamic Analysis of a Fatigue Test Bench for High Precision Flexure Hinges" in: L. Zentner et al. (eds.), *Microactuators and Micromechanisms*, Mechanisms and Machine Science 45, DOI 10.1007/978-3-319-45387-3_15

The author name in the original version of this chapter was published with incorrect initial as 'C. Corves'. It has now been updated with the correct initial as 'B. Corves'.

The updated original online version for this chapter can be found at 10.1007/978-3-319-45387-3_15

D. Schoenen (✉) · M. Hüsing · B. Corves
Department of Mechanism Theory and Dynamics of Machines, RWTH Aachen University, Aachen, Germany
e-mail: schoenen@igm.rwth-aachen.de

M. Hüsing
e-mail: huesing@igm.rwth-aachen.de

B. Corves
e-mail: corves@igm.rwth-aachen.de

L. Zentner et al. (eds.), *Microactuators and Micromechanisms*, Mechanisms and Machine Science 45, DOI 10.1007/978-3-319-45387-3_23

MIX
Papier aus verantwortungsvollen Quellen
Paper from responsible sources
FSC® C105338

If you have any concerns about our products,
you can contact us on
ProductSafety@springernature.com

In case Publisher is established outside the EU,
the EU authorized representative is:
Springer Nature Customer Service Center GmbH
Europaplatz 3, 69115 Heidelberg, Germany

Printed by Libri Plureos GmbH
in Hamburg, Germany